# PHYSICS

## SIXTH EDITION

## Paul E. Tippens
## Paul Zitzewitz
## Graig Kramer

Glencoe
McGraw-Hill

New York, New York   Columbus, Ohio   Woodland Hills, California   Peoria, Illinois

**Cover photo:** Richard Magna/Fundamental Photographs

Physics, Sixth Edition     ISBN 0-07-820340-6
Activities Manual     ISBN 0-07-820341-4
Instructor's Manual     ISBN 0-07-820342-2

*Glencoe/McGraw-Hill*

*A Division of The McGraw·Hill Companies*

**Activities Manual: Study Guide and Experiments for Physics, Sixth Edition**

Send all inquiries to:
Glencoe/McGraw-Hill
8787 Orion Place
Columbus, OH 43240

ISBN 0-07-820341-4

1 2 3 4 5 6 7 8 9   045   06 05 04 03 02 01 00

# Contents

# Preface

This volume provides study guide material and laboratory experiments for use with *Physics, Sixth Edition,* a basic noncalculus physics textbook. The book effectively combines student support and course enhancement material under one cover. This approach allows students to review classroom lectures and their homework assignments, and to prepare for assigned laboratory activities.

## UNIT A

Unit A, the first part of this book, is the study guide, which carefully follows the contents of each chapter in *Physics, Sixth Edition.* Each chapter in Unit A includes the following items:

- A contents listing major chapter topics
- Definitions of important terms
- Concepts and examples
- A concise chapter summary
- Key equations listed according to chapter section
- True-false questions
- Multiple-choice questions
- Answers to all questions and problems

All the material has been tested in the classroom and revised to effect maximum learning. Unit A should be considered as an aid to the students and as a supplement to the text. Worked-out solutions to typical problems and opportunities to test general understanding through thoughtful questions are especially valuable to the student. Answers are provided for all questions and for the supplementary problems.

The following approach has been found effective for the use of Unit A:

1. First, glance over the contents for the unit, and then read the objectives at the beginning of each chapter to see where emphasis is placed.

2. Read text material first and work through each example given in the text. Then review the appropriate definitions from the study guide.

3. After the lecture, return to Unit A to review the section on Concepts and Examples.

4. Test your understanding of basic principles by answering true-false, multiple-choice, and completion questions from Unit A. (Make a note to ask your instructor about answers you do not understand.)

5. The most important part of any physics course is the application of physics in the solution of problems.

Remember that a study guide does not replace a text, and it does not replace the instructor. It should be thought of as a companion that supplements all other experiences of a typical beginning physics course. It is most useful for a quick review of major concepts and as a resource for additional practice, which is necessary for a good understanding of basic physics.

Unit B consists of 29 experiments for the beginning student of physics. The experiments illustrate the concepts found in this introductory course. Both qualitative and quantitative experiments, requiring manipulation of apparatus, observation, and collection of data, are included. The experiments are designed to help you utilize the processes of science to interpret data and draw conclusions.

The laboratory report is an important part of the laboratory experience. It helps students learn to communicate their observations and conclusions to others. The laboratory report pages are included with each experiment to aid students in recording their data. Students are encouraged to record their results on separate graph paper and to draw diagrams of the experimental apparatus. Likewise, instructors may wish to increase or decrease the number of experimental observations based upon lab time and the equipment available. Thus, each experiment provides an outline for preparing a lab report.

Although accuracy is always desirable, other goals are of equal importance in laboratory work that accompanies early courses in science. A high priority is given to how well laboratory experiments introduce, develop, or make the physics theories learned in the classroom realistic and understandable and to how well laboratory investigations illustrate the methods used by scientists. The investigations in Unit B place more emphasis on the implications of laboratory work and its relationship to general physics principles, rather than to how closely results compare with accepted quantitative values.

## Processes of Science

The scientifically literate person uses the processes of science in making decisions, solving problems, and expanding an understanding of nature. The Laboratory Manual in Unit B utilizes many processes of science in all of the lab activities. Throughout this unit, you are asked to collect and record data, plot graphs, make and identify assumptions, perform experiments, and draw conclusions. In addition, the following processes of science are included in Unit B.

**Observing**   Using the senses to obtain information about the physical world.

**Classifying**   Imposing order on a collection of items or events.

**Communicating**   Transferring information from one person to another.

**Measuring**   Using an instrument to find a value, such as length or mass, that quantifies an object or event.

**Using numbers**   Using numbers to express ideas, observations, and relationships.

**Controlling variables**   Identifying and managing various factors that may influence a situation or event, so that the effect of any given factor may be learned.

**Designing experiments**   Performing a series of data-gathering operations that provide a basis for testing an hypothesis or answering a specific question.

**Defining operationally**   Producing definitions of an object, concept, or event in terms that give it a physical description.

**Formulating models**   Devising a mechanism or structure that describes, acts, or performs as if it were a real object or event.

**Inferring**   Explaining an observation in terms of previous experience.

**Interpreting data**   Finding a pattern or meaning inherent in a collection of data, which leads to a generalization.

**Predicting**   Making a projection of future observations based on previous information.

**Questioning**   Expressing uncertainty or doubt that is based on the perception of a discrepancy between what is known and what is observed.

**Hypothesizing**   Explaining a relatively large number of events by making a tentative generalization, which is subject to testing, either immediately or eventually, with one or more experiments.

## The Experiment

Experiments are organized into several sections. Each experiment opens with a statement describing the purpose of the activity. The Purpose is followed by a Concept and Skill Check, which reviews or introduces relevant physics concepts and background information. Numerous illustrations are provided to help you assemble the apparatus and understand the principles under investigation.

The Materials section lists the equipment used in the experiment and allows you to assemble the required materials quickly and efficiently. All of the equipment listed is either common to the average physics lab or can be readily obtained from local sources or a science supply company. Slight variations of equipment can be made and will not affect the basic integrity of the experiments in Unit B.

The Procedure section contains step-by-step instructions to perform the experiment. This format helps you take advantage of limited laboratory time. Safety symbols alert you to potential dangers in the laboratory investigation, and caution statements are provided where appropriate.

The Observations and Data section helps organize the lab report. All tables are outlined and properly labeled. In more qualitative experiments, questions are provided to guide your observations.

In the Analysis section, you relate observations and data to the general principles outlined in the purpose of the experiment. Graphs are drawn and interpreted and conclusions concerning data are made. Questions relate laboratory observations and conclusions to basic physics principles studied in the text and in classroom discussions.

The Application section allows you to apply some aspects of the physics concept investigated. Often this section illustrates a current application of the concept.

Many of the experiments include an Extension section. *Extensions* are supplemental procedures or problems that expand the scope of the experiment. They are designed to further the investigation and to challenge the more interested student.

## Purpose of Laboratory Experiments

The laboratory work in physics is designed to help you to better understand basic principles of physics. You will, at the same time, gain a familiarity with the scientific methods and techniques employed in the laboratory. In each experiment, you will be seeking a definite goal, investigating a specific principle, or solving a definite problem. To find the answer to the problem, you will make measurements, list your measurements as data, and then interpret the data to find the results of your measurements.

The values you obtain may not always agree with accepted values. Frequently this result is to be expected because your laboratory equipment is usually not sophisticated enough for precision work, and the time allowed for each experiment is not extensive. The relationships between your observations and the broad general laws of physics are of much more importance than strict numerical accuracy.

## Preparation of Your Lab Report

One very important aspect of laboratory work is the communication of your results obtained during the investigation. Unit B is designed so that laboratory report writing is as efficient as possible. In most of the laboratory experiments in this book, you will write your report on the report sheets placed immediately following each experimental procedure. All tables are outlined and properly labeled for ease in recording data and calculations. For instructions on writing a formal laboratory report, see Appendix E.

### Using Significant Digits

When making observations and calculations, you should stay within the limitations imposed upon you by your equipment and measurements. Each time you make a measurement, you will read the scale to the smallest calibrated unit and then obtain one smaller unit by estimating. The doubtful or estimated figure is significant because it is better than no estimate at all and should be included in your written values.

It is easy, when making calculations using measured quantities, to indicate a precision greater than your measurements actually allow. To avoid this error, use the following guidelines.

- When adding or subtracting measured quantities, round off all values to the same number of significant decimal places as the quantity having the least number of decimal places.
- When multiplying or dividing measured quantities, retain in the product or quotient the same number of significant digits as in the least precise quantity.

### Accuracy and Precision
Whenever you measure a physical quantity, there is some degree of uncertainty in the measurement. Error may come from a number of sources, including the type of measuring device, how the measurement is made, and how the measuring device is read. How close your measurement is to the accepted value refers to the accuracy of a measurement. In several of these laboratory activities, experimental results will be compared to accepted values.

When you make several measurements, the agreement, or closeness, of those measurements refers to the precision of the measurement. The closer the measurements are to each other, the more precise is the measurement. It is possible to have excellent precision, but to have inaccurate results. Likewise, it is possible to have poor precision, but to have accurate results when the average of the data is close to the accepted value. Ideally, the goal is to have good precision and good accuracy.

### Relative Error
Although absolute error is the absolute value of the difference between your experimental value and the accepted value, relative error is the percentage deviation from an accepted value. The relative error is calculated according to the following relationship:

$$\text{Relative error} = \frac{|\text{accepted value} - \text{experimental value}|}{\text{accepted value}} \times 100\%$$

### Graphs
Frequently an experiment involves finding out how one quantity is related to another. The relationship is found by keeping constant all quantities except the two in question. One quantity is then varied, and the corresponding change in the other is measured. The quantity that is deliberately varied is called the independent variable. The quantity that changes due to the variation in the independent variable is called the dependent variable. Both quantities are then listed in a table. It is customary to list the values of the independent variable in the first column of the table and the corresponding values of the dependent variable in the second column.

More often than not, the relationship between the dependent and independent variables cannot be ascertained simply by looking at the written data. But if one quantity is plotted against the other, the resulting graph gives evidence of what sort of relationship, if any, exists between the variables. When plotting a graph, use the following guidelines:

- Plot the independent variable on the horizontal *x*-axis (abscissa).
- Plot the dependent variable on the vertical *y*-axis (ordinate).
- Draw the smooth line that best fits the most plotted points.

It is the hope of the author and the publisher that both students and instructors will enjoy this combined economical volume, which allows students to review chapter assignments and prepare for laboratory exercises. The author welcomes comments from students and instructors. Comments may be directed to the physics department at Southern Technical Institute, Marietta, Georgia, or to the publisher.

**Paul E. Tippens**

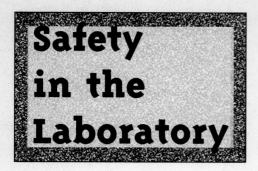

# Safety in the Laboratory

If you follow instructions exactly and understand the potential hazards of the equipment and the procedure used in an experiment, the physics laboratory is a safe place for learning and applying your knowledge. You must assume responsibility for the safety of yourself, your fellow students, and your teacher. Here are some safety rules to guide you in protecting yourself and others from injury and in maintaining a safe environment for learning.

1. The physics laboratory is to be used for serious work.

2. Never bring food, beverages, or make-up into the laboratory. Never taste anything in the laboratory. Never remove lab glassware from the laboratory, and never use this glassware for eating or drinking.

3. Do not perform experiments that are unauthorized. Always obtain your teacher's permission before beginning an activity.

4. Study your laboratory assignment before you come to the lab. If you are in doubt about any procedure, ask your teacher for help.

5. Keep work areas and the floor around you clean, dry, and free of clutter.

6. Use the safety equipment provided for you. Know the location of the fire extinguisher, safety shower, fire blanket, eyewash station, and first-aid kit.

7. Report any accident, injury, or incorrect procedure to your teacher at once.

8. Keep all materials away from open flames. When using any heating element, tie back long hair and loose clothing. If a fire should break out in the lab, or if your clothing should catch fire, smother it with a blanket or coat or use a fire extinguisher. *NEVER RUN.*

9. Handle toxic, combustible, or radioactive substances only under the direction of your teacher. If you spill acid or another corrosive chemical, wash it off immediately with water.

10. Place broken glass and solid substances in designated containers. Keep insoluble waste material out of the sink.

11. Use electrical equipment only under the supervision of your teacher. Be sure your teacher checks electric circuits before you activate them. Do not handle electric equipment with wet hands or when you are standing in damp areas.

12. When your investigation is completed, be sure to turn off the water and gas and disconnect electrical connections. Clean your work area. Return all materials and apparatus to their proper places. Wash your hands thoroughly after working in the laboratory.

# First Aid in the Laboratory

Report all accidents, injuries, and spills to your teacher immediately.
*You Must Know:*

- Safe laboratory techniques
- Where and how to report an accident, injury, or spill
- The location of first aid equipment, fire alarm, and telephone

| Situation | Safe response |
| --- | --- |
| Burns | Flush with cold water. |
| Cuts and bruises | Treat as directed by instructions included in your first aid kit. |
| Electric shock | Provide person with fresh air, have person recline in a position such that the head is lower than the body; if necessary, perform CPR. |
| Fainting or collapse | See electric shock. |
| Fire | Turn off all flames and gas jets; wrap person in fire blanket; use fire extinguisher to put out fire. *DO NOT* use water to extinguish fire, as water may react with the burning substance and intensify the fire. |
| Foreign matter in eyes | Flush with plenty of water; use eye bath. |
| Poisoning | Note the suspected poisoning agent, contact your teacher for antidote; if necessary, call poison control center. |
| Severe bleeding | Apply pressure or a compress directly to the wound and get medical attention immediately. |
| Spills, general acid burns | Wash area with plenty of water; use safety shower; use sodium hydrogen carbonate, $NaHCO_3$ (baking soda). |
| Base burns | Use boric acid, $H_3BO_3$. |

# Safety Symbols

In Unit B you will find several safety symbols that alert you to possible hazards and dangers in a laboratory activity. Be sure that you understand the meaning of each symbol before you begin an experiment. Take necessary precautions to avoid injury to yourself and others and to prevent damage to school property.

| | | | |
|---|---|---|---|
|  | **DISPOSAL ALERT**<br>This symbol appears when care must be taken to dispose of materials properly. |  | **LASER SAFETY**<br>This symbol appears when care must be taken to avoid staring directly into the laser beam or at bright reflections. |
|  | **BIOLOGICAL HAZARD**<br>This symbol appears when there is danger involving bacteria, fungi, or protists. |  | **RADIOACTIVE SAFETY**<br>This symbol appears when radioactive materials are used. |
|  | **OPEN FLAME ALERT**<br>This symbol appears when use of an open flame could cause a fire or an explosion. |  | **CLOTHING PROTECTION SAFETY**<br>This symbol appears when substances used could stain or burn clothing. |
|  | **THERMAL SAFETY**<br>This symbol appears as a reminder to use caution when handling hot objects. |  | **FIRE SAFETY**<br>This symbol appears when care should be taken around open flames. |
|  | **SHARP OBJECT SAFETY**<br>This symbol appears when a danger of cuts or punctures caused by the use of sharp objects exists. |  | **EXPLOSION SAFETY**<br>This symbol appears when the misuse of chemicals could cause an explosion. |
|  | **FUME SAFETY**<br>This symbol appears when chemicals or chemical reactions could cause dangerous fumes. |  | **EYE SAFETY**<br>This symbol appears when a danger to the eyes exists. Safety goggles should be worn when this symbol appears. |
|  | **ELECTRICAL SAFETY**<br>This symbol appears when care should be taken when using electrical equipment. |  | **POISON SAFETY**<br>This symbol appears when poisonous substances are used. |
| | **PLANT SAFETY**<br>This symbol appears when poisonous plants or plants with thorns are handled. |  | **CHEMICAL SAFETY**<br>This symbol appears when chemicals used can cause burns or are poisonous if absorbed through the skin. |

# Activities Manual for Physics

**UNIT**

**A**

# Chapter
# 1

# Introduction

**DEFINITIONS OF KEY TERMS**

**Physics**   The science that investigates the fundamental concepts of matter, energy, and space and the relationships among them.

**Statics**   The study of the physical phenomena associated with bodies at rest.

**Dynamics**   The description of motion and the treatment of its causes.

**Hypothesis**   A statement or an equation that is a provisional conjecture to guide further investigation.

**Scientific theory**   An experimentally verified hypothesis that is consistently repeatable and therefore capable of prediction.

**True-False Questions**

T   F   1.  Physics is a field of science that is distinguished easily from other fields such as chemistry or biology.

T   F   2.  Mathematics serves many purposes, but its main value is to provide a professional tool for the scientist or engineer.

T   F   3.  All quantities in physics should have specific, measurable definitions.

T   F   4.  *Physics* and *science* are different terms frequently used to describe the same thing, and so they may be used interchangeably.

T   F   5.  Dynamics is a branch of physics devoted to the study of objects at rest or in motion with constant speed.

T   F   6.  If a problem or definition discussed by an instructor is also in the text, it is better to jot down a reference rather than to take detailed notes.

T   F   7.  Physics is a quantitative science.

T   F   8.  Since some information in the text might be misunderstood, it is better to wait until after the lecture to read the material.

T   F   9.  If a workable hypothesis fails only one time in many trials, it still may be incorporated into a scientific law.

T   F   10.  A good background in mathematics, though desirable, is not really helpful in understanding basic concepts.

T   F   11.  All physics learning takes place in the classroom.

T   F   12.  It is better to wait until the weekend to study physics than to study daily.

T   F   13.  Learning most often comes from working problems.

# Chapter 2

# Technical Mathematics

**CONTENTS**

## DEFINITIONS OF KEY TERMS

**Base**   A number raised to an exponent power.

**Exponent**   A superscript number representing the number of times a quantity will be multiplied by itself.

**Formula**   An equation with letters representing quantities in the physical world.

**Parallel**   Two coplanar lines that will never touch.

**Perpendicular**   Two lines that meet at right angles.

**Radical**   A fractional exponent, for example, $a^{1/2} = \sqrt{a}$; $a^{1/3} = \sqrt[3]{a}$

**Right triangle**   A triangle in which one of the angles is a right angle or 90° angle.

**Scientific notation**   A number expressed with one nonzero digit to the left of the decimal point, all the remaining significant digits to the right of the decimal point, and multiplied by a power of 10.

**Triangle**   A figure made by three lines with three angles.

## CONCEPTS AND EXAMPLES

To solve problems, you need to be able to move the unknown quantity to only one side of the equation. This is a simple procedure that you can learn to do by working carefully and checking afterward for mistakes.

**EXAMPLE 2-1**

Solve the following equation for $c$:

$$9c + 6 = 2c - 8$$

*Solution*
First, subtract 6 from both sides.

$$9c = 2c - 14$$

Next, subtract $2c$ from both sides.

$$7c = -14$$

Now divide both sides by 7.

$$c = -2$$

Exponents and radicals follow clear rules that you can use reliably. Be sure that you recognize whether exponents should be added, subtracted, multiplied, or divided to come up with the correct answer. Whenever you are in doubt, use examples with powers of 2, which you can figure out in your head, to reassure yourself that you are using the appropriate operation.

**EXAMPLE 2-2**

Give two other ways of showing the quantity $(a^3)^{-5}$

*Solution*

When a power is raised to a power, the two exponents are multiplied.

$$(a^3)^{-5} = a^{(3)(-5)} = a^{-15}$$

Another way of expressing a negative exponent is by changing its sign and placing it in the bottom part of a fraction.

$$a^{-15} = \frac{1}{a^{15}}$$

**EXAMPLE 2-3**

Simplify the expression $\sqrt{2^4}$

*Solution*

A radical or square root is equivalent to raising the quantity inside the radical to a power of $^1/_2$.

$$2^4 = (2^4)^{1/2} = 2^{(4)(1/2)} - 2^{4/2} = 2^2 = 4$$

When working a problem, often you will be able to use what you have learned about geometry to help solve it. Two geometry facts will help: (1) Alternate interior angles are always equal. (2) For any triangle, the sum of the interior angles is equal to 180 degrees.

**EXAMPLE 2-4**    Find the angles equal to *a* in each figure.

**1.**

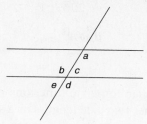

Fig. 2-1

**2.**

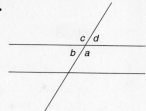

Fig. 2-2

**3.**

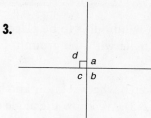

Fig. 2-3

**4.**

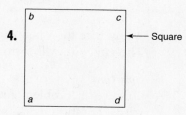

Fig. 2-4

*Solution*

1. *a* and *b* are alternate interior angles; *b* and *d* are opposite angles, and therefore also equal.
2. *c* is the opposite angle to *a*, so *c* is equal to *a*.
3. Since *d* is a right angle, all four angles must be right angles; therefore *b*, *c*, and *d* are all equal to *a*.
4. Since the figure is a square, all the angles must be 90°; therefore *b*, *c*, and *d* are all equal to *a*.

The pythagorean theorem states that the square of the hypotenuse equals the sum of the squares of the short sides of a right triangle.

$$R^2 = x^2 + y^2$$

You can use this theorem to find the length of the third side of any right triangle if you know the lengths of the other two sides.

**EXAMPLE 2-5**    A person leans a 20-ft ladder against a wall. The ladder touches the wall 15 ft from the ground. How long is the distance from the wall to the base of the ladder?

*Solution*

The pythagorean theorem states that $R^2 = x^2 + y^2$. Since the ladder is 20 ft long, $R = 20$. The height of the wall is 15 ft, so $y = 15$. To solve for *x*, we first move the terms around.

$$x^2 = R^2 - y^2$$
$$x = \sqrt{R^2 - y^2}$$

Now we substitute the known values for the ladder length and the height of the wall.

$$x = \sqrt{20^2 - 15^2}$$

$$x = 13.2 \text{ ft}$$

You also learned about the trigonometric functions sine, cosine, and tangent. Using these functions, you have either two side lengths or a small angle and one side length. You can learn all the other angles and lengths of the triangle.

**EXAMPLE 2-6**

A guy wire for a new high-tech windmill is to be placed 50 m away from the base. The guy wire will form a 68° angle with the ground. How long will the wire be?

*Solution*

To solve the problem, first we sketch what we know and label what we wish to know with a question mark (see Fig. 2-5). We know the angle and the length of the adjacent side. We need to use a formula that includes the information we know and only one unknown.

$$\cos \theta = \text{adj/hyp}$$

Now solve for the unknown quantity.

$$\text{hyp} = \text{adj/cos } \theta$$

Now substitute the known values and calculate the answer.

$$\text{hyp} = 50 \text{ m/cos } 68 = 50 \text{ m/0.3746} = 133.5 \text{ m}$$

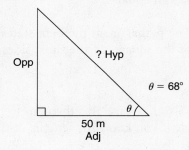

**Fig. 2-5**

---

**Summary**

This chapter covered the mathematics concepts you will need to successfully complete the text *Physics*.

- To solve a problem, use this process.
  **a.** Write the formula with only one unknown, the quantity you wish to find.
  **b.** Rearrange the formula so that your unknown is alone on one side of the equation.
  **c.** Substitute the values in the equation to calculate your unknown quantity.
  **d.** As a check, ask yourself whether the number makes sense.

- To multiply two numbers whose exponents have the same base, add the exponents.
- To divide two numbers whose exponents have the same base, subtract the exponent of the bottom part from the exponent of the top part.
- To express a number in scientific notation, move the decimal point one digit to the right of the leftmost nonzero digit. If this process moves the decimal to the left, count up to find the power of 10; if the decimal moves to the right, count down to find the power of 10.
- When two straight lines intersect, they form opposing right angles which are equal.
- When a straight line intersects two parallel lines, the alternate interior angles are equal.
- For any triangle, the sum of the interior angles is 180°; for a right triangle, the sum of the smaller angles is 90°.

## Key Equations by Section

| Section | Topic | Equation | Equation No. |
|---|---|---|---|
| 2-2 | Volume of a solid | $V = lwh$ | 2-1 |
| 2-2 | Volume of a cone | $V = \pi r^2 h/3$ | 2-2 |
| 2-3 | Multiplication of numbers with exponents | $(a^m)(a^n) = a^{m+n}$ | 2-3 |
| 2-3 | Division of numbers with exponents | $a^m/a^n = a^{m-n}$ | 2-6 |
| 2-3 | Numbers with exponents raised to a power | $(a^m)^n = a^{mn}$ | 2-7 |
| 2-3 | Radicals | $\sqrt[n]{b} = b^{1/n}$ | |
| 2-7 | Pythagorean theorem | $R^2 = x^2 + y^2$ | 2-12 |
| 2-7 | Definitions of sin, cos, tan | $\sin \theta = \dfrac{\text{opp } \theta}{\text{hyp}}$ | 2-16 |
| | | $\cos \theta = \dfrac{\text{adj } \theta}{\text{hyp}}$ | |
| | | $\tan \theta = \dfrac{\text{opp } \theta}{\text{adj}}$ | |

## True-False Questions

T  F  **1.** When multiplying two exponents with the same base, add the exponents.

T  F  **2.** When multiplying two exponents with the same base, multiply the exponents.

T  F  **3.** On a cold winter day, the temperature rose from −5°C to 20°C. The increase in temperature was equal to 25°C.

**T  F   4.** Three boys have three marbles, and each marble has three blue dots. The total number of dots is $3^3$ or 27.

**T  F   5.** Thirty-two apples, or $2^5$, are to be divided equally among 16 people, or $2^4$. Since $2^5/2^4 = 2^{5-4}$, or $2^1$, each person gets two apples.

**T  F   6.** If a straight line crosses another straight line at a 90° angle, the adjacent angles are equal.

**T  F   7.** To multiply two numbers in scientific notation, you must subtract the second exponent from the first exponent.

**T  F   8.** The side of a right triangle opposite the right angle can be equal to one of the other sides.

**T  F   9.** If you know the length of one side of a right triangle and the size of the right angle, you can find the other two sides and two angles.

**T  F  10.** If two variables increase in proportion to one another, they are said to have an inverse relationship.

## Multiple-Choice Questions

1. A positive number multiplied by an even number of negative numbers will always result in which type of number?
   (a) positive          (b) negative          (c) exponent    (d) fraction

2. Which of the following formulas is solved for the unknown quantity $a$?
   (a) $a = 1/2a + b$          (b) $a = 3^a - b^n$    (c) $a = \sin\theta/50$
   (d) $\theta = \sin a/50$

3. Which of the following is equivalent to the quantity $a^{11}/a^{15}$?
   (a) $a^{-4}$          (b) $a^{26}$          (c) $a^4$          (d) $a^{1/4}$

4. Which of the following is equivalent to the fourth root of $a$?
   (a) $a^{-4}$          (b) $a^4$          (c) $a^0$          (d) $a^{1/4}$

5. The speed and distance of a certain train are directly related. This means that when the speed increases, the distance does what?
   (a) increases          (b) decreases          (c) remains the same
   (d) may increase or decrease

6. A ladder base is 3 m away from a house, and the fully extended ladder is 7 m long when it leans against the house. How high does the ladder reach on the house?
   (a) 58 m          (b) 4 m          (c) about 6.3 m
   (d) about 7.6 m

7. An anchor holding a boat stationary on a river forms a 30° angle with the river bottom. The anchor line is 100 m long. Assuming that the river has a uniform depth, how deep is the river directly below the boat?
   (a) 0.005 m          (b) $5 \times 10^3$ m    (c) 86 m    (d) 50 m

8. To make a bridge across the swift-flowing upper Niagara river, it was necessary to fly a kite across in order to span the river for the first time. If 300 m of kite string was let out as the kite was directly over the opposite bank, and the kite string formed a 50° angle with the level ground, how wide was the river span?
   (a) about $6.4 \times 10^{-1}$ m    (b) 40°          (c) about 230 m
   (d) about 193 m

9. Two parallel crosswalk lines cross a street at a 30° angle to connect a pedestrian mall. What is the largest angle formed by the intersection of the crosswalk lines and the street?
   (a) 90°          (b) 30°          (c) 60°          (d) 150°

**10.** At 5 minutes after the hour, a minute arm forms a 30° angle. At 15 minutes, it forms a 90° angle, and at 20 minutes it forms a 120° angle. What angle will it form at 30 minutes?

(a) 30°                (b) 180°                (c) 0°                (d) 270°

## Completion Questions

**1.** When an odd number of negative numbers is multiplied, the answer will always be _____.

**2.** The square of a number is the same as the number raised to a power of _____.

**3.** A number raised to a fractional power can also be expressed as a _____.

**4.** As one quantity increases, another decreases proportionally. These quantities have an _____ relationship.

**5.** A relationship using letters to represent variable events in the physical world is a _____.

**6.** A superscript number to the right of a variable is its _____ or _____.

**7.** The longest side of a right triangle is its _____.

**8.** The side of the triangle next to the angle in question is called the _____ angle.

**9.** The remaining angle in a right triangle with one 25° angle is _____.

**10.** When two parallel lines are crossed by a single straight line, the alternate interior angles are _____.

# Chapter 3

# Technical Measurement and Vectors

**DEFINITIONS OF KEY TERMS**

**Unit** A standard by which a physical quantity is measured, such as a unit of length (meter) or a unit of force (newton).

**Vector quantity** A quantity that is specified completely by its magnitude and direction, for example, 40 m, north or 50 N, 40°.

**Scalar quantity** A quantity that is specified completely by its magnitude only, for example, 24 m or 144 lb.

**Displacement** A vector quantity indicating the straight-line distance from a reference point to a known position.

**Resultant force** A resultant force is the single-force vector whose effect is the same as that of a number of vectors acting simultaneously.

**Component** The projection of a vector on a particular coordinate axis, such as the $x$ or $y$ components of a force vector.

**Concurrent forces** Forces that intersect at a common point or have the same point of application.

**Accuracy** The degree to which a measurement truly reflects the quantity measured.

**Component method** A graphical method of vector addition that uses components to calculate the resultant.

**Dimensions** Any measurable quantity, such as length, mass, time, or velocity.

**Fundamental quantity** A dimension with unreducible units such as meters, kilograms, or seconds. Fundamental quantities are combined to make derived quantities such as area ($m^2$), density ($kg/m^3$), and velocity ($m/s$).

**Parallelogram method** A graphical addition method for finding the resultant of two vectors. Both vectors are drawn from the same origin, and a parallelogram is formed by completing the parallel sides. The resultant is the diagonal of the parallelogram.

**Polygon method** A method for the graphical addition of vectors by placing the head of each vector serially on the tail of the previous vector. The resultant is found by drawing the vector from the origin to the last vector head.

**Precision**   The careful rendering of a measurement, achieved by reporting the smallest graduations of measure.

**SI units**   The metric system of measurement, whose units are the meter, the kilogram, and the second. *SI* is the French abbreviation of International System of Units.

**Significant figures**   All digits taken in a measurement. The answer to a problem can only be as precise as the measurements taken to achieve the result.

**Unit Conversions**   When converting units, first we write the quantity to be converted, including the number and the unit(s). Next, we write the desired unit(s) and try to recall a definition of the desired unit(s) in terms of the given unit(s). From each definition we form conversion factors and multiply by those factors that cancel non-desired units.

**EXAMPLE 3-1**

Convert a pressure of 14 lb/in.$^2$ to newtons per square meter, given that 1 N = 0.2248 lb and 1 m = 100 cm.

*Solution*
We write the quantity to be converted and then multiply by the appropriate conversion factors. Units are canceled algebraically as follows:

$$\left(14\frac{\text{lb}}{\text{in.}^2}\right)\left(\frac{1\text{ N}}{0.2248\text{ lb}}\right)\left(\frac{1\text{ in.}}{2.54\text{ cm}}\right)^2\left(\frac{100\text{ cm}}{1\text{ m}}\right)^2$$

$$\frac{(14)(100)^2}{(2.2248)(2.54)^2}\frac{\text{N}}{(\text{m}^2)} = \boxed{\textbf{96,530 N/m}^2}$$

**Vector Components**   Finding the components of vectors requires that we draw the vector and label its components opposite and adjacent to the given angle. If the vector is $(R, \theta)$, then the $x$ and $y$ components are found from

$$\mathbf{R}_x = R\cos\theta \qquad \mathbf{R}_y = R\sin\theta$$

**EXAMPLE 3-2**

What are the $x$ and $y$ components of a 60-N force acting at an angle of 330°? Refer to Fig. 3-1.

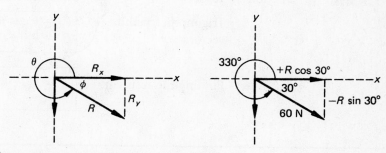

**Fig. 3-1**   Finding the components of a vector.

*Solution*
The components are found as follows:

$$\mathbf{R}_x = +\,(60\text{ N})(\cos 30°) = \boxed{\textbf{+52.0 N}}$$

$$\mathbf{R}_y = -\,(60\text{ N})(\sin 30°) = \boxed{\textbf{−30.0 N}}$$

Even if you decide to use the polar angle $\theta$ instead of the reference angle $\phi$, you should still draw the figure as shown so that you can verify the signs of the components. It's just as easy, and many times safer, to work with the reference angle ($\phi = 30°$) to find the magnitude of the components and then refer to the figure for the signs. In this example, $R_y$ is negative because it points downward.

**Helpful Hints**   Most beginners make sign errors. It's a good idea to check your signs twice by looking at the figure. (Measure twice and cut once.) Rounding errors also give us trouble if we are not careful. Always carry one more significant digit in your calculations than you ultimately round off to. For example, if you want your final answer to be accurate to three significant digits, you should carry at least four significant digits throughout all calculations. In all problems in this text, we assume an accuracy to three significant digits.

## Summary

- To convert one unit to another
  **a.** Write the quantity to be converted (number and unit).
  **b.** Recall the necessary definitions.
  **c.** Form two conversion factors for each definition.
  **d.** Multiply the quantity to be converted by the conversion factors that cancel all but the desired units.

- In the polygon method of vector addition, the resultant vector is found by drawing each vector to scale, placing the tail of one vector to the tip of another until all vectors are drawn. The resultant is the straight line drawn from the starting point to the tip of the last vector.

- In the parallelogram method of vector addition, the resultant of two vectors is the diagonal of a parallelogram formed by the two vectors as adjacent sides.

## Key Equations by Section

| Section | Topic | Equation | Equation No. |
|---------|-------|----------|--------------|
| 3-6 | Trigonometry and vectors | $F_x = F \cos \theta$ <br> $F_y = F \sin \theta$ | 3-1 |
| 3-2 | Finding the resultant | $R = \sqrt{F_x^2 + F_y^2} \quad \tan \theta = \dfrac{F_y}{F_x}$ | 3-2 |

## True-False Questions

T   F   **1.** The fundamental unit for length is the mile.

T   F   **2.** The kilogram is equivalent to a mass of 1000 g.

T   F   **3.** Only vectors having the same dimensions may be added.

T   F   **4.** The difference between two vectors is obtained by adding one vector to the negative of the other.

T   F   **5.** The order in which two or more vectors are added does not affect their resultant.

| T | F | **6.** | Given the $x$ component of one vector and the $y$ component of another vector, it is possible to find the resultant of the two vectors mathematically. |
|---|---|---|---|
| T | F | **7.** | The graphical methods of vector addition are not as accurate as the mathematical method. |
| T | F | **8.** | Vector addition can be performed only for concurrent vectors. |
| T | F | **9.** | If a boat travels upstream with a speed of 8 km/h in a current whose speed is 3 km/h, the boat's speed relative to the shore is 5 km/h. |
| T | F | **10.** | The $x$ component of the resultant vector is equal to the sum of the $x$ components of the individual vectors. |

## Multiple-Choice Questions

**1.** Which of the following is not a fundamental quantity?
  (a) Length  (b) Force  (c) Mass  (d) Time

**2.** The resultant of 10 and 15 lb acting in opposite directions on an object is
  (a) 150 lb  (b) 25 lb  (c) 5 lb  (d) 20 lb

**3.** Which is a scalar quantity?
  (a) Velocity  (b) Force  (c) Speed  (d) Displacement

**4.** A force of 3 N acts perpendicularly to a force of 4 N. Their resultant has a magnitude of
  (a) 12 N  (b) 7 N  (c) 5 N  (d) 1 N

**5.** Which is a vector quantity?
  (a) Volume  (b) Time  (c) Distance  (d) Displacement

**6.** Given that the units of $s$, $v$, $a$, and $t$ are feet, feet per second, feet per second squared, and seconds, respectively, which of the following equations is dimensionally incorrect?
  (a) $s = vt + \frac{1}{2}at$       (b) $2as = v_f^2 - v_0^2$
  (c) $v = v_0 + at$       (d) $s = vt$

**7.** A force of 16 N is directed 30° north of east. The $y$ component of the force is
  (a) 8 N  (b) 13.8 N  (c) 12 N  (d) 4.8 N

**8.** A man walks 9 km east and then 12 km north. The magnitude of his resultant displacement is
  (a) 21 km  (b) 15 km  (c) 13 km  (d) 3 km

**9.** The resultant of the forces in Fig. 3-2 is
  (a) 66.5 N, 222°       (b) 66.5 N, 132°
  (c) 66.5 N, 228°       (d) none of these

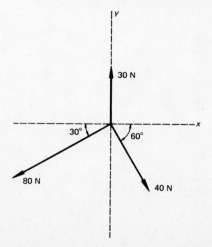

**Fig. 3-2**

**10.** A 20-lb force acts to the left while an 80-lb force acts upward and to the right at an angle of 37°. The magnitude of the resultant force is
(a) 80 lb          (b) 70 lb     (c) 65 lb          (d) 100 lb

**Completion Questions**

**1.** If two quantities are to be added or subtracted, they must be of the same _____.

**2.** Three examples of vector quantities are _____, _____, and _____.

**3.** A physical quantity that is specified completely by a number and a unit is called a(n) _____.

**4.** Every vector can be resolved into two perpendicular vectors called its _____.

**5.** A physical quantity that is specified completely by a number, a unit, and a direction is called a(n) _____.

**6.** In the _____ method of vector addition, the tail of one vector is connected to the tip of another until all vectors have been represented.

**7.** The _____ is a single force whose effect is the same as that of a given set of concurrent forces.

**8.** In the component method, the $x$ component of the resultant vector is equal to the sum of the _____ of each vector.

**9.** Forces that intersect at a common point or have the same point of application are said to be _____.

**10.** The difference between two vectors is obtained by adding one vector to the _____ of the other.

# Chapter 4

# Translational Equilibrium and Friction

## DEFINITIONS OF KEY TERMS

**Inertia** A property of matter that represents its resistance to any change in its state of rest or motion.

**Reaction force** The equal and opposite force that, according to Newton's third law, results when any force is applied to an object.

**Equilibrium** A condition in which all forces or tendencies are exactly counterbalanced or neutralized by equal and opposite forces and tendencies. Translational equilibrium exists when the sum of all forces acting on a body is zero.

**Equilibrant** A force vector that is equal in magnitude to the resultant force but opposite in direction.

**Free-body diagram** A vector diagram depicting every force that acts on a body and labeling the components of each force opposite and adjacent to known angles.

**Friction force** Resistive forces between two surfaces in contact that oppose actual or impending motion. The force of static friction exists when motion is impending but no sliding occurs. The force of kinetic friction exists between two surfaces in relative motion.

**Coefficient of friction** The ratio of the magnitude of a friction force to the normal force between two surfaces when motion is actual or impending.

**Normal force** A supporting force exerted perpendicular to the plane of motion or support.

**Angle of repose** The angle of inclination for which static friction is just overcome. The angle whose tangent is equal to the coefficient of static friction.

## CONCEPTS AND EXAMPLES

**Equilibrium** The first step in any equilibrium problem is to read the problem and draw a rough sketch to help understand the forces involved. Then, from the sketch, draw a free-body diagram, being careful to label all forces and their components. Letters should be assigned to unknown forces, and all given information should be placed on the diagram. Two equations can be written for translational equilibrium:

$$\sum F_x = 0 \qquad \sum F_y = 0$$

Solving these equations for the unknown forces usually provides the information required by the problem.

**EXAMPLE 4-1**    Determine the force $A$ exerted by the rope and the force $B$ of compression in the light boom for Fig. 4-1.

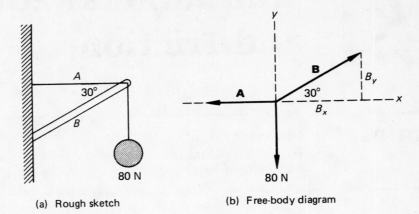

(a) Rough sketch          (b) Free-body diagram

**Fig. 4-1**

*Solution*
The free-body diagram for this situation is shown in Fig. 4-1b. We neglect the weight of the boom. Note that force **B** is directed up and outward instead of along the structure of the boom. The reason is that we must have a system of concurrent forces, each acting *on* the same object (in this case, the end of the boom). Applying the first condition for equilibrium gives

$$F_x = B \cos 30° - A = 0 \qquad \text{or} \quad A = B \cos 30°$$
$$F_y = B \sin 30° - 80 \text{ N} = 0 \quad \text{or} \quad B \sin 30° = 80 \text{ N}$$

Solving the second equation for $B$ gives  **160 N** ; then substitution into the first equation gives  **139 N**  for $A$.

**Friction**    There are two kinds of friction forces. The static friction force exists when motion is impending, and the kinetic friction force occurs during relative motion. Each is proportional to the normal force and is given by

$$\mathscr{F}_s = \mu_s \mathscr{N} \qquad \mathscr{F}_k = \mu_k \mathscr{N}$$

In friction problems, usually the $x$ axis should be chosen along the plane of motion. This places the normal force and the friction force perpendicular to each other. If the system is at rest or if an object moves with constant speed, the first condition for equilibrium is satisfied and the sum of forces along either axis must be zero. In the rest condition, motion may be impending and static friction applies. Kinetic friction is operative when motion is occurring.

**EXAMPLE 4-2**    What push **P** up the plane is necessary just to hold the 40-lb block on the 50° incline of Fig. 4-2?

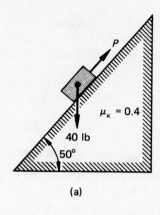
(a)

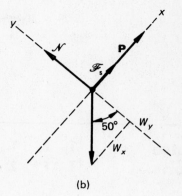
(b)

**Fig. 4-2**

*Solution*

From the free-body diagram (Fig. 4-2b), we obtain two equations by summing forces along the plane and perpendicular to the plane. Note that the friction force is directed up the plane since motion is impending down the plane.

$$\sum F_x = P + \mathscr{F}_s - W_x = 0 \qquad \text{or} \qquad P = W_x - \mu_s \mathscr{N}$$
$$\sum F_y = \mathscr{N} - W_y = 0 \qquad \text{or} \qquad \mathscr{N} = W_y = 40 \cos 50°$$

Thus, $\mathscr{N} = 25.7$ lb, and the friction force is

$$\mathscr{F}_s = \mu_s \mathscr{N} = (0.4)(25.7 \text{ lb}) \qquad \text{or} \qquad \mathscr{F}_s = 10.3 \text{ lb}$$

It is also seen from Fig. 4-2 that $W_x = 40 \sin 50° = 30.6$ lb. Then, we obtain $P$ by substituting into the first equation:

$$P = W_x - \mathscr{F}_s = 30.6 \text{ lb} - 10.3 \text{ lb} \qquad \text{or} \qquad \boxed{P = 20.3 \text{ lb}}$$

Therefore, a push of 20.3 lb up the plane will just stop the block from sliding down the plane.

## Summary

- Newton's first law of motion states that an object at rest and an object in motion with constant speed will maintain the state of rest or constant motion unless acted on by a resultant force.

- Newton's third law of motion states that every action must produce an equal and opposite reaction. The action and reaction forces do not act on the same body.

- From the conditions of the problem, a neat sketch is drawn and all known quantities are labeled. Then a force diagram indicating all forces and their components is constructed. This is a free-body diagram.

- A body in translational equilibrium has no resultant force acting on it.

- Static friction exists between two surfaces when motion is impending; kinetic friction occurs when the two surfaces are in relative motion.

## Key Equations by Section

| Section | Topic | Equation | Equation No. |
|---|---|---|---|
| 4-3 | Equilibrium | $\sum F_x = 0 \quad \sum F_y = 0$ | 4-1 |
| 4-6 | Coefficient of static friction | $\mathscr{F}_s = \mu_s \mathcal{N}$ | 4-10 |
| 4-6 | Coefficient of kinetic friction | $\mathscr{F}_k = \mu_k \mathcal{N}$ | 4-11 |

## True-False Questions

T  F  **1.** A moving body cannot be in equilibrium.

T  F  **2.** In equilibrium problems, the normal force will always equal the weight of a body.

T  F  **3.** According to Newton's first law of motion, every action force has an equal but opposite reaction force.

T  F  **4.** Action and reaction forces do not neutralize each other because they act on different objects.

T  F  **5.** Newton's first law of motion is also referred to as the *law of inertia.*

T  F  **6.** Under a condition of equilibrium, any one of the forces acting on an object is the equilibrant of the vector sum of all the other forces.

T  F  **7.** Static friction does not exist until motion is impending.

T  F  **8.** The maximum force of static friction is directly proportional to the weight of a body.

T  F  **9.** The parallel force required to move an object over a surface at constant speed is equal to the magnitude of the force of kinetic friction.

T  F  **10.** The higher the coefficient of friction between two surfaces, the higher will be the angle of repose for those surfaces.

## Multiple-Choice Questions

**1.** If two vertical ropes are used to support a 20-N weight, the tension in each rope must be
(a) 20 N        (b) 40 N      (c) 10 N        (d) 30 N

**2.** A rope attached to a post in the ground is pulled horizontally with a force of 80 lb. The pole pulls back with a force of
(a) 40 lb        (b) 80 lb      (c) 100 lb        (d) 160 lb

**3.** Which of the following cannot apply for a body in translational equilibrium?
(a) $\sum F_x = 0$                (b) Constant speed
(c) Increasing speed              (d) $\sum F_y = 0$

**4.** Find the tension in the cable for the arrangement in Fig. 4-3. Neglect the weight of the board.

(a) 72.1 N      (b) 67.3 N      (c) 45.0 N      (d) 37.2 N

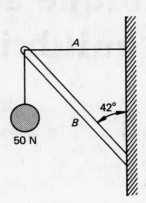

**Fig. 4-3**

**5.** A 40-N block is supported by two ropes. One rope is horizontal, and the other makes an angle of 30° with the ceiling. The tension in the rope attached to the ceiling is approximately

(a) 80 N      (b) 34.6 N      (c) 40 N      (d) 46.2 N

**6.** A 200-lb block is at rest on a 37° inclined plane. The normal force exerted by the plane is approximately

(a) 120 lb      (b) 200 lb      (c) 160 lb      (d) 100 lb

**7.** A 60-lb block rests on a smooth horizontal surface. The horizontal force that just starts the block moving is found to be 30 lb. Which of the following is true for this situation?

(a) $\mu_k = 0.5$      (b) $\mu_s = 0.5$   (c) $\mu_s = 0.18$   (d) $F_k = 30$ lb

**8.** A 40-N block slides down a board with constant speed when the angle of elevation is 30°. Which of the following is *not* true for this situation?

(a) $\tan 30° = \mu_k$         (b) $F_k = 20$ N
(c) $F_k = 34.6$ N          (d) $\mu_k = 0.577$

**9.** It is easier to pull a sled at an angle than it is to push a sled at an angle because of a

(a) lower normal force        (b) lower weight
(c) higher normal force      (d) mechanical advantage

**10.** If $\mu_k = 0.1$, the force required to push a 20-N sled up a 40° incline with constant speed is approximately

(a) 26.1 N      (b) 14.4 N      (c) 15.3 N      (d) 31.1 N

# Chapter 5

# Torque and Rotational Equilibrium

**Moment arm**   The perpendicular distance from the line of action of a force to the axis of rotation.

**Axis of rotation**   A chosen point about which forces tend to cause rotation. Once the axis is chosen, the sum of torques about the axis will be zero for rotational equilibrium.

**Torque**   The product of a force and its moment arm. It is a vector quantity that is negative for clockwise tendencies and positive for counterclockwise tendencies.

**Rotational equilibrium**   An object is said to be in rotational equilibrium when the sum of all the torques about a chosen axis is zero.

**Center of gravity**   The point at which all the weight of an object may be considered to act without changing the resultant torque on the object. For regularly shaped objects of uniform density, this point is at the geometrical center.

**Line of action**   An imaginary line of a force extended indefinitely along the vector in both directions.

**Resultant Torque**   To find the resultant torque, we draw a free-body diagram and label all known forces and distances. By extending the line of action of each force, we can determine the moment arms as the perpendicular distance from a line of action of the axis of rotation. The torque due to each force is calculated and given a positive sign for counterclockwise rotation or a negative sign for clockwise rotation. The resultant torque is the algebraic sum of these torques.

**EXAMPLE 5-1**

Determine the resultant torque about axis *A* in Fig. 5-1.

*Solution*
The lines of action of each force are drawn as shown in Fig. 5-1b, and the torque contributed by each force is calculated and given a positive or negative sign. We see from the figure that $\tau_1$ will be positive and that $\tau_2$ will be negative. The moment arm for $F_3$ is zero, and the other two are found as follows:

$$r_1 = (6 \text{ m})(\sin 50°) = 4.596 \text{ m}$$
$$r_2 = (4 \text{ m})(\sin 30°) = 2.00 \text{ m}$$

Since there is no torque due to $F_3$, the resultant torque is

$$\tau_R = \tau_1 + \tau_2 = (50 \text{ N})(4.596 \text{ m}) - (80 \text{ N})(2.0 \text{ m})$$
$$= 229.8 \text{ N} \cdot \text{m} - 160 \text{ N} \cdot \text{m} = \boxed{+69.8 \text{ N} \cdot \text{m}}$$

The resultant is 69.8 N · m counterclockwise.

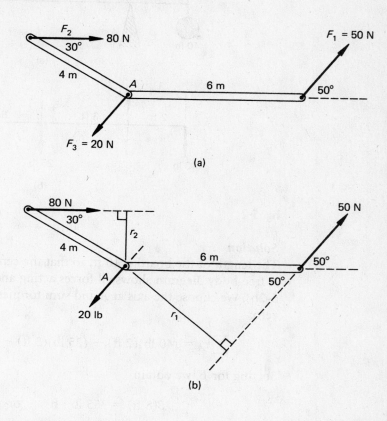

**Fig. 5-1**

## Rotational Equilibrium

If a system is in total equilibrium, the sum of all forces and the sum of all torques must be zero. Thus, three equations may be written for that system:

$$\sum F_x = 0 \qquad \sum F_y = 0 \qquad \sum \tau_A = 0$$

In summing torques, we have the advantage of choosing the axis anywhere we wish. Since there is no resultant torque, the sum of all torques will be zero about *any* axis. It makes sense, however, to choose an axis at the point of application of an unknown force. The zero moment arm of this force will eliminate it from the resulting sum of torques.

**EXAMPLE 5-2**    A 25-lb board is laid across two supports as shown in Fig. 5-2. Determine the upward forces exerted by the supporters $A$ and $B$.

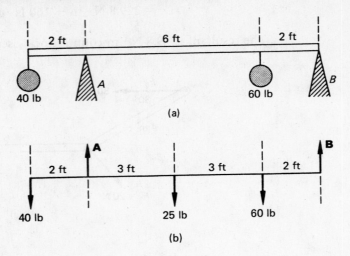

(a)

(b)

**Fig. 5-2**

*Solution*
The length of the board is 10 ft, so that the center of gravity is 5 ft from each end. A free-body diagram shows all forces acting and the distances between forces (Fig. 5-2b). We choose the axis at $A$ and sum torques about that axis. The result is set to zero:

$$\sum \tau_A = (40 \text{ lb})(2 \text{ ft}) - (25 \text{ lb})(3 \text{ ft}) - (60 \text{ lb})(6 \text{ ft}) + B(8 \text{ ft}) = 0$$

Solving for $B$, we obtain

$$B(8 \text{ ft}) = 355 \text{ lb} \cdot \text{ft} \qquad \text{or} \qquad \boxed{B = 44.4 \text{ lb}}$$

Summing vertical forces and setting to zero give

$$A + B - 40 \text{ lb} - 25 \text{ lb} - 60 \text{ lb} = 0 \qquad \text{or} \qquad A + B = 125 \text{ lb}$$

The force $A$ is now found by substitution:

$$A = 125 \text{ lb} - 44.4 \text{ lb} \qquad \text{or} \qquad \boxed{A = 80.6 \text{ lb}}$$

As a check, you could sum the torques about $B$ to verify the value found for $A$.

**Summary**

- The moment arm of a force is the perpendicular distance from the line of action of the force to the axis of rotation.
- The torque about a given axis is defined as the product of the magnitude of a force and its moment arm.
- The resultant torque $\tau_R$ about a particular axis $A$ is the algebraic sum of the torques produced by each force.

- A body in rotational equilibrium has no resultant torque acting on it. In such cases, the sum of all the torques about any axis must equal zero. The axis may be chosen anywhere because the system is not tending to rotate about any point.
- Total equilibrium exists when a body has translational equilibrium and rotational equilibrium.
- The center of gravity of a body is the point through which the resultant weight acts, regardless of how the body is oriented. For applications involving torque, the entire weight of the object may be considered as acting at this point.

## Key Equations by Section

| Section | Topic | Equation | Equation No. |
|---|---|---|---|
| 5-3 | Torque | $\tau = Fr$ | 5-1 |
| 5-4 | Resultant torque | $\tau_R = \sum \tau = \tau_1 + \tau_2 + \tau_3 + \cdots$ | 5-2 |
| 5-5 | Translational equilibrium | $\sum F_x = 0 \qquad \sum F_y = 0$ | 5-3 |
| 5-5 | Rotational equilibrium | $\sum \tau = \tau_1 + \tau_2 + \tau_3 + \cdots = 0$ | 5-4 |

## True-False Questions

T   F   **1.** When considering only concurrent forces, we do not need to calculate torques.

T   F   **2.** If an object is suspended from the ceiling by a rope attached to its center of gravity, the system will attain equilibrium.

T   F   **3.** The moment arm is the distance from the point of application of a force to the axis of rotation.

T   F   **4.** When a force tends to cause clockwise rotation, it produces a positive torque by convention.

T   F   **5.** In applying the second condition for equilibrium, the axis of rotation may be chosen for convenience.

T   F   **6.** The center of gravity of a body is always at the geometrical center of the body.

T   F   **7.** In applying the second condition for equilibrium, it is acceptable to equate the clockwise torques about any axis to the counterclockwise torques about the same axis.

T   F   **8.** In computing the resultant torque, the axis of rotation may be chosen for convenience.

T   F   **9.** The unit of torque in the metric system is the newton-meter.

T   F   **10.** The net torque will be the same if a force is doubled but moved to a point halfway to the axis of rotation.

## Multiple-Choice Questions

1. To apply the second condition for equilibrium, the axis of rotation
   (a) must be at the center of the body
   (b) must be at one end
   (c) may be chosen anywhere
   (d) must be at the center of gravity

**2.** To compute torque, one must know
    (a) the force magnitude        (b) the moment arm
    (c) both of these                (d) neither of these

**3.** A rope is wrapped in a clockwise direction around a 12-in.-diameter axle. If a force of 20 lb is exerted on the rope, what is the resultant torque?
    (a) $-20$ lb · ft    (b) $+20$ lb · ft    (c) $-10$ lb · ft    (d) $10$ lb · ft

**4.** A 60-lb boy sits on the left end of a 12-ft seesaw, and a 40-lb boy sits on the right end. The resultant torque about the center is approximately
    (a) $-120$ lb · ft    (b) $+20$ lb · ft    (c) $+120$ lb · ft    (d) $240$ lb · ft

**5.** Which of the following is not a unit of torque?
    (a) Pound-foot
    (b) Kilogram-newton
    (c) Newton-meter
    (d) Pound-inch

**6.** If the resultant of a set of concurrent forces is zero, the sum of the torques
    (a) is also zero                (b) may or may not be zero
    (c) is not zero               (d) depends on the axis of rotation

**7.** A 20-lb uniform board is 16 ft long and is supported at a point 6 ft from the right end. The upward force that must be exerted at the left end to balance the system is approximately
    (a) 4 lb         (b) 2 lb         (c) 3.6 lb        (d) 5 lb

**8.** A rope is wrapped around a 16-in.-diameter drum in a counterclockwise direction. What force must be exerted on the rope to give a torque of 120 lb · ft?
    (a) 90 lb        (b) 75 lb        (c) 200 lb       (d) 180 lb

**9.** A girl and a boy carry a sack weighing 40 N on a pole between them. If the pole is 6 m long and the load is 2 m from the girl, what force does the boy exert?
    (a) 13.3 N      (b) 26.7 N      (c) 40 N       (d) 80 N

**10.** The boom in Fig. 5-3 is 2 m long and weighs 100 N. If the 800-N load is suspended 50 cm from the end, the cable tension is approximately
    (a) 310 N      (b) 352 N      (c) 424 N      (d) 512 N

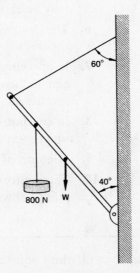

**Fig. 5-3** Find the cable tension.

**Completion Questions**

1. The _____ distance from the line of action of a force to the point of rotation is known as the _____.

2. The _____ of a body is that point at which all the weight of the body may be considered to be concentrated.

3. The second condition for equilibrium states that the sum of all the _____ acting on an object must be _____.

4. By convention, torques that tend to cause clockwise motion are considered _____, and those that tend to cause counterclockwise motion are considered _____.

5. If the line of action of a force passes through the axis of rotation, the moment arm is _____.

6. In problems involving nonconcurrent forces, both the resultant _____ and the resultant _____ must be zero.

7. The unit for moment of force in the SI system is the _____.

8. The center of gravity of a regular solid of uniform density, such as a sphere or cube, is located at _____.

9. An alternate way of computing the torque due to a single force is to resolve the force into parallel and perpendicular components and _____.

10. Three tools that utilize the principle of torque are _____, _____, and _____.

# Chapter 6

# Uniform Acceleration

**DEFINITIONS OF KEY TERMS**

**Constant speed**   An object that covers the same distances in each successive interval of time is said to move with constant speed. The magnitude is the ratio of distance to time.

**Average speed**   The average speed for a specific distance is calculated by dividing that distance by the time elapsed.

**Instantaneous speed**   A scalar quantity representing the speed at the instant an object is at a specific point.

**Velocity**   A vector quantity denoting the time rate of change in displacement. A statement of velocity includes a scalar magnitude, expressed in units of length divided by time, and a direction relative to some point of reference.

**Instantaneous velocity**   The velocity of an object at a particular instant and at a particular point in its path.

**Acceleration**   A vector quantity representing the time rate of change in velocity. The scalar magnitude has units of length per unit of time squared.

**Uniform acceleration**   Motion in a straight line in which the speed changes at a constant rate. Since there is no change in direction, such constant acceleration can be computed by a scalar equation that determines the change in velocity per unit of time.

**Acceleration due to gravity**   The acceleration with which an object falls at a specific location in a gravitational field. At sea level on the earth, the value is 9.8 m/s$^2$ (32.17 ft/s$^2$).

**Projectile**   An object that is given an initial velocity and then allowed to move freely under the influence of gravity.

**Trajectory**   The path followed by a projectile.

**Range**   The horizontal distance traveled by a projectile from the point of release to the point of final impact.

## Velocity and Acceleration

Definitions of these terms lead to four equations involving the five parameters: $a$, $v_f$, $v_0$, $s$, and $t$. By determining three given quantities, the others may be determined from one of the four equations listed:

$$(1) \quad s = \frac{v_0 + v_f}{2} t$$

$$(2) \quad v_f = v_0 + at$$

$$(3) \quad s = v_0 t + \tfrac{1}{2} at^2$$

$$(4) \quad 2as = v_f^2 - v_0^2$$

Draw a figure, choose a positive direction, list three givens and two unknowns, select an equation, and solve for the unknown.

**EXAMPLE 6-1**

A car traveling with a speed of 20 m/s slows to a speed of 5 m/s in a distance of 30 m. What was the acceleration?

*Solution*

We draw a sketch and determine the givens first.

Given: $v_0 = 20$ m/s

$v_f = 5$ m/s

$s = 30$ m

Find $a$ and $t$. The fourth equation is used because it contains $a$ and not $t$:

$$2as = v_f^2 - v_0^2$$

$$a = \frac{v_f^2 - v_0^2}{2s} = \frac{(5 \text{ m/s})^2 - (20 \text{ m/s})^2}{2(30 \text{ m})}$$

$a = -6.25 \text{ m/s}^2$     (directed opposite to initial speed)

## Acceleration Due to Gravity

The only force acting on a body that is falling freely above the earth is its weight. Near the surface of the earth, all objects fall with constant acceleration g equal approximately to 9.8 m/s² (32 ft/s²). The choice of a positive direction and appropriate use of positive and negative signs are essential for solving gravity problems. Remember:

1. Velocity $v$ is plus or minus depending on the direction of motion.
2. The displacement $s$ is plus or minus dependent on position (right, left, up, down) with respect to the starting position.
3. Acceleration $a$ is plus or minus depending on the direction of the force that causes the velocity to change. The acceleration due to gravity is always downward because the weight is directed downward. If up is positive, then g must be negative.

Neither the sign of position $s$ nor the sign of acceleration $a$ is determined from the direction in which an object is moving. Only the signs of the initial and final velocities are a function of the direction of motion.

**Horizontal Projection**  The key to all projectile problems is to recognize that the horizontal and vertical motion can be treated separately. The vertical motion of an object projected horizontally is identical to that of an object dropped from rest. The horizontal motion is the same as that of an object traveling at a constant speed. You should review the equations listed in the chapter summary.

---

**EXAMPLE 6-2**  An object is projected horizontally with an initial speed of 23 m/s from a position 6 m from the earth. What are the range and the time to strike the ground?

*Solution*
The range is determined by the horizontal speed and the time, so first we need to know the time in the air:

$$s = \tfrac{1}{2}at^2 \qquad \text{or} \qquad t^2 = \frac{2s}{a}$$

Thus,

$$t = \sqrt{\frac{2(-6 \text{ m})}{(-9.8 \text{ m/s}^2)}} = \boxed{1.107 \text{ s}}$$

Now, we can find the horizontal range:

$$x = v_x t = (23 \text{ m/s})(1.107 \text{ s}) = \boxed{25.5 \text{ m}}$$

---

**General Trajectories**  The projection of an object at an angle $\theta$ is handled by working with horizontal and vertical components separately. It is the vertical component of the initial velocity that determined vertical positions and velocities; the horizontal component affects horizontal parameters.

---

**EXAMPLE 6-3**  An arrow is fired with an initial speed of 40 m/s at an angle of 52°. Find the position and velocity 2 *s* later.

*Solution*
First, we find the horizontal and vertical components of the initial velocity:

$$v_{0x} = (40 \text{ m/s})(\cos 52°) = 24.6 \text{ m/s}$$
$$v_{0y} = (40 \text{ m/s})(\sin 52°) = 31.5 \text{ m/s}$$

Now, the horizontal and vertical position $(x,y)$ is found:

$$x = (24.6 \text{ m/s})(2 \text{ s}) = \boxed{49.3 \text{ m}}$$
$$y = (31.5 \text{ m/s})(2 \text{ s}) + \tfrac{1}{2}(-9.8 \text{ m/s}^2)(2 \text{ s})^2 = \boxed{43.4 \text{ m}}$$

The horizontal component of the final velocity is 24.6 m/s. The vertical component is given by

$$v_{fy} = 31.5 \text{ m/s} + (-9.8 \text{ m/s}^2)(2 \text{ s}) = \boxed{11.9 \text{ m/s}}$$

From the $x$ and $y$ components of the final velocity, we can show that the final velocity is 27.3 m/s at 25.8°.

## Summary

- Average velocity is the distance traveled per unit of time, and average acceleration is the change in velocity per unit of time.
- To solve acceleration problems, read the problem with a view to establishing the three given parameters and the two that are unknown.
- Problems involving gravitational acceleration can be solved like other acceleration problems.
- The key to problems involving projectile motion is to treat the horizontal motion and the vertical motion separately.
- In applying the equations, remember to be consistent throughout with units and sign conversion.

## Key Equations by Section

| Section | Topic | Equation | Equation No. |
|---|---|---|---|
| 6-1 | Average speed of a moving object | $Average\ speed = \dfrac{distance\ traveled}{time\ elapsed}$ $\boxed{\bar{v} = \dfrac{s}{t}}$ | 6-1 |
| 6-2 | Acceleration | $Acceleration = \dfrac{change\ in\ velocity}{time\ interval}$ $\mathbf{a} = \dfrac{\mathbf{v}_f - \mathbf{v}_0}{t}$ | 6-2 |
| 6-3 | Constant acceleration | $a = \dfrac{v_f - v_0}{t}$ | 6-3 |
| 6-3 | Final velocity | $Final\ velocity = initial\ velocity + change\ in\ velocity$ $v_f = v_0 + at$ | 6-4 |
| 6-3 | Average velocity | $\bar{v} = \dfrac{v_f + v_0}{2}$ | 6-5 |
| 6-3 | Distance traveled | $\boxed{s = \bar{v}t = \dfrac{v_f + v_0}{2}t}$ | 6-6 |
| 6-4 | Distance traveled in terms of initial velocity, time, and acceleration | $s = v_0 t + \frac{1}{2}at^2$ | 6-7 |
| 6-4 | Distance traveled in terms of final velocity, time, and acceleration | $s = v_f t - \frac{1}{2}at^2$ | 6-8 |
| 6-4 | Distance traveled in terms of acceleration, final velocity, and initial velocity | $2as = v_f^2 - v_0^2$ | 6-9 |

T  F  **1.**  An object in motion can have a constant velocity only if it moves in a straight path.

T  F  **2.**  An object falling freely from rest near the surface of the earth falls a distance of 32 ft by the end of the first second.

T  F  **3.**  In a vacuum, all bodies fall with the same velocity.

T  F  **4.**  If an object's velocity is decreasing constantly, it will always have a negative acceleration.

T  F  **5.**  An object thrown downward in a gravitational field has the same acceleration as one dropped from rest.

T  F  **6.**  If an object has an acceleration of 8 m/s$^2$, its distance will increase by 8 m every second.

T  F  **7.**  For a ball thrown vertically upward, its upward motion with respect to position and velocity is just the reverse of its downward motion.

T  F  **8.**  The velocity and position of a free-falling body after 2 s are numerically the same.

T  F  **9.**  In the absence of friction, all bodies, large or small, heavy or light, fall to the earth with the same acceleration.

T  F  **10.**  If any two of the parameters $v_0$, $v_f$, $a$, $s$, and $t$ are given, the other three can be calculated from derived equations.

T  F  **11.**  The motion of a projectile fired at an angle is an example of uniformly accelerated motion.

T  F  **12.**  The resultant force acting on a projectile is its weight.

T  F  **13.**  A projectile fired horizontally will strike the ground in the same time as one dropped vertically from the same position if we neglect the effect of air resistance.

T  F  **14.**  A projectile launched into space at any angle will have a constant horizontal velocity.

T  F  **15.**  The horizontal range is greatest when the angle of projection is 45°.

T  F  **16.**  The vertical motion of a projectile is uniformly accelerated motion.

T  F  **17.**  The range of a projectile depends only on its initial speed.

T  F  **18.**  A horizontally fired projectile will drop 32 ft during the first second.

T  F  **19.**  If an apple drops from a tree at the same instant a projectile is fired toward it from the ground, they will still collide in the air.

T  F  **20.**  When a projectile is fired at a 45° angle, its maximum height will equal its range.

**Multiple-Choice Questions**

**1.**  An object falls freely from rest. Its position after 2 s is how far below the point of release?

(a) 32 ft      (b) 64 ft      (c) 96 ft      (d) 48 ft

**2.**  If the initial velocity, the distance traveled, and the time elapsed are known, which equation would you use to calculate the acceleration?

(a) $s = vt$ 　　　　　　　(b) $s = v_0 t + \frac{1}{2}at^2$

(c) $v_f = v_0 + at$ 　　　　　(d) $2as = v_f - v_0$

3. The algebraic sign of acceleration depends on
   (a) the direction on the force of the accelerated object
   (b) whether an object is speeding up or slowing down
   (c) the sign of the final velocity
   (d) the position of the object

4. An object traveling initially at 24 ft/s slows to 12 ft/s in 3 s. Its acceleration is
   (a) 4 ft/s$^2$    (b) −4 ft/s    (c) −4 ft/s$^2$    (d) −8 ft/s$^2$

5. The distance traveled by the object in Question 4 is
   (a) 48 ft    (b) 66 ft    (c) 98 ft    (d) 54 ft

6. An object is thrown downward with an initial velocity of 32 ft/s. Its velocity after 3 s is
   (a) 102 ft/s    (b) 96 ft/s    (c) 80 ft/s    (d) 128 ft/s

7. A car accelerates from rest at 4 m/s$^2$. How far will it travel in 4 s?
   (a) 32 m    (b) 19.6 m    (c) 78.4 m    (d) 94.5 m

8. An arrow is shot vertically upward with an initial velocity of 96 ft/s. It will first come to a stop in
   (a) 4 s    (b) 2 s    (c) 3 s    (d) 6 s

9. An object is projected upward with an initial velocity of 64 ft/s. What will be its position above the point of release after 3 s?
   (a) 48 ft    (b) 16 ft    (c) 32 ft    (d) 64 ft

10. A car accelerates for 10 s at 6 m/s$^2$. What is its final velocity if its initial velocity was 4 m/s?
    (a) 60 m/s    (b) 64 m/s    (c) 34 m/s    (d) 30 m/s

11. The acceleration of a projectile is
    (a) $g$                    (c) 0
    (b) $-g$                    (d) dependent on its initial velocity

12. A projectile is fired horizontally with an initial velocity of 20 m/s. Its horizontal velocity 3 s later is
    (a) 20 m/s    (b) 60 m/s    (c) 6.67 m/s    (d) 29.4 m/s

13. The vertical velocity of the projectile in Question 12 after 3 s is approximately
    (a) 60 m/s    (b) 9.8 m/s    (c) 29.4 m/s    (d) 20 m/s

14. A cannonball is projected horizontally with a velocity of 1200 ft/s from the top of a cliff 128 ft high. It will strike the water below in approximately
    (a) 8 s    (b) 2.83 s    (c) 0.1 s    (d) 9.38 s

15. In Question 14, the horizontal range is approximately
    (a) 3396 ft    (b) 1200 ft    (c) 938 ft    (d) 9600 ft

16. A projectile is fired at an angle of 30° with an initial velocity of 640 ft/s. The time to reach its maximum height is approximately
    (a) 17.3 s    (b) 20 s    (c) 5 s    (d) 10 s

17. In Question 16, the horizontal range is
    (a) 5542 ft    (b) 11,084 ft    (c) 3200 ft    (d) 6400 ft

18. A projectile is fired at an angle of 37° with an initial velocity of 100 m/s. What is the approximate vertical component of its velocity after 2 s?
    (a) 60 m/s    (b) 40 m/s    (c) 80 m/s    (d) 100 m/s

19. In Question 18, the position above the ground after 3 s is approximately
    (a) 200 m    (b) 140 m    (c) 136 m    (d) 120 m

20. Which of the following projection angles will result in the greatest range?
    (a) 20°    (b) 37°    (c) 48°    (d) 60°

## Completion Questions

1. The total distance traveled divided by the time elapsed is a measure of
_____.

2. _____ is motion in a straight line in which the speed changes
at a constant rate.

3. At least _____ of the following parameters must be known to
find the other two: _____, _____,
_____, _____, and
_____.

4. The final velocity is equal to the initial velocity plus _____.

5. In the absence of friction, all objects fall to the earth with the same
_____, independent of size or weight.

6. A body that has a continually increasing negative velocity has a(n)
_____ acceleration.

7. The acceleration due to gravity near the earth is 32 or _____
$m/s^2$.

8. If the upward direction is chosen as positive, a negative distance $s$ indicates that
the final position is _____.

9. The distances traveled by an object dropped from rest after 1, 2, and 3 s are
_____ ft, _____ ft, and
_____ ft, respectively.

10. An acceleration of 4 $ft/s^2$ means that every _____ the velocity
increases by _____.

11. A projectile is an object launched into space under the influence of
_____ only.

12. In working with trajectories, it is easier to treat the _____ and
the _____ motions separately.

13. The vertical component of the velocity of a projectile as a function of time is
calculated from _____.

14. The vertical position of a projectile as a function of time is given by
_____.

15. The range of a projectile can be calculated from _____.

16. Projectiles fired upward, downward, or at an angle all have the same
_____.

17. The only force acting on a projectile is its _____.

**18.** The maximum height of a projectile may be found by dividing the

_____ component of the initial _____ by

the _____.

**19.** For projectile motion, the _____ component of the

_____ is constant.

**20.** The vertical position of a projectile fired horizontally can be calculated from

_____.

# Chapter 7

# Newton's Second Law

**DEFINITIONS OF KEY TERMS**

**Newton's second law of motion**    When a resultant force acts on a body, it produces an acceleration in the direction of the force that is directly proportional to the force and inversely proportional to the mass of the body.

**Mass**    A physical measure of the inertia of a body. It is the constant ratio of an applied force to the acceleration it produces. In a gravitational field, the mass of an object is the constant ratio of its weight to the gravitational acceleration.

**Weight**    The gravitational force with which a body is attracted to the earth. Its magnitude is equal to the product of the mass of the body and the acceleration due to gravity. Weight, therefore, varies with the acceleration due to gravity.

**Newton**    The SI unit for force that is equal in magnitude to the force requires to give a 1-kg mass an acceleration of 1 m/s$^2$.

**Slug**    An outdated unit of mass still used in the United States; it represents the mass to which a force of 1 lb will impart an acceleration of 1 ft/s$^2$.

**CONCEPTS AND EXAMPLES**

**Single-Body Problems**    In all applications of Newton's second law, it is important not to confuse mass with weight. Remember that weight is a vector quantity that has direction, and mass is a scalar measure of a body's inertia. Weight is gravity-dependent and is always measured in newtons (or pounds). Mass is a universal constant that is independent of location. The SI unit of mass is the kilogram (kg). In U.S. Customary System (USCS) units, the slug is the mass unit and the pound (lb) is the weight unit.

---

**EXAMPLE 7-1**

What is the weight of a 40-kg mass? What is the mass of a 60-lb ball of cord?

*Solution*
Recall from Sec. 7-3 in the textbook that weight is given by $W = mg$ and that mass is found from $m = W/g$. Substitution gives

$$W = mg = (40 \text{ kg})(9.8 \text{ m/s}^2) = \boxed{392 \text{ N}}$$

and

$$m = \frac{W}{g} = \frac{60 \text{ lb}}{32 \text{ ft/s}^2} = \boxed{1.88 \text{ slugs}}$$

---

Remember that objects are usually described by their mass (kilograms) in the metric system and by their weight (pounds) in USCS units. You can tell whether a number represents mass or weight by looking at the unit.

You should review the problem-solving strategy in Sec. 7-4 of the text. Draw detailed free-body diagrams, choose a positive axis along the direction of motion, and then apply Newton's second law.

---

**EXAMPLE 7-2**

A 4-kg block is dragged by a horizontal pull **P** against a friction force of 12 N. What pull **P** is required to produce an acceleration of 5 m/s²?

*Solution*

The resultant force on the block is the difference between the pull **P** and the 12-N drag of friction. Applying Newton's second law, we obtain

$$\text{Resultant force} = \text{mass} \times \text{acceleration}$$
$$P - 12 \text{ N} = (4 \text{ kg})(5 \text{ m/s}^2)$$
$$P - 12 \text{ N} = 20 \text{ N} \qquad \text{or} \qquad \boxed{P = 32 \text{ N}}$$

Working a similar problem in USCS units would be different because of the description given the object. For example, you might be asked what pull **P** is required to drag a 64-lb block against a 12-lb friction force resulting in an acceleration of 5 ft/s². You must recognize that 64 lb is the *weight* of the block and that the mass must be determined as 64/32, or 2 slugs. As an additional exercise you should show that $P = 22$ lb.

---

## Applications to Systems of Bodies

When a resultant force accelerates two or more bodies, you must consider the unbalanced force on the entire system and set it equal to the mass of the entire system times the acceleration of the system.

---

**EXAMPLE 7-3**

Determine the acceleration of the system and the tension in the cord for Fig. 7-1.

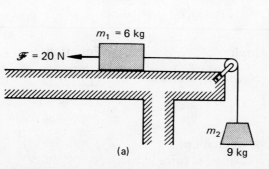
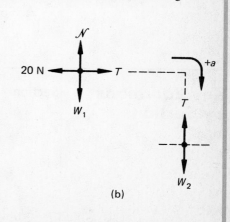

**Fig. 7-1**

*Solution*

First, draw a free-body diagram for each body in the system, as shown in Fig. 7-1b. Note that the direction of positive acceleration is consistent and labeled on both diagrams. The masses are given, and the weights are calculated:

$$m_1 = 6 \text{ kg} \qquad W_1 = (6 \text{ kg})(9.8 \text{ m/s}^2) = 58.8 \text{ N}$$
$$m_2 = 9 \text{ kg} \qquad W_2 = (9 \text{ kg})(9.8 \text{ m/s}^2) = 88.2 \text{ N}$$

Now, the resultant force on the entire system is the difference between the weight $W_2$ and the 20-N friction force:

$$\text{Resultant force} = \text{total mass} \times \text{acceleration}$$
$$88.2 \text{ N} - 20 \text{ N} = (6 \text{ kg} + 9 \text{ kg})(a)$$

So

$$a = \frac{68.2 \text{ N}}{15 \text{ kg}} \qquad \text{or} \qquad \boxed{a = 4.55 \text{ m/s}^2}$$

Now the tension in the cord must be found by looking at either body independently. Let's look at the 6-kg mass:

$$\text{Resultant force} = \text{mass} \times \text{acceleration}$$
$$T - 20 \text{ N} = (6 \text{ kg})(4.55 \text{ m/s}^2)$$

So

$$T - 20 \text{ N} = 27.3 \text{ N} \qquad \text{or} \qquad \boxed{T = 47.3 \text{ N}}$$

## Summary

- Application of Newton's second law of motion
  **a.** Construct a free-body diagram for each body undergoing an acceleration. Indicate on this diagram the direction of positive acceleration.
  **b.** Determine an expression for the net resultant force on a body or a system of bodies.
  **c.** Set the resultant force equal to the total mass of the system multiplied by the acceleration of the system.
  **d.** Solve the resulting equation for the unknown quantity.

## Key Equations by Section

| Section | Topic | Equation | Equation No. |
|---------|-------|----------|--------------|
| 7-1 | Newton's second law | $F = ma$<br>*Resultant force =<br>mass × acceleration* | 7-1 |
| 7-1 | A resultant or unbalanced force | $\sum F_x = ma_x$ | 7-2 |

| Section | Topic | Equation | Equation No. |
|---------|-------|----------|--------------|
| 7-2 | Relationship between a body's weight and its mass | $W = mg \quad \text{or} \quad m = \dfrac{W}{g}$ | 7-3 |
| 7-3 | Relationship for constant force and no acceleration in the $y$ direction | $\sum F_x = ma_x$ <br> $\sum F_y = ma_y = 0$ | 7-4 |

## True-False Questions

T   F   **1.** The kilogram is the metric unit of weight.

T   F   **2.** The mass of an object on the moon is the same as its mass on earth.

T   F   **3.** The British unit for mass is the slug.

T   F   **4.** Newton's second law of motion is strictly true only in the absence of friction forces.

T   F   **5.** The mass of a body is dependent on the acceleration due to gravity.

T   F   **6.** The acceleration an object receives is proportional to the applied force, the constant of proportionality being the reciprocal of its mass.

T   F   **7.** The weight of an object is equal to the product of its mass and the gravitational acceleration.

T   F   **8.** Newton's second law of motion applies either to a system of bodies moved together as a whole or to any body that is a part of such a system.

T   F   **9.** The weight of a body whose mass is 1 slug is equal to approximately 32 lb.

T   F   **10.** The weight of a 9.8-kg body is 1 N.

## Multiple-Choice Questions

**1.** The mass of an 80-lb sled is
(a) 5 slugs   (b) 2.5 slugs   (c) 2.5 kg   (d) 256 slugs

**2.** The weight of a 10-kg block is
(a) 98 N   (b) 9.8 N   (c) 10 N   (d) 0.98 N

**3.** What acceleration will a force of 40 N impart to a mass of 5 kg?
(a) 8 m/s   (b) 200 m/s$^2$   (c) 8 m/s$^2$   (d) 8 cm/s$^2$

**4.** What resultant force will give a 64-lb object an acceleration of 6 ft/s$^2$?
(a) 384 lb   (b) 10.7 lb   (c) 12 slugs   (d) 12 lb

**5.** When a 1-N force acts on a 1-kg body, the body receives
(a) an acceleration of 9.8 m/s$^2$       (b) a speed of 1 m/s
(c) an acceleration of 980 m/s$^2$       (d) an acceleration of 1 m/s$^2$

**6.** A force of 20 N gives an object an acceleration of 5 m/s$^2$. What force will give the same object an acceleration of 16 m/s$^2$?
(a) 64 N   (b) 32 N   (c) 4 N   (d) 256 N

**7.** A 10-kg mass is suspended by a rope. What is the acceleration if the tension in the rope is 118 N while the mass is in motion?
(a) 4 ms/$^2$   (b) 2 m/s$^2$   (c) 11.8 m/s$^2$   (d) 20 m/s$^2$

**8.** A 65-lb body is suspended by a rope and accelerated upward with an acceleration of 8 ft/s$^2$. The tension in the rope is approximately
(a) 96 lb   (b) 16 lb   (c) 81 lb   (d) 48 lb

9. The mass $m$ in Fig. 7-2 required to give the system an acceleration of 6 m/s² is approximately
(a) 31.0 kg    (b) 31.6 kg    (c) 32.6 kg    (d) 41.9 kg

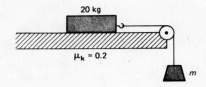

**Fig. 7-2**

10. A 30-kg girl steps to shore from a 60-kg boat, giving the boat an acceleration of 2 m/s². If water resistance is neglected, the girl receives an acceleration of
(a) −4 ms/²    (b) −15 m/s²    (c) −1 m/s²    (d) −8 m/s²

---

**Completion Questions**

1. The acceleration an object receives is directly proportional to the _____ and inversely proportional to the _____.

2. The mass of a particle is equal to _____ divided by the _____.

3. In USCS units, the mass unit, called a _____, is derived from the chosen unit of _____ for force.

4. In the SI system, the mass unit is the _____, and the derived unit is the _____.

5. Weight has the same units as the units of _____.

6. In USCS units, the mass of an object is found by dividing its _____ by _____.

7. In the metric system, objects are usually described by giving their _____. In the USCS, objects are described by giving their _____.

8. _____ is the force of gravitational attraction and is quite dependent on the _____.

9. If the force is held constant, an increase in _____ produces a proportionate _____ in acceleration.

10. If the unit of mass is the kilogram, the unit of force will be the _____ and the unit for acceleration will be _____.

# Chapter

# 8

# Work, Energy, and Power

| **CONTENTS** | 8-1 Work | 8-5 Potential Energy |
|---|---|---|
| | 8-2 Resultant Work | 8-6 Conservation of Energy |
| | 8-3 Energy | 8-7 Energy and Friction Forces |
| | 8-4 Work and Kinetic Energy | 8-8 Power |

## DEFINITIONS OF KEY TERMS

**Work**  A scalar quantity equal to the product of the displacement and the component of the force in the direction of the displacement.

**Potential energy**  Energy possessed by a system by virtue of position or condition.

**Kinetic energy**  Energy possessed by a body by virtue of its motion.

**Joule**  The SI unit of work or energy. It is equivalent to the work done by a force of 1 N when it displaces an object through a parallel distance of 1 m.

**Conservation of energy**  A principle that states that energy cannot be created or destroyed, only changed in form.

**Power**  The time rate at which work is accomplished. The SI unit of power is the watt.

**Watt**  Work or energy being expended at the rate of 1 J/s. The kilowatt (1 kW) is equal to 1000 watts (1000 W).

**Horsepower**  A USCS unit of power equal to 550 ft · lb/s, 33,000 ft · lb/min, or 746 W.

**Kilowatt**  One thousand watts.

## CONCEPTS AND EXAMPLES

**Work**  When work is calculated, it is important to distinguish between resultant work and the work of an individual force. For example, if we drag a block across the floor at constant speed, our pull does work, and the force of friction does negative work, but the resultant work is zero. By definition, work is always the product of force and displacement, but resultant work is the work done by the resultant force.

## EXAMPLE 8-1

A block is pulled across the floor by a pull of 400 N at an angle of 35° with the horizontal. The force of the kinetic friction is 150 N, and the block is moved a horizontal distance of 23 m. What is the resultant work?

*Solution*

A rough sketch such as Fig. 8-1 is drawn with a free-body diagram and all the given information shown. Note that the vertical forces are balanced, and the resultant force is the difference between $P_x$ and the friction force:

$$\text{Work} = (P_x - \mathcal{F}_k)(s) = (P \cos\theta - 150 \text{ N})(23 \text{ m})$$
$$= (400 \text{ N} \cos 35° - 150 \text{ N})(23 \text{ m}) = (177.7 \text{ N})(23 \text{ m})$$
$$= 4086 \text{ J}$$

In this case the block will be accelerated since resultant work implies the existence of a resultant force.

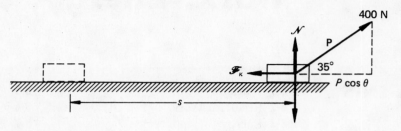

**Fig. 8-1** Finding resultant work.

## Conservation of Mechanical Energy

The total energy of a system is constant. If we lift a 4-N weight to a height of 10 m, it will have a potential energy of 40 J and a kinetic energy of 0. The total energy of the system $E_p + E_k$ is 40 J. If there is no friction, the object will lose potential energy and gain kinetic energy as it falls, but the sum will always be 40 J. When the object reaches the bottom of its fall, it will have 40 J of kinetic energy and zero potential energy. In the case of friction, some of the energy initially available to the system is lost in doing work against friction. Example 8-2 illustrates this point.

**EXAMPLE 8-2**

A 40-kg sled rests at the top of a 34° incline that is 500 m long. If the coefficient of sliding friction is 0.2, what will be the speed at the bottom of the slope? The 40-kg mass includes the mass of the man.

*Solution*

The speed at the bottom will be determined by the kinetic energy at that point. Since energy is conserved, the kinetic energy at the bottom is equal to the potential energy at the top minus the work done against friction. A sketch and free-body diagram as shown in Fig. 8-2 are essential to understand the problem. First, we need to know the total energy available to the system—the initial potential energy. This energy is a function of only the weight $mg$ and the height $h$. The height is found from the angle:

$$h = (500 \text{ m})(\sin 34°) = 280 \text{ m}$$

Now the potential energy (also the total energy) at the top is

$$E_p = mgh = (40 \text{ kg})(9.8 \text{ m/s}^2)(280 \text{ m}) = 109,602 \text{ J}$$

This is the total energy available to the system. Some of this energy will be lost in doing work against friction:

$$(\text{Work})_{\mathcal{F}} = \mathcal{F}_k s = (\mu_k \mathcal{N})(s) \qquad \text{and} \qquad \mathcal{N} = mg \sin 34°$$
$$(\text{Work})_{\mathcal{F}} = (\mu_k)(mg \sin 34°)(s)$$
$$= (0.2)(40 \text{ kg})(9.8 \text{ m/s}^2)(\sin 34°)(500 \text{ m}) = 21,920$$

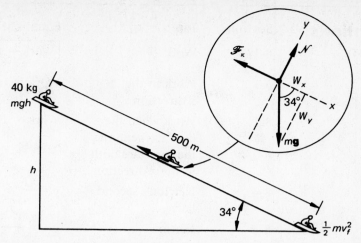

**Fig. 8-2** The kinetic energy $\frac{1}{2} mv^2$ at the bottom is equal to the potential energy $mgh$ at the top less the work $\mathscr{F}_k \cdot s$ done against the friction force.

J

Now the kinetic energy at the bottom must be equal to the potential energy at the top minus the work done against friction:

$$E_k = 109{,}602 \text{ J} - 21{,}920 \text{ J} = 87{,}762 \text{ J}$$

Finally, the velocity is found as follows:

$$\tfrac{1}{2}mv^2 = 87.762 \text{ J}$$

So

$$v^2 = \frac{2(87{,}762 \text{ J})}{(40 \text{ kg})} = 4388 \text{ m}^2/\text{s}^2$$

Taking the square root of both sides gives

**Power**   Power is the rate at which work is done or the rate at which energy is expended. It is the product of force and distance divided by time, or the product of force and velocity.

**Summary**

- Kinetic energy $E_k$ is energy in the form of a mass in motion.
- Gravitational potential energy is the energy that results from the position of an object relative to the earth.
- Potential energy $E_p$ has the same units as work.
- Net work is equal to the change in kinetic energy.
- Power is the rate at which work is done.

| | Section | Topic | Equation | Equation No. |
|---|---|---|---|---|
| | 8-1 | Work | $\boxed{\text{Work} = F_x s}$ | 8-1 |
| | 8-1 | Work in the $x$ direction | $\text{Work} = (F\cos\theta)s$ | 8-2 |
| | 8-1 | Work when the displacement and the force are both in the $x$ direction | $\text{Work} = Fs$ | 8-3 |
| | 8-4 | Newton's second law | $a = \dfrac{F}{m}$ | 8-4 |
| | 8-4 | Motion | $\dfrac{F}{m} = \dfrac{v_f^2 - v_0^2}{2s}$ | 8-5 |
| | 8-4 | Kinetic energy | $\boxed{E_k = \tfrac{1}{2}mv^2}$ | 8-6 |
| | 8-5 | Potential energy | $\boxed{E_p = Wh = mgh}$ | 8-7 |
| | 8-6 | Conservation of energy | $(E_p + E_k)_{\text{BEG}} = (E_p + E_k)_{\text{END}}$ $mgh_o + \tfrac{1}{2}mv_o^2 = mgh_f + \tfrac{1}{2}mv_f^2$ | 8-8 |
| | 8-7 | Conservation of energy with energy losses | $(E_p + E_k)_{\text{BEG}} =$ $(E_p + E_k)_{\text{END}} +$ $|\text{energy losses}|$ | 8-9 |
| | 8-7 | Conservation of energy with friction | $mgh_o + \tfrac{1}{2}mv_o^2 =$ $mgh_f + \tfrac{1}{2}mv_f^2 +$ $|\mathscr{F}_k s|$ | 8-10 |
| | 8-7 | Conservation of energy, with friction, of an object starting at rest | $mgh_o = \tfrac{1}{2}mv_f^2 + |\mathscr{F}_k s|$ | 8-11 |
| | 8-7 | Power | $P = \dfrac{\text{work}}{t}$ | 8-12 |
| | 8-7 | Power | $P = \dfrac{\text{work}}{t} = \dfrac{Fs}{t}$ | 8-13 |
| | 8-7 | Power | $P = F\dfrac{s}{t} = Fv$ | 8-14 |

**True-False Questions**

T  F  **1.** In the absence of friction, air resistance, or other dissipative forces, the total kinetic energy remains constant.

T  F  **2.** The total energy of a system of bodies in isolation may be defined as the sum of the individual kinetic and potential energies.

T  F  **3.** All moving bodies possess kinetic energy.

T  F  **4.** The work of a resultant force on a body is equal to the change in kinetic energy.

T  F  **5.** As a body falls, its potential energy increases with its speed.

T  F  **6.** A 1-hp engine will do work at a faster rate than a 1-kW engine.

**T  F  7.** The kilowatthour is a unit of energy.

**T  F  8.** As an object falls freely from the top of a building, its *total* energy remains constant.

**T  F  9.** If we consider friction, the potential energy at the top of an inclined plane is less than it would be in the absence of friction.

**T  F  10.** According to convention, negative work means that the direction of the work is downward or to the left.

---

## Multiple-Choice Questions

**1.** Which of the following is not necessary for work to be done?
(a) An applied force
(b) A force component along the displacement
(c) A displacement
(d) A constant speed

**2.** Which of the following is not a unit of work or energy?
(a) Newton-meter          (b) Joule
(c) Foot-pound per second (d) Kilowatthour

**3.** The largest unit of power is the
(a) kilowatt      (b) watt      (c) foot-pound per second      (d) horsepower

**4.** A force of 20 N moves a 10-kg block through a distance of 400 cm. The work done by the 20-N force is
(a) 8000 J      (b) 80,000 J      (c) 80 J      (d) 80 N · m

**5.** A 10-kg block is lifted 20 m above the ground in a gravitational field. The work done by the field is
(a) negative                      (b) positive
(c) equal to the final potential  (d) a vector quantity
energy

**6.** In Question 5, the gravitational potential energy of the block after it has been lifted 20 m is
(a) 196 J      (b) 1960 J      (c) −1960 J      (d) 200 J

**7.** A 1-hp motor will lift a 200-lb block to what height in 2 s?
(a) 0.01 ft      (b) 0.727 ft      (c) 100 ft      (d) 5.5 ft

**8.** A block of mass $m$ slides down an inclined plane of height $h$ and slope distance $s$. The kinetic energy at the bottom will be equal to
(a) $mgh$      (b) $\frac{1}{2}mv^2 - Fs$   (c) $mgh - Fs$      (d) $mgh + Fs$

**9.** A 2-kg ball has a potential energy of 6400 J at a point $A$ above the ground. What will its velocity be when it strikes the ground after being released from point $A$?
(a) 80 m/s      (b) 28.3 m/s      (c) 6400 m/s      (d) 800 m/s

**10.** A bullet whose initial kinetic energy is 400 J strikes a block where an 8000-N resistive force brings it to a stop. The depth of penetration into the wood is approximately
(a) unknown for lack of        (b) 0.2 m
information
(c) 0.5 m                      (d) 0.05 m

## Completion Questions

1. Three examples of potential energy are _____,

   _____, and _____.

2. The work of a resultant external force on a body is equal to the change in

   _____ of the body.

3. The total mechanical energy of a body is the sum of its

   _____. This total is _____ in the absence of

   friction.

4. A force of 1 N acting through a distance of 1 m represents

   _____ equal to 1 _____.

5. _____ is the rate at which work is done.

6. The net work done by a number of forces acting on the same object is equal to

   the work of _____ force.

7. The kilowatthour is a unit of _____.

8. When the resultant force on an object is opposite to the direction of displace-

   ment, the work is considered _____.

9. Two things that are necessary in the performance of work are

   _____ and _____.

10. The product of force and velocity is a measure of _____.

**Chapter 9**

# Impulse and Momentum

## DEFINITIONS OF KEY TERMS

**Impulse**    A vector quantity equal in magnitude to the product of a force and the time interval in which it acts ($F\,\Delta t$). The direction of the impulse is the same as the direction of the applied force.

**Momentum**    A vector quantity equal in magnitude to the product of a mass $m$ and its velocity $v$. The direction is the same as that for the velocity.

**Conservation of momentum**    The total linear momentum of colliding bodies before impact is equal to their total momentum after impact. Thus, momentum is said to be conserved.

**Elastic impact**    A collision in which the total kinetic energy remains constant. No energy is lost because of heat or deformation.

**Inelastic impact**    A collision in which some of the initial energy possessed by the colliding bodies is lost to heat or deformation.

**Coefficient of restitution**    The negative ratio of the relative velocity after collision to the relative velocity before collision. For elastic impacts, the coefficient is equal to unity; for inelastic impacts, it is zero.

## CONCEPTS AND EXAMPLES

**Impulse**    Impulse is the product of a force and the time through which it acts ($F\,\Delta t$). It is equal in magnitude to the change in linear momentum that it produces, and its direction is the same as that of the applied force.

### EXAMPLE 9-1

A 3-kg sledgehammer is moving at 15 m/s as it strikes a steel spike. It is brought to a stop in 0.025 s. Find the average driving force on the spike.

*Solution*
Since the final velocity of the hammer is zero, the change in momentum is equal to the initial momentum of the hammer. The force causing this change in momentum is directed upward by the stake on the block:

$$F\,\Delta t = mv_f - mv_0 = -mv_0 \quad v_f = 0$$

Let's choose the upward direction as positive, making the initial velocity $-15$ m/s. Substitution yields

$$F(0.025 \text{ s}) = -(3 \text{ kg})(-15 \text{ m/s}) = +45 \text{ kg} \cdot \text{m/s}$$

From this we solve for $F$:

$$F = \frac{45 \text{ kg} \cdot \text{m/s}}{0.025 \text{ s}} = \boxed{1800 \text{ N}}$$

The driving force of the hammer on the stake is equal and opposite to the force we calculated ($-1800$ N).

## Conservation of Momentum

When two bodies collide as in Fig. 9-1, their total momentum before impact is equal to their total momentum after impact. This knowledge allows us to determine unknown masses or velocities for such collisions.

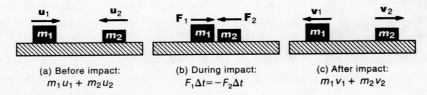

(a) Before impact:
$m_1 u_1 + m_2 u_2$

(b) During impact:
$F_1 \Delta t = -F_2 \Delta t$

(c) After impact:
$m_1 v_1 + m_2 v_2$

**Fig. 9-1**   Head-on collision of two masses.

**EXAMPLE 9-2**

A 54-Mg freight car traveling at a speed of 16 km/h collides with a stationary 27-Mg freight car. The two lock and move off together after impact. What is the velocity of the two-car train after impact?

*Solution*
The total momentum before collision must equal the total momentum after collision, and they have the same velocity $v$ after collision. Thus,

$$m_1 u_1 + m_2 u_2 = (m_1 + m_2)(v)$$

The consistent unit for mass is kilograms (kg) and for speed is meters per second (m/s). Recall that 1 Mg = 1000 kg and 1 km/h = 0.2778 m/s. Making the conversions and substituting yield

$$(54{,}000 \text{ kg})(4.444 \text{ m/s}) + m_2(0 \text{ m/s}) = (54{,}000 \text{ kg} + 27{,}000 \text{ kg})(v)$$

The combined velocity is found by solving for $v$:

$$\boxed{v = 2.96 \text{ m/s}}$$

## Elastic and Inelastic Collisions

In a perfectly elastic collision, no energy is lost on impact. In this ideal situation, the relative velocity is the same after impact as it was before impact except for the sign. When a ball rebounds from an elastic collision with the floor, it leaves with the same speed at which it arrived, but in the opposite direction. A completely inelastic collision represents the other extreme. In these instances the relative velocity is zero after impact, which really means that they remain together with the same velocity. For example, when a bullet lodges into a block of wood, the system moves off together after impact.

The most general type of collision is somewhere in between the two extremes. In such cases the coefficient of restitution is a useful concept, and it is derived from energy and momentum considerations. It is the negative ratio of the relative velocities before and after impact. For perfectly elastic collisions it is equal to 1, and for completely inelastic collisions it is equal to 0. By using this concept and the conservation of momentum, even general problems can be solved. You should study Example 9-5 in the text. When working with USCS units, you must remember to convert the given weights to masses before substituting into the momentum equations ($m = W/g$).

## Summary

- The *impulse* is the product of the average force **F** and the time interval $\Delta t$ through which it acts.
- The *momentum* of a particle is its mass times its velocity.
- The impulse is equal to the change in momentum.
- *Conservation of momentum* occurs when the total momentum before impact is equal to the total momentum after impact.
- The *coefficient of restitution* or elastic coefficient is found from relative velocities before and after collision or from the rebound height.

## Key Equations by Section

| Section | Topic | Equation | Equation No. |
|---|---|---|---|
| 9-1 | Problems involving impact | $\boxed{\mathbf{F}\,\Delta t = m\mathbf{v}_f - m\mathbf{v}_0}$ | 9-1 |
| 9-2 | Law of conservation of momentum | $\boxed{\begin{array}{l} m_1 u_1 + m_2 u_2 = \\ m_1 v_1 + m_2 v_2 \end{array}}$ | 9-2 |
| | | *Total momentum before impact = total momentum after impact* | |
| 9-3 | Velocity relationship for completely elastic collisions | $\begin{array}{l} v_1 - v_2 = u_2 - u_1 = \\ -(u_1 - u_2) \end{array}$ | 9-3 |
| 9-3 | Coefficient of restitution (elastic coefficient) | $\boxed{e = \dfrac{v_2 - v_1}{u_1 - u_2}}$ | 9-4 |
| 9-3 | Coefficient of restitution for a single bouncing object | $\boxed{e = \sqrt{\dfrac{h_2}{h_1}}}$ | 9-5 |

## True-False Questions

**T F 1.** The total linear energy of colliding bodies before impact must always equal the total energy after impact.

**T F 2.** The vigor with which a body restores itself to its original shape after deformation is a measure of its elasticity or restitution.

**T F 3.** A coefficient of restitution equal to 0.5 is an indication that the collision is inelastic.

**T F 4.** The units for impulse are equivalent to the units for momentum.

**T F 5.** An object that has linear momentum also has kinetic energy.

**T F 6.** The foot-pound per second is a unit of linear momentum.

**T F 7.** An object bouncing against a floor with perfectly elastic collisions will continue to bounce at the same height.

**T F 8.** Momentum is a scalar quantity.

**T F 9.** Momentum is conserved in all collisions.

**T F 10.** The kilogram-meter per second is a larger unit of momentum than the slug-foot per second.

## Multiple-Choice Questions

**1.** Which of the following is *not* a unit of impulse?
(a) Pound-second      (b) Newton-second
(c) Pound-hour      (d) Newton-meter

**2.** When the velocity of a body is doubled,
(a) its kinetic energy is doubled
(b) its momentum is doubled
(c) its acceleration is doubled
(d) its potential energy is doubled

**3.** A 100-kg astronaut releases 1 g of gas from a special pistol at a speed of 50 m/s. As a result, he moves in the opposite direction at
(a) 5 cm/s    (b) 50 cm/s    (c) 0.5 cm/s    (d) 0.05 cm/s

**4.** If a car is to gain momentum, it must
(a) move rapidly      (b) accelerate
(c) lose inertia      (d) lose weight

**5.** The face of a golf club exerts an average force of 4000 N while it is in contact with the golf ball. If the impulse is 80 N · s, the time of contact is
(a) 0.2 s      (b) 0.02 s      (c) 2 s      (d) 0.002 s

**6.** In a completely elastic collision, which of the following quantities need *not* be conserved?
(a) Kinetic energy      (b) Momentum
(c) Mass      (d) Potential energy

**7.** Eight steel balls are suspended from equal heights by threads so that they are all in contact. If two balls are pulled away to the left and released,
(a) one ball will leave from the right with twice the speed
(b) two balls will leave from the right at approximately the same speed
(c) either (a) or (b) might apply
(d) neither (a) nor (b) applies

**8.** A 40-kg cart moving at 3 m/s makes a head-on collision with a 20-kg cart at rest, as illustrated in Fig. 9-2. If the collision is completely inelastic, the speed with which the carts leave the impact is approximately
(a) 2 m/s      (b) 6 m/s      (c) 20 m/s      (d) 0.2 m/s

          

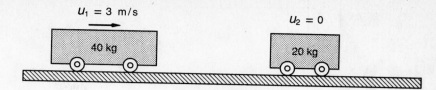

**Fig. 9-2**

9. If the collision in Question 8 is completely elastic, the speed of the 20-kg cart after collision will be approximately
   (a) 6 m/s    (b) 3 m/s    (c) 4 m/s    (d) 2 m/s

10. A 10-g bullet is fired into a 2-kg block of wood suspended from a cord. If the block and bullet rise to a height of 10 cm after impact, the initial velocity of the bullet was approximately
    (a) 2.8 m/s    (b) 281 m/s    (c) 235 m/s    (d) 28.1 m/s

---

**Completion Questions**

1. An elastic collision is one in which both _____ and _____ are conserved.

2. _____ is the product of mass and velocity and has the metric units _____.

3. The negative ratio of the relative velocities after impact to the relative velocities before impact is known as the _____.

4. The product of the average force acting on a body and the time of action is called _____, and it is equal to the _____ of the body.

5. The coefficient of restitution has a maximum value of _____ for a perfectly _____ impact and a minimum value of _____ for a perfectly _____ impact.

6. If two colliding bodies stick together after impact and move off with the same velocity, the impact is said to be _____.

7. In USCS, the unit for impulse is _____, and the unit for momentum is _____.

8. When a rifle is fired, the _____ of the bullet is the same as the recoil _____ of the rifle.

9. The square root of the ratio of the height to which a ball bounces to its original height is a measure of the _____.

10. The vigor with which an object restores itself to its original shape after deformation is a measure of its _____.

# Chapter 10

# Uniform Circular Motion

## DEFINITIONS OF KEY TERMS

**Uniform circular motion**   Circular motion in which the speed is constant and only the direction changes.

**Centripetal acceleration**   The acceleration toward the center that produces circular motion. The change in velocity may be only a change in direction, as is true for uniform circular motion. Its magnitude is equal to $v^2/R$.

**Centripetal force**   The inward force necessary to impart centripetal acceleration and therefore maintain circular motion. Its magnitude is equal to $mv^2/R$.

**Universal law of gravitation**   Each particle attracts each other particle with a force that is directly proportional to the product of their masses and inversely proportional to the square of the distance between them.

**Linear speed**   Distance covered per unit of time, whether the distance is along a straight line or along a curved path.

**Period**   In circular motion, the time for one complete revolution.

**Frequency**   In circular motion, the number of complete revolutions occurring per unit of time. Units are revolutions per minute and revolutions per second (1 Hz = 1 rev/s).

**Critical speed**   For motion in a vertical circle, the minimum speed at the bottom of the path for which the motion will remain circular.

**Conical pendulum**   A mass $m$ revolving in a horizontal circle with constant speed $v$ at the end of a cord of length $L$.

**Gravitational constant**   On earth, gravity has a constant value of 9.8 m/s.

## CONCEPTS AND EXAMPLES

**Centripetal Acceleration and Centripetal Force**   In uniform circular motion, a central force causes the path of an object moving at constant linear speed to be a circle. The acceleration, which represents a constant change in the direction of the linear velocity, is also directed toward the center. Its magnitude is the ratio of the square of the linear speed to the radius of revolution.

**EXAMPLE 10-1**    The governor on a steam engine consists of two 1.6-kg balls that swing in a horizontal circle. At the instant that the rotational speed is 300 rpm, the radius of the circular path is 300 mm. Find the centripetal acceleration and the centripetal force on each ball.

*Solution*

The centripetal acceleration is determined by the linear speed $v$ and the radius of the path. First, we find $v$:

$$v = 2\pi fR = (2\pi \text{ rad/rev})(300 \text{ rev/min})(1 \text{ min/60 s})(0.3 \text{ m}) = 9.42 \text{ m/s}$$

The centripetal acceleration is, therefore,

$$a_c = \frac{v^2}{R} = \frac{(9.42 \text{ m/s})^2}{0.3 \text{ m}} = \boxed{296 \text{ m/s}^2}$$

Finally, we calculate the centripetal force:

$$F_c = ma_c = (1.6 \text{ kg})(296 \text{ m/s}^2) = \boxed{474 \text{ N}}$$

Of course, we could plug directly into the equation that expresses centripetal force as a function of frequency. (See the summary at the end of Chap. 10 in the text.)

**Friction and the Banking of Curves**    The centripetal force is often provided by other forces such as friction between the tires of an automobile and the road or by the normal force exerted by the road on the car. In the case of banking curves, the entire centripetal force should be provided as an inward component of the normal force. This, in theory, would eliminate the need for friction as would be required on horizontal roadways. The necessary banking angle is found from

$$\tan \theta = \frac{v^2}{gR}$$

**Conical Pendulum**    The same equation applies for problems on the conical pendulum except that the centripetal force is provided by the tension in a cord instead of by a normal force.

**EXAMPLE 10-2**    What frequency of revolution (in revolutions per minute) is required to cause the pendulum of length 3 ft to swing in a horizontal circle so that the cord makes a 30° angle with the vertical?

*Solution*

An analysis of Fig. 10-1 and computations such as those given in the text for a conical pendulum given the following relation:

$$\tan \theta = \frac{v^2}{gR} \qquad \text{where} \qquad v = 2\pi fR$$

Substituting $v^2 = (2\pi fR)^2$, we have

$$\tan \theta = \frac{4\pi^2 f^2 R}{g} \qquad \text{or} \qquad f^2 = \frac{g \tan \theta}{4\pi^2 R}$$

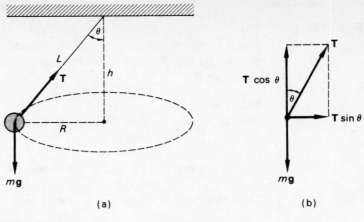

**Fig. 10-1** The conical pendulum.

So

$$f^2 = \frac{(32 \text{ ft/s}^2)(\tan 30°)}{(4\pi^2)(3 \text{ ft})} = 0.156 \text{ s}^{-2}$$

Taking the square root of both sides, we obtain

$$f = 0.394 \text{ rev/s} \quad or \quad \boxed{f = 23.7 \text{ rpm}}$$

## Motion in a Vertical Circle

For motion in a vertical circle, the centripetal force necessary for circular motion is affected significantly by the weight of the object. Consider, for example, a mass $m$ that is tied to a cord and swung in a vertical circle. At the top of the path, both the tension $T$ in the rope and the weight $mg$ add to give the centripetal force:

$$F_c = \frac{mv^2}{R} = T + mg$$

At the bottom of the path, the centripetal force is provided by the difference between the tension and the weight $mg$:

$$F_c = \frac{mv^2}{R} = T - mg$$

Equations such as these are useful for determining unknown forces. It is also possible to determine the critical speed an object must have as it passes its lowest point so that circular motion will not be aborted at the top of its path. This is done by setting $T = 0$ in the first equation listed above and then solving for $v$:

$$v_c = \sqrt{gR} \quad \text{critical speed}$$

You should study Example 10-6 in the text to see how to apply these concepts.

## Newton's Universal Law of Gravitation

Every particle in the universe attracts every other particle with a force that is directly proportional to the product of the masses and inversely proportional to their separation. (See Fig. 10-2).

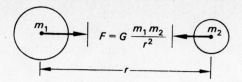

**Fig. 10-2** The universal law of gravitation.

Doubling the separation of such objects will result in one-fourth the gravitational force; tripling the separation results in one-ninth the original force. Study Examples 10-7 and 10-8 in the text as a guide to applications of this fundamental law.

## Summary

- The linear speed $v$ of an object in uniform circular motion can be calculated from the period $T$ or frequency $f$.
- The centripetal acceleration $a_c$ is found from the linear speed, the period, or the frequency.
- The centripetal force $F_c$ is equal to the product of the mass $m$ and the centripetal acceleration $a_c$.
- The force of a banking car's static friction must be less than or equal to the centripetal force for the no-slip condition.
- The areas swept by the radius of a moving planet are equal if the sweeping times are equal. This is Kepler's second law.

## Key Equations by Section

| Section | Topic | Equation | Equation No. |
|---------|-------|----------|--------------|
| 10-2 | Centripetal acceleration | $\mathbf{a} = \dfrac{\Delta \mathbf{v}}{\Delta t} = \dfrac{\mathbf{v_2} - \mathbf{v_1}}{\Delta t}$ | 10-1 |
| 10-2 | Centripetal acceleration graphical proportion | $\dfrac{\Delta v}{v} = \dfrac{s}{R}$ | 10-2 |
| 10-2 | Centripetal acceleration | $a_c = \dfrac{v^2}{R}$ | 10-3 |
| 10-2 | Speed of rotation based on period and radius | $v = \dfrac{2\pi R}{T}$ | 10-4 |
| 10-2 | Frequency of rotation | $f = \dfrac{1}{T}$ | 10-5 |
| 10-2 | Linear speed | $v = 2\pi f R$ | 10-6 |
| 10-3 | Centripetal force | $F_c = ma_c = \dfrac{mv^2}{R}$ | 10-7 |
| 10-3 | Centripetal force based on frequency | $F_c = \dfrac{mv^2}{R} = 4\pi^2 f^2 m R$ | 10-8 |
| 10-3 | Maximum speed without sliding | $v = \sqrt{\mu_s g R}$ | 10-9 |

| Section | Topic | Equation | Equation No. |
|---------|-------|----------|--------------|
| 10-3 | Banking angle for the no-slip condition | $\tan \theta = \dfrac{v^2}{Rg}$ | 10-10 |
| 10-5 | The conical pendulum | $h = \dfrac{gR^2}{v^2}$ | 10-11 |
| 10-5 | Frequency independent of radius | $f = \dfrac{1}{2\pi} \sqrt{\dfrac{g}{h}}$ | 10-12 |
| 10-6 | Motion in a vertical circle, the mass at the highest point | $T_1 + mg = \dfrac{mv_1^2}{R}$ | 10-13 |
| 10-6 | Motion in a vertical circle, the mass at the lowest point | $T_2 - mg = \dfrac{mv_2^2}{R}$ | 10-14 |
| 10-6 | Motion in a vertical circle, the minimum high point velocity needed to keep moving in a circle | $v_1 = \sqrt{gR}$ | 10-15 |
| 10-7 | Gravitation between two bodies | $F = G\dfrac{m_1 m_2}{r^2}$ | 10-16 |
| 10-8 | Earth's gravitation | $g = \dfrac{Gm_e}{R_e^2}$ | 10-17 |
| 10-8 | Earth's gravitation | $g = \dfrac{F_g}{m} = \dfrac{Gm_e}{R^2}$ | 10-18 |
| 10-9 | Satellites in circular orbits, finding the velocity | $v = \sqrt{\dfrac{Gm_e}{R}}$ | 10-19 |
| 10-9 | Satellites in circular orbits, finding the period | $T^2 = \left(\dfrac{4\pi^2}{Gm_e}\right)R^3$ | 10-20 |
| 10-10 | Kepler's third law | $T^2 = \dfrac{4\pi a^3}{Gm_s}$ | 10-21 |

## True-False Questions

T  F   **1.** When a rock is made to move in a circular path at the end of a string, an outward force is exerted on the rock.

T  F   **2.** If the string breaks in Question 1, centripetal acceleration will cause the rock to move inward.

T   F   **3.** Centripetal and centrifugal forces have the same magnitudes but are opposite in directions; they don't cancel because they act on different objects.

T   F   **4.** When a car moves in a horizontal circle on a level road, the centripetal force is exerted by the tires.

T   F   **5.** The maximum speed with which a car can negotiate a curve does not depend on the weight of the car.

T   F   **6.** The period of revolution may be thought of as the number of complete revolutions per second.

T   F   **7.** Just as a resultant force is needed to change the speed of a body, a resultant force is necessary to change its direction.

T   F   **8.** Uniform circular motion is motion in which the linear speed changes at a constant rate.

T   F   **9.** The formula for computing the banking angle also applies for the angle that the cord makes with the vertical for a conical pendulum.

T   F   **10.** If two masses are separated by twice the distance, they will experience one-half the gravitational attraction.

---

**Multiple-Choice Questions**

**1.** A ball is tied to a string and swung in a horizontal circle. When the string breaks, the ball will follow a path that is
(a) toward the center      (b) away from the center
(c) at a tangent to its circular path      (d) none of these

**2.** The force exerted on the ball in Question 1 is a
(a) centrifugal force      (b) centripetal force
(c) gravitational force      (d) fictitious force

**3.** A body traveling in a circular path at constant speed
(a) has an outward acceleration      (b) has inward acceleration
(c) has constant velocity      (d) is not accelerated

**4.** An object makes one complete revolution in 0.5 s. Its frequency is
(a) 2 s      (b) 5 rev/s      (c) 2 rev/s      (d) 0.5 rev/s

**5.** An object swings at the end of a string in uniform circular motion. Which of the following changes would not cause an increased centripetal force?
(a) A longer string      (b) A shorter string
(c) A greater linear speed      (d) A larger mass

**6.** The linear speed of an object swinging in a circular path of radius 2 m with a frequency of 5 rev/s is
(a) $2\pi$ m/s      (b) $20\pi$ m/s      (c) $4\pi$ m/s      (d) $10\pi$ m/s

**7.** The centripetal acceleration of a 2-kg mass swinging in a 0.4-m radius with a linear speed of 4 m/s is
(a) 4 m/s$^2$      (b) 40 m/s$^2$      (c) 10 m/s$^2$      (d) 20 m/s$^2$

**8.** An 8-kg ball is swung in a horizontal circle by a cord of length 2 m. If the period is 0.5 s, the tension in the cord is approximately
(a) 264 N      (b) 202 N      (c) 1200 N      (d) 2527 N

**9.** The banking angle for a curve of radius 400 ft for a speed of 60 mi/h should be approximately
(a) 31°      (b) 16°      (c) 26°      (d) 37°

**10.** The gravitational constant is $6.67 \times 10^{-11}$ N $\cdot$ m$^2$/kg$^2$. What is the gravitational force between two 4-kg balls separated by 0.2 m?

(a) $5.34 \times 10^{-7}$ N

(b) $2.67 \times 10^{-8}$ N

(c) $1.33 \times 10^{-8}$ N

(d) $6.67 \times 10^{-8}$ N

---

**Completion Questions**

**1.** Uniform circular motion is motion in which there is no change in _____ but only a change in _____.

**2.** The _____ is the number of revolutions per unit of time, whereas the _____ is the time for one revolution.

**3.** For a body moving in a horizontal circle, doubling the linear speed has the effect of increasing the centripetal acceleration by a factor of

_____.

**4.** The gravitational force between two particles is directly proportional to the product of their _____ and inversely proportional to the _____ of the _____ between them.

**5.** The force required to keep mass moving in a circular path at constant speed $v$ is called the _____ force and has a magnitude equal to _____, where $R$ is the radius of the circle.

**6.** The proper banking angle $\theta$ to eliminate the necessity for a frictional force is given by the relation _____, where $v$ is the velocity, $R$ is the radius of the curve, and $g$ is the acceleration due to gravity.

**7.** Whenever the centripetal force acts on an object, there is an equal and opposite _____ force exerted by the object.

**8.** When a curve is not banked, the centripetal force is provided by

_____.

**9.** The _____ speed is the minimum speed needed to maintain circular motion in a vertical plane.

**10.** The correct set of USCS units for the gravitational constant is

_____.

# Chapter 11

# Rotation of Rigid Bodies

## DEFINITIONS OF KEY TERMS

**Angular displacement**   The amount of rotation a body describes, denoted by an angle measured in degrees, radians, or revolutions.

**Angular velocity**   The time rate of change of the angular displacement, usually expressed in radians per second.

**Angular acceleration**   The time rate of change in angular velocity, normally expressed in radians per second squared (rad/s$^2$).

**Moment of inertia**   For rotational motion, the moment of inertia $I$ of a body about a particular axis is the rotational analog of mass in linear motion. Sometimes it is described as rotational inertia, and it depends on how the mass is distributed in a body.

**Rotational kinetic energy**   Energy possessed by a body by virtue of its moment of inertia $I$ and its angular speed $\omega$.

**Radius of gyration**   The radial distance from the center of rotation of a rigid body to a point at which the total mass of the body might be concentrated without changing its moment of inertia.

**Rotational work**   The scalar product of rotational torque and the angular displacement it produces.

**Angular momentum**   With reference to a fixed axis of rotation, the angular momentum is equal to the product of a body's angular velocity and its moment of inertia.

**Radian**   A measure of angle size. There are $2\pi$ radians in a circle.

**Tangential acceleration**   Acceleration in the direction tangent to the circular motion of a mass.

## CONCEPTS AND EXAMPLES

**Rotational Acceleration**   In rotational motion, we are concerned with angular displacement $\theta$, angular velocity $\omega$, and the angular acceleration $\alpha$. The definitions result in formulas similar to those used in Chap. 6 for linear acceleration. Tables 6-1 and 6-3 in the textbook summarize these equations:

$$\theta = \frac{\omega_f + \omega_0}{2}t$$

$$\omega_f = \omega_0 + \alpha t$$

$$\theta = \omega_0 t + \tfrac{1}{2}\alpha t^2$$

$$2\alpha\theta = \omega_f^2 - \omega_0^2$$

When any three of the five parameters $\theta$, $\alpha$, $t$, $\omega_f$, and $\omega_0$ are given, the other two can be found from one of these equations. Choose a direction of rotation as being positive throughout your calculations.

In applying the equations for constant angular acceleration, it is important to choose the appropriate units; the angle must be in radians. It is also important to choose a direction (clockwise or counterclockwise) as positive and to follow through consistently in giving the correct size to each quantity.

**EXAMPLE 11-1**  A pulley 30 cm in diameter is rotating initially at 5 rev/s. It then undergoes a constant acceleration of 6 rad/s². (a) What is its angular speed after the pulley has completed 50 rev? (b) What is the linear speed of a belt attached to the pulley at this instant?

**Solution**

(a)  First we write the given information.

Given: $\omega_0 = 5$ rev/s $= 31.4$ rad/s       Find: $\omega_f = ?$

$\alpha = 6$ rad/s²       $t = ?$

$\theta = (50 \text{ rev})(2\pi \text{ rad/rev}) = 314.2$ rad

Now $\omega_f^2 - \omega_0^2 = 2\alpha\theta$     or     $\omega_f = \sqrt{\omega_0^2 + 2\alpha\theta}$
So

$$\omega_f = \sqrt{(31.4 \text{ rad/s})^2 + 2(6 \text{ rad/s}^2)(314.2 \text{ rad})} = \boxed{69.0 \text{ rad/s}}$$

(b)  Now we find the linear speed $v$ from $v = \omega R$:

$$v = (69.0 \text{ rad/s})(0.15 \text{ m}) = \boxed{10.3 \text{ m/s}}$$

## Rotational Work, Energy and Power

The best way to approach a study of rotational motion is by comparing it with the linear cases with which we already have experience. Consider the analogies given in this table, for example:

| Concept | Linear motion | Rotational motion |
|---|---|---|
| Displacement | Distance $s$ (m) | Angle $\theta$ (rad) |
| Inertia | Mass $m$ (kg) | Moment of inertia $I$ |
| Effort | Force $F$ (N) | Torque $\tau$ (N · m) |
| Newton's law | $F = ma$ | $\tau = I a$ |
| Work | Work $= Fs$ | Work $= \tau\theta$ |
| Kinetic energy | $E_k = \tfrac{1}{2}mv^2$ | $E_k = \tfrac{1}{2}I\omega^2$ |
| Power | $P = Fv$ | $P = \tau\omega$ |

**EXAMPLE 11-2**    A solid 94-kg turbine wheel has a radius of gyration equal to 1.4 m. What torque must be applied to the wheel to change its rotational velocity from rest to 600 rpm in 60 rev of the wheel?

*Solution*

We must first determine the moment of inertia of the wheel from its radius of gyration:

$$I = mk^2 = (94 \text{ kg})(1.4 \text{ m})^2 = 184.2 \text{ kg} \cdot \text{m}^2$$

Now remember that the work done by any effort must be equal to the change in kinetic energy produced. This means that the rotational work $\tau\theta$ must equal the final rotational kinetic energy. (The initial kinetic energy was zero.)

$$\tau\theta = \tfrac{1}{2}I\omega^2 \qquad \text{or} \qquad \tau = \frac{I\omega^2}{2\theta}$$

The angular velocity and displacement are now calculated:

$$\omega = 2\pi f = (2\pi)(600 \text{ rev/min})(1 \text{ min}/60 \text{ s}) \qquad \text{or} \qquad \omega = 62.8 \text{ rad/s}$$

$$\theta = (60 \text{ rev})(2\pi \text{ rad/rev}) = 377 \text{ rad}$$

Finally we determine the torque by substitution:

$$\tau = \frac{(184.2 \text{ kg} \cdot \text{m}^2)(62.8 \text{ rad/s})^2}{2(377 \text{ rad})} = \boxed{964 \text{ N} \cdot \text{m}}$$

Of course, the same result could be obtained by calculating the angular acceleration and then applying Newton's second law. You should do it this way as a check on the above result.

---

**Summary**

- The angle in radians is the ratio of the arc distance $s$ to the radius $R$ of the arc.
- The radian is a unitless ratio of two lengths.
- Angular velocity, which is the rate of angular displacement, can be calculated from $\theta$ or from the frequency of rotation.
- Angular acceleration is the time rate of change in angular speed.

---

**Key Equations by Section**

| Section | Topic | Equation | Equation No. |
|---|---|---|---|
| 11-1 | Angular displacement in radians | $\theta = \dfrac{s}{R}$ | 11-1 |
| 11-2 | Angular velocity in radians per second | $\overline{\omega} = \dfrac{\theta}{t}$ | 11-2 |
| 11-2 | Conversion factor between radians and frequency | $\omega = 2\pi f$ | 11-3 |

| Section | Topic | Equation | Equation No. |
|---|---|---|---|
| 11-3 | Final angular velocity based on angular acceleration | $\omega_f = \omega_0 + \alpha t$ | 11-4 |
| 11-3 | Angular displacement | $\theta = \overline{\omega}t = \dfrac{\omega_f + \omega_0}{2}t$ | 11-5 |
| 11-4 | Linear velocity based on rotational velocity | $v = \omega R$ | 11-6 |
| 11-4 | Tangential acceleration based on rotational acceleration | $a_T = \alpha R$ | 11-7 |
| 11-4 | Centripetal acceleration based on linear velocity | $a_c = \dfrac{v^2}{R}$ | 11-8 |
| 11-5 | Rotational inertia | $I = \Sigma mr^2$ | 11-9 |
| 11-5 | Kinetic energy in terms of angular momentum and angular velocity | $E_k = \frac{1}{2}I\omega^2$ | 11-10 |
| 11-5 | Rotational inertia in terms of the radius of gyration | $I = mk^2$ | 11-11 |
| 11-6 | Newton's second law for rotational motion | $\tau = I\alpha$ | 11-12 |
| 11-7 | Rotational work | $\text{Work} = \tau\theta$ | 11-13 |
| 11-7 | Rotational power | $\text{Power} = \dfrac{\text{work}}{t} = \dfrac{\tau\theta}{t}$ | 11-14 |
| 11-7 | Rotational power in terms of average rotational velocity | $\text{Power} = \tau\overline{\omega}$ | 11-15 |
| 11-8 | Angular momentum of a particle | $L\,(\text{particle}) = mvr$ | 11-16 |
| 11-9 | Angular momentum of a particle in terms of moment of inertia and angular momentum | $L = I\omega$ | 11-17 |
| 11-9 | Conservation of angular momentum where an external torque $\tau$ is acting on a body | $\tau t = I\omega_f - I\omega_0$<br>Angular impulse = change in angular momentum | 11-18 |

**True-False Questions**

T   F   **1.** If the sum of the external torques acting on a body or system of bodies is zero, the angular momentum is also zero.

T   F   **2.** An angle of 1 rad is an angle whose arc distance $s$ is equal in length to the radius $R$.

T   F   **3.** Angular frequency and angular velocity are representations of the same physical quantity.

T   F   **4.** Both angular acceleration and tangential acceleration represent a rate of change in angular velocity.

T   F   **5.** The rotational kinetic energy of an object on a rotating platform decreases as the object moves toward the center of rotation.

T   F   **6.** A solid disk of mass $M$ will roll to the bottom of an incline quicker than a circular hoop of the same mass and diameter.

T   F   **7.** Torque is the rotational analog of linear force.

T   F   **8.** A resultant torque will produce an angular acceleration directly proportional to the applied torque and inversely proportional to the total mass of the object.

T   F   **9.** The angular impulse is equal to a change in rotational kinetic energy.

T   F   **10.** For a rotating rigid body, the ratio of linear speed to angular speed is equal to the radius of revolution.

**Multiple-Choice Questions**

**1.** Increasing the angular speed of a rotating body will not cause an increase in
(a) rotational kinetic energy    (b) angular momentum
(c) the momentum of inertia    (d) linear speed

**2.** Which of the following would not necessarily cause angular acceleration?
(a) Change in frequency
(b) Change in angular displacement
(c) Change in angular velocity
(d) Change in linear speed

**3.** A point on the edge of a rotating disk of radius 8 m moves through an angle of 2 rad. The length of the arc described by the point is
(a) 4 m      (b) 0.25 m      (c) 16 m      (d) $4\pi$ rad

**4.** If the frequency is 2 rev/s, the angular speed is
(a) $4\pi$ rad/s    (b) $2\pi$ rad/s    (c) $\pi$ rad/s    (d) 4 rad/s

**5.** A flywheel increases its angular speed from 4 to 12 rad/s in 4 s. Its angular acceleration is
(a) 3 rad/s$^2$    (b) 12 rad/s$^2$    (c) 4 rad/s$^2$    (d) 2 rad/s$^2$

**6.** A drive shaft has a frequency of rotation of 1200 rpm. The linear speed of flyweights positioned 2 ft from the axis is approximately
(a) $80\pi$ ft/s    (b) $40\pi$ ft/s    (c) 40 ft/s    (d) 20 ft/s

**7.** A circular disk has a moment of inertia of 2 kg · m$^2$ and a rotational kinetic energy of 400 J. The angular speed must be approximately
(a) 200 rad/s    (b) 40 rad/s    (c) 400 rad/s    (d) 20 rad/s

**8.** Which of the following objects has the largest moment of inertia, assuming they all have the same mass and the same radius?
(a) A solid sphere          (b) A solid disk
(c) A circular hoop         (d) A solid cylinder

9. A circular grinding disk of radius 0.5 ft has a moment of inertia of 16 slug · ft$^2$ and is rotating at 600 rpm. The frictional force applied to the edge of the disk, to stop the wheel in 10 s, is approximately
   (a) 201 lb      (b) 100 lb      (c) 402 lb      (d) 32 lb

10. A wheel with angular momentum of 10 kg · m$^2$/s has a moment of inertia equal to 0.5 kg · m$^2$. Its angular velocity is
    (a) 40 rad/s      (b) 20 rad/s      (c) 5 rad/s      (d) 0.5 rad/s

---

**Completion Questions**

1. The angular displacement in _____ is the ratio of the length of arc to its _____.

2. If the sum of the external _____ acting on a body or system of bodies is zero, the angular _____ remains unchanged. This is a statement of the conservation of _____.

3. Rotational work is the scalar product of _____ and _____.

4. A resultant torque applied to a rigid body will always result in a(n) _____ that is directly proportional to the applied _____ and inversely proportional to the body's _____.

5. The USCS units of the moment of inertia are _____.

6. The ratio of the tangential acceleration to the _____ is equal to the radius of revolution.

7. In rotational motion, the final angular velocity is equal to the initial _____ plus the product of the angular _____ plus the product of the angular _____ and _____.

8. The _____ is the radial distance from the center of rotation to a point at which the total mass of the body might be concentrated without changing its _____.

9. The linear speed divided by the radius of rotation is the _____ of the body.

10. Rotational power is equal to the product of _____ and angular _____.

# Chapter 12

# Simple Machines

## DEFINITIONS OF KEY TERMS

**Machine**   Any device that transmits the application of a force into useful work.

**Efficiency**   The ratio of the work output by a machine to the work input, usually expressed as a percentage.

**Lever**   A simple machine consisting of any rigid bar pivoted at a certain point called the fulcrum. The principle of the lever is used in many simple machines, such as pulley systems and gears.

**Inclined plane**   A simple machine that uses a gradual slope to give the desired mechanical advantage (the wedge and the screw).

**Actual mechanical advantage**   The ratio of an output force to the input force under actual conditions, in other words, with friction.

**Ideal mechanical advantage**   The ratio of the distance the input force moves to the distance the output force moves under ideal conditions, in other words, in the absence of friction.

**Pitch**   The distance between two adjacent threads on a screw.

**Speed ratio**   In pulleys or gears, the ratio of the angular velocity of the input pulley or gear to the angular velocity of the output pulley or gear.

**Belt drive**   A simple machine application of the pulley with two different-sized pulleys coordinated by a belt.

**Gears**   Two or more notched wheels meshed together to transmit torque.

**Pulley**   A simple machine that transmits the direction of force. Pulleys can be used to reduce the applied force if they are allowed to move or if they are used in combination with other pulleys as a block and tackle.

**Screw**   A simple machine application of the wedge wrapped around a cylinder.

**Wedge**   A simple machine application of the inclined plane.

**Wheel and axle**   A simple machine application of the lever principle that allows the continued action of the input force.

**Mechanical Advantage**   A simple machine is a device that converts a single input force $F_i$ into a single output force $F_o$. The input force moves through a distance $s_i$, and the output force moves through a distance $s_o$. The actual mechanical advantage is the ratio $F_o/F_i$; the ideal mechanical advantage is the ratio $s_i/s_o$.

---

**EXAMPLE 12-1**   A steel bar 1.2 m long is used to lift a 30-kg box. Where should the fulcrum be placed to give an ideal mechanical advantage of 8? What input force is required to lift the box?

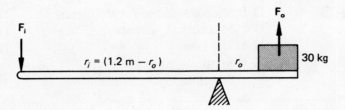

**Fig. 12-1**

**Solution**

In Figure 12-1, the fulcrum is placed a distance $r_o$ from the 30-kg mass. This means that $r_i = 1.2$ m $- r_o$, and since the ideal mechanical advantage is $r_o/r_i$, we have

$$\frac{r_i}{r_o} = \frac{1.2 \text{ m} - r_o}{r_o} = 8$$

We now solve for $r_o$ as follows:

$$8r_o = 1.2 \text{ m} - r_o \qquad \text{or} \qquad r_o = 0.133 \text{ m}$$

Thus, the fulcrum should be placed 133 mm from the 30-kg mass. If we neglect friction, the actual mechanical advantage $F_o/F_i$ is also equal to 8. The output force is the weight $mg$. Solving for $F_i$ gives

$$F_i = \frac{F_o}{8} = \frac{mg}{8} \qquad F_i = \frac{(30 \text{ kg})(9.8 \text{ m/s}^2)}{8} = \boxed{36.8 \text{ N}}$$

---

**Transmission of Torque**   Torque is usually transmitted from a driving gear to an output gear. The ideal mechanical advantage is the ratio of the output diameter to the input diameter, which is also equivalent to the ratio of the number of teeth in the case of gears.

---

**EXAMPLE 12-2**   A particular 6-kW engine that is 80 percent efficient, the input gear has 18 teeth and operates at 400 rpm during normal use. If the driven gear has 12 teeth, calculate the torque and rpm delivered to the output gear.

### Solution

First, we determine the ideal mechanical advantage:

$$M_I = \frac{N_o}{N_i} = \frac{12}{18} = 0.667$$

Now, since the machine is only 80 percent efficient, $M_a$ is given by

$$M_A = (0.80)(0.667) = 0.533$$

We recall that the actual mechanical advantage is equal to the ratio of the output torque to the input torque. Thus, we must determine the input torque. Remember that $P = \tau\overline{\omega}$ and that $\omega = 2\pi f$. Therefore,

$$P_i = \tau_i\omega_i \qquad \omega_i = (2\pi)(400 \text{ rev/min})(1 \text{ min/60 s}) = 41.9 \text{ rev/s}$$

Solving for $\tau_i$, we have

$$\tau_i = \frac{P_i}{\omega_i} = \frac{6000 \text{ W}}{41.9 \text{ rev/s}} = \boxed{143 \text{ N} \cdot \text{m}}$$

To find the output rpm, we must recall that the ideal mechanical advantage 0.667 is equal to the speed ratio $\omega_i/\omega_o$. Solving for $\omega_o$ and substituting yield

$$\omega_o = \frac{\omega_i}{M_I} = \frac{400 \text{ rpm}}{0.667} = \boxed{600 \text{ rpm}}$$

---

## Summary

- A simple machine is a device that converts a single input force $F_i$ into a single output force $F_o$.
- The input force moves through a distance $s_i$, and the output force moves a distance $s_o$.
- The efficiency of a machine is a ratio of output to input work.
- Efficiency is normally expressed as a percentage.

---

## Key Equations by Section

| Section | Topic | Equation | Equation No. |
|---------|-------|----------|--------------|
| 12-1 | Efficiency of a machine | $E = \dfrac{\text{work output}}{\text{work input}}$ | 12-1 |
| 12-1 | Alternate expression for efficiency | $E = \dfrac{\text{power output}}{\text{power input}} = \dfrac{P_o}{P_i}$ | 12-2 |
| 12-2 | Actual mechanical advantage | $M_A = \dfrac{\text{output force}}{\text{input force}} = \dfrac{F_o}{F_i}$ | 12-3 |
| 12-2 | Ideal mechanical advantage of a simple machine | $M_I = \dfrac{F_o}{F_i} = \dfrac{s_i}{s_o}$ | 12-4 |

| Section | Topic | Equation | Equation No. |
|---------|-------|----------|--------------|
| 12-2 | Efficiency is equal to the actual mechanical advantage divided by the ideal mechanical advantage | $E = \dfrac{M_A}{M_I}$ | 12-5 |
| 12-3 | Ideal mechanical advantage of the lever | $M_I = \dfrac{F_o}{F_i} = \dfrac{r_i}{r_o}$ | 12-6 |
| 12-4 | Applications of the lever principle | $M_I = \dfrac{F_o}{F_i} = \dfrac{R}{r}$ | 12-7 |
| 12-4 | Ideal mechanical advantage for a single fixed pulley | $M_I = \dfrac{F_o}{F_i} = 1$ | 12-8 |
| 12-4 | Ideal mechanical advantage for a single movable pulley | $M_I = \dfrac{F_o}{F_i} = 2$ | 12-9 |
| 12-5 | Ideal mechanical advantage of a belt drive based on the input and output diameters | $M_I = \dfrac{D_o}{D_i}$ | 12-10 |
| 12-5 | Ideal mechanical advantage of a belt drive or two gears based on input and output angular velocities | $M_I = \dfrac{D_o}{D_i} = \dfrac{\omega_i}{\omega_o}$ | 12-11 |
| 12-5 | Ideal mechanical advantage of two gears based on the number of teeth or the diameters of the gears | $M_I = \dfrac{N_o}{N_i} = \dfrac{D_o}{D_i}$ | 12-13 |
| 12-6 | Ideal mechanical advantage of the inclined plane | $M_I = \dfrac{W}{F_i} = \dfrac{s}{h}$ | 12-14 |
| 12-7 | Ideal mechanical advantage of the wedge | $M_I = \dfrac{L}{t}$ | 12-15 |
| 12-7 | Ideal mechanical advantage of the screw jack | $M_I = \dfrac{s_i}{s_o} = \dfrac{2\pi R}{p}$ | 12-16 |

## True-False Questions

T  F  **1.** The efficiency of a machine is defined as the ratio of the work input to the work output.

T  F  **2.** The actual mechanical advantage of a machine can never be less than unity.

T  F  **3.** The wheel and axle is an application of the principle of the lever.

T  F  **4.** A single fixed pulley offers no mechanical advantage; it changes only the direction of an input force.

T  F  **5.** In a pulley system with a speed ratio greater than 1, the output torque is less than the input torque.

T  F  **6.** In a machine using two gears in which the number of teeth on the input gear is greater than the number of teeth on the output gear, there will be a reduction in the output torque.

T  F  **7.** The speed ratio $\omega_i/\omega_o$ represents the ideal mechanical advantage and not the actual mechanical advantage.

T  F  **8.** In a belt drive, maximum efficiency is obtained for minimum belt tension.

T  F  **9.** For an inclined plane, the actual mechanical advantage is the ratio of the slope distance to the height of the plane.

T  F  **10.** For a screw jack, a larger pitch results in a smaller mechanical advantage.

## Multiple-Choice Questions

**1.** Which of the following ratios does *not* represent the mechanical advantage of a machine?
(a) $D_o/D_i$  (b) $\omega_o/\omega_i$  (c) $N_o/N_i$  (d) $F_o/F_i$

**2.** Which of the following is *not* an indication of the efficiency of a machine?
(a) $M_A/M_I$  (b) $P_o/P_i$  (c) $F_o r_o/(F_i r_i)$  (d) $L_o \omega_o/(L_i \omega_i)$

**3.** The actual mechanical advantage may be increased for a given machine by
(a) increasing $F_i$  (b) increasing $F_o$
(c) lubrication  (d) none of these

**4.** Which of the following machines is *not* an application of the lever principle?
(a) Screw jack  (b) Wheelbarrow
(c) Wheel and axle  (d) Pliers

**5.** An 80-hp motor has an efficiency of 60 percent. The output power is approximately
(a) 133 hp  (b) 75 hp  (c) 60 hp  (d) 48 hp

**6.** An iron pipe 10 ft long is used to lift a 400-lb weight. If the fulcrum is placed 2 ft from the weight, the force that must be exerted at the end of the pipe is approximately
(a) 200 lb  (b) 100 lb  (c) 40 lb  (d) 80 lb

**7.** The ideal mechanical advantage of a belt drive is 4.0. If the angular speed of the input pulley is 8 rad/s, the angular speed of the output pulley is
(a) 2 rad/s  (b) 32 rad/s  (c) 0.5 rad/s  (d) 4 rad/s

**8.** A 100-lb box is pushed up a 200-ft ramp to a height of 20 ft. The ideal mechanical advantage is approximately
(a) 4  (b) 10  (c) 5  (d) 0.1

**9.** Consider a belt drive in which the diameter of the driving pulley is 8 cm and the diameter of the driven pulley is 24 cm. If the efficiency is 50 percent, what is the actual mechanical advantage?
(a) 15  (b) 6  (c) 1.5  (d) 3

**10.** In the operation of a screw jack of pitch 0.1 in., the input force of 4 lb turns through a circle of radius 2 ft, lifting an 80-lb weight. The efficiency is approximately
(a) 48 percent  (b) 14.4 percent
(c) 20 percent  (d) 1.3 percent

**Completion Questions**

1. The efficiency of a machine is defined as the ratio of the _____ to the _____.

2. In general, for a simple machine the _____ is defined as the ratio of the output force to the input force. The _____ is the ratio of the distance the _____ force moves to the distance the _____ force moves.

3. Three applications of the lever principle are demonstrated by the following tools: _____, _____, and _____.

4. If the speed ratio is greater than 1, the machine produces an output torque _____ than the input torque.

5. Three common types of gears are _____, _____, and _____.

6. The ideal mechanical advantage of a wedge is equal to its _____ plus the _____.

7. For a simple machine, the work input is equal to the _____ plus the _____.

8. The ideal mechanical advantage of a set of gears is equal to the ratio of the _____ on the output gear to the _____ on the input gear.

9. For a screw jack, the reciprocal of the number of threads per inch is the _____ of the screw.

10. In a single movable pulley, the input force moves through _____ the distance that the output force moves, resulting in a mechanical advantage of _____.

# Chapter 13

# Elasticity

## DEFINITIONS OF KEY TERMS

**Elasticity**   The property whereby a body, when deformed, automatically recovers its normal shape when the deforming forces are removed.

**Hooke's law**   Provided that the elastic limit is not exceeded, an elastic deformation (strain) is directly proportional to the magnitude of the applied force per unit area (stress).

**Strain**   The relative change in the dimensions or shape of a body as a result of an applied stress, say, $(l - l_0)/l_0$.

**Elastic limit**   The maximum stress a body can experience without becoming permanently deformed.

**Ultimate strength**   The greatest stress a material can withstand without rupturing.

**Young's modulus**   The modulus of elasticity that relates to the change in a single dimension caused by a parallel stress. For a rod or wire, it is the ratio of force per unit area to the resultant change in length per unit length.

**Shear modulus**   The property of a material defined as the ratio of an applied stress to the resulting angle of shear, expressed in radians.

**Bulk modulus**   The negative ratio of an applied force per unit area to the resultant change in volume per unit volume.

**Compressive stress**   The force per square area that causes compressive deformation, that is, crushing deformation.

**Shear stress**   The force per square area that causes a shear deformation, that is, a deformation caused by two noncoaxial forces.

**Spring constant**   The proportionality constant $k$, which describes how much elastic deformation will occur for an applied force.

**Tensile stress**   The force per square area that causes a tensile deformation, that is, pulling deformation.

## CONCEPTS AND EXAMPLES

**Modulus of Elasticity**   The ratio of a stress $F/A$ to a strain is referred to as the modulus of elasticity. Young's modulus applies to the change in length of wires or rods, the shear modulus refers to an angular deformation, and the bulk modulus applies to a change in volume. Values for these constants are determined experimentally and may be obtained from tables. Our knowledge of these properties of materials allows us to use certain metals with more confidence or suggests the appropriate building material for a given structure.

**EXAMPLE 13-1**    An 8-kg mass is supported by a steel wire that is 1.5 m long and 0.8 mm in diameter. How much will the wire stretch under this load? What mass would result in permanent deformation of the wire?

*Solution*

We must use the values of Young's modulus and the elastic limit for steel from Table 13-1 given in the text. Then,

$$Y = \frac{F/A}{\Delta l/l_0} = \frac{Fl_0}{\Delta l\, A} \qquad \text{or} \qquad \Delta l = \frac{Fl_0}{YA}$$

Now, we must find $F = mg$ and $A = \pi R^2$.

$$F = (8 \text{ kg})(9.8 \text{ m/s}^2) = 78.4 \text{ N}$$

and

$$A = \pi R^2 = \pi(0.0004 \text{ m})^2 = 5.03 \times 10^{-7} \text{ m}^2$$

The change in length is determined by substitution:

$$\Delta l = \frac{(78.4 \text{ N})(1.5 \text{ m})}{(2.07 \times 10^{11} \text{ Pa})(5.03 \times 10^{-7} \text{ m}^2)} = \boxed{1.13 \times 10^{-3} \text{ m}}$$

Therefore, the wire will stretch by 1.13 mm.

The maximum load that can be hung will occur when $F/A$ exceeds the elastic limit (248 MPa). Solving for $F$ gives

$$F = (248 \times 10^6 \text{ Pa})(5.03 \times 10^{-7} \text{ m}^2) = 124.7 \text{ N}$$

Since this force represents the maximum weight, we determine the mass by dividing by gravity:

$$m = \frac{124.7 \text{ N}}{9.8 \text{ m/s}^2} = \boxed{12.7 \text{ kg}}$$

## Summary

- An elastic body will deform or elongate an amount $s$ multiplied by a factor $k$ under the application of a force $F$ according to Hooke's law.
- The proportionality constant $k$ is the *spring constant*.
- The ratio of an applied force to the area over which it acts is called *stress*.
- The relative change in dimensions that results from the stress is the *strain*.
- The modulus of elasticity is the constant ratio of stress to strain.
- A shearing strain occurs when an angular deformation $\phi$ is produced.

## Key Equations by Section

| Section | Topic | Equation | Equation No. |
|---------|-------|----------|--------------|
| 13-1 | Hooke's law | $F = ks$ | 13-1 |
| 13-1 | Definition of modulus | *Modulus of elasticity* $= \dfrac{stress}{strain}$ | 13-2 |

| Section | Topic | Equation | Equation No. |
|---------|-------|----------|--------------|
| 13-2 | Young's modulus | $Y = \dfrac{F/A}{\Delta l/l} = \dfrac{Fl}{A\,\Delta l}$ <br><br> *Young's modulus = longitudinal stress ÷ longitudinal strain* | 13-3 |
| 13-3 | Shear modulus | $S = \dfrac{\text{shearing stress}}{\text{shearing strain}} = \dfrac{F/A}{\phi}$ | 13-4 |
| 13-3 | Shear modulus | $S = \dfrac{F/A}{\tan \phi} = \dfrac{F/A}{d/l}$ | 13-5 |
| 13-4 | Bulk modulus | $B = \dfrac{\text{volume stress}}{\text{volume strain}} = \dfrac{F/A}{\Delta V/V}$ | 13-6 |
| 13-4 | Bulk modulus for liquids, in terms of pressure | $B = \dfrac{-P}{\Delta V/V}$ | 13-7 |
| 13-4 | Definition of compressibility $k$ | $k = \dfrac{1}{B} = -\left(\dfrac{1}{P}\right)\dfrac{\Delta V}{V_0}$ | 13-8 |

## True-False Questions

**T F 1.** A material having a large spring constant will experience a greater change in length for a given applied force than a material with a small spring constant.

**T F 2.** Stress is the relative change in the dimensions or shape of a body as the result of an applied force.

**T F 3.** The ultimate strength of a material is the greatest stress it can withstand without rupturing.

**T F 4.** The smaller the bulk modulus of a material, the higher its compressibility.

**T F 5.** The radian may be a unit of strain.

**T F 6.** The modulus of elasticity has the same units as stress.

**T F 7.** Young's modulus is constant for a particular material but varies with the length and cross-sectional area of a wire.

**T F 8.** Water has a greater compressibility than steel.

**T F 9.** Usually the shearing strain can be approximated by the tangent of the shearing angle.

**T F 10.** Young's modulus applies for compressive stresses as well as tensile stresses.

## Multiple-Choice Questions

**1.** Two wires $A$ and $B$, are made of the same material and are subjected to the same loads. The strain is greater for $A$ when
(a) $A$ is twice as long as $B$
(b) $A$ has twice the diameter of $B$

(c) *A* has twice the length and half the diameter

(d) *A* has twice the diameter and half the length

2. A shearing stress acting on a body changes its
   - (a) shape
   - (b) volume
   - (c) length
   - (d) area

3. A unit for strain is the
   - (a) inch
   - (b) pound per square inch
   - (c) newton per centimeter
   - (d) radian

4. The volume strain for a constant applied force increases directly with an increase in
   - (a) surface area
   - (b) compressibility
   - (c) volume
   - (d) bulk modulus

5. According to Hooke's law, an applied force will result in an elongation equal to
   - (a) *ks*
   - (b) *s/k*
   - (c) *F/s*
   - (d) *F/k*

6. The modulus of rigidity is another name for
   - (a) compressibility
   - (b) bulk modulus
   - (c) shear modulus
   - (d) Young's modulus

7. When a force of 20 N produces an elongation of 0.4 cm, the spring constant is
   - (a) 5 N/cm
   - (b) 50 N/cm
   - (c) 8 N · cm
   - (d) 0.02 cm/N

8. The cross-sectional area of a 20-in. copper wire is 0.001 in.$^2$. A force of 400 lb causes a stress of
   - (a) $4 \times 10^5$ lb/in.$^2$
   - (b) 0.4 lb/in.$^2$
   - (c) 20 lb/in.
   - (d) 50 lb/in.

9. In Question 8 the Young's modulus for copper is $17 \times 10^6$ lb/in.$^2$. The resultant elongation of the wire is approximately
   - (a) 0.012 in.
   - (b) 0.12 in.
   - (c) 0.47 in.
   - (d) 0.0047 in.

10. A mechanical press contains 2 m$^3$ of oil ($B = 1700$ MPa). If the volume of the oil decreases by $2 \times 10^{-8}$ m$^3$, the applied pressure must be approximately
    - (a) 17 Pa
    - (b) 68 Pa
    - (c) 0.06 Pa
    - (d) $1.7 \times 10^{17}$ Pa

---

**Completion Questions**

1. Three types of stress are _____, _____ and

   _____.

2. The constant ratio of stress to strain is called the _____.

3. _____ is the relative change in the dimensions or shape of a

   body as a result of an applied force.

4. Whenever a shearing stress is applied, the strain can be approximated by

   _____.

5. The reciprocal of the bulk modulus is called _____ and is

   denoted by the symbol _____.

6. The stress on a body is the ratio of _____ to

   _____.

**7.** The ultimate strength of a material is always _____ its elastic limit.

**8.** Provided that the _____ is not exceeded, an elastic strain is directly proportional to the _____. This is a statement of _____ law.

**9.** The term *elastic limit* refers to the maximum _____ a body can experience without becoming permanently deformed.

**10.** The only elastic modulus that applies for liquids is the _____.

# Chapter 14

# Simple Harmonic Motion

**DEFINITIONS OF KEY TERMS**

**Simple harmonic motion (SHM)**   Periodic motion, in the absence of friction, produced by a restoring force that is directly proportional to the displacement and oppositely directed.

**Restoring force**   The elastic force that always tends to restore a vibrating body to its equilibrium position.

**Reference circle**   An imaginary circle drawn to relate simple harmonic motion to analogous circular motion.

**Period**   The time required for one complete oscillation.

**Frequency**   The number of complete oscillations per unit of time. The frequency is the reciprocal of the period.

**Displacement**   In SHM the displacement from the equilibrium position at a particular instant.

**Amplitude**   In SHM the maximum displacement in either direction from the equilibrium position.

**Simple pendulum**   An ideal pendulum for which all the mass is concentrated at the center of gravity of the bob and the restoring force acts at a single point.

**Torsion pendulum**   A solid disk or cylinder supported at the end of a thin rod or wire.

**Hertz**   A unit of frequency equal to one oscillation per second.

**CONCEPTS AND EXAMPLES**

**Simple Harmonic Motion**   In simple harmonic motion, the restoring force is directly proportional to the displacement. Both the frequency $f$ and the amplitude $A$ remain constant. Displacement, velocity, and acceleration undergo periodic changes as a function of time. In general

$$x = A \cos 2\pi ft \qquad v = -2\pi fA \sin 2\pi ft \qquad a = -4\pi^2 f^2 x$$

The maximum values for these quantities occur when the arguments of the trigonometric functions are equal to $+1$ or $-1$. Thus, the maximum displacement is $A$, the maximum velocity is $2\pi fA$, and the maximum acceleration is $4\pi^2 f^2 A$.

**EXAMPLE 14-1**   A body vibrates in SHM with an amplitude of 15 cm and a period of 1.5 s. What are its position, velocity, and acceleration 5 s after it passes its equilibrium position, moving from the left to the right?

*Solution*

The frequency $f$ is equal to $1/T$. Therefore, the frequency must be 1/1.5 s, or 0.667 Hz. The amplitude is given as 15 cm. The displacement after 5 s is found by direct substitution:

$$x = A\cos\pi ft = (15\text{ cm})\{\cos[2\pi(0.667\text{ Hz})(5\text{ s})]\}$$
$$= (15\text{ cm})(\cos 20.95\text{ rad})$$

Make sure that your calculator is set to work with angles in *radians*. Then the displacement of the body becomes:

$$x = (15\text{ cm})(\cos 20.95\text{ rad}) = \boxed{-7.58\text{ cm}}$$

The meaning of the negative sign is that the position is to the left of the equilibrium position.

Now the velocity is a function of the same angle, and we substitute accordingly:

$$v = -2\pi fA \sin 20.95\text{ rad} = (-2\pi)(0.667\text{ Hz})(15\text{ cm})(0.866)$$

$$\boxed{v = -54.2\text{ cm/s}} \qquad \text{motion is to the left}$$

The acceleration is a function of the displacement $x$. Thus,

$$a = -4\pi^2 f^2 x = (-4\pi^2)(0.667\text{ Hz})^2(-7.58\text{ cm})$$

$$\boxed{a = 133\text{ cm/s}^2} \qquad \text{positive acceleration}$$

The positive sign of acceleration means that the restoring force is directed to the right at this instant, which is what we would expect.

---

## Summary

- Simple harmonic motion is produced by a *restoring force* **F**.
- A convenient way to study simple harmonic motion is to use the *reference* circle.
- A torsion pendulum consists of a solid disk or cylinder of moment of *inertia I* suspended at the end of a thin rod.

---

## Key Equations by Section

| Section | Topic | Equation | Equation No. |
|---------|-------|----------|--------------|
| 14-1 | Force as a function of displacement for an object in simple harmonic motion | $F = -kx$ | 14-1 |

| Section | Topic | Equation | Equation No. |
|---------|-------|----------|--------------|
| 14-2 | Newton's second law defining acceleration for an object in simple harmonic motion | $a = -\dfrac{k}{m}x$ | 14-2 |
| 14-2 | Displacement as a function of rotational angle or of frequency and time | $x = A \cos \theta = A \cos \omega t$ | 14-3 |
| 14-2 | Displacement as a function of time | $\boxed{x = A \cos 2\pi f t}$ | 14-4 |
| 14-3 | Velocity of an object in simple harmonic motion based on the tangential velocity and either the angle or the frequency and the time | $v = -v_T \sin \theta = -v_T \sin \omega t$ | 14-5 |
| 14-3 | Velocity of a body in simple harmonic motion based on the frequency $f$, the radius of the reference circle $A$, and the time $t$ | $\boxed{v = -2\pi f A \sin 2\pi f t}$ | 14-6 |
| 14-4 | Acceleration of an object in simple harmonic motion based on the centripetal acceleration and either the angle or the frequency and the time | $a = -a_c \cos \theta = -a_c \cos \omega t$ | 14-7 |
| 14-4 | Acceleration of a body in simple harmonic motion based on the frequency $f$, the radius of the reference circle $A$, and the time $t$ | $a = -4\pi^2 f^2\, A \cos 2\pi f t$ | 14-8 |
| 14-4 | Acceleration of a body in simple harmonic motion based on the frequency and the displacement | $\boxed{a = -4\pi^2 f^2 x}$ | 14-9 |

| Section | Topic | Equation | Equation No. |
|---|---|---|---|
| 14-5 | Frequency of an object in simple harmonic motion based on the acceleration and the displacement | $f = \dfrac{1}{2\pi}\sqrt{-\dfrac{a}{x}}$ | 14-10 |
| 14-5 | Period of an object in simple harmonic motion based on the acceleration and the displacement | $T = 2\pi\sqrt{-\dfrac{x}{a}}$ | 14-11 |
| 14-5 | Frequency of a body in simple harmonic motion under the influence of an elastic restoring force based on the spring constant and the mass of the body | $f = \dfrac{1}{2\pi}\sqrt{\dfrac{k}{m}}$ | 14-12 |
| 14-5 | Period of a body in simple harmonic motion under the influence of an elastic restoring force based on the mass on the spring constant | $T = 2\pi\sqrt{\dfrac{m}{k}}$ | 14-13 |
| 14-6 | The restoring force of a simple pendulum | $F = -kx = -kl\theta$ | 14-14 |
| 14-6 | The restoring force of a simple pendulum given by the tangential component of the weight | $F = -mg\sin\theta$ | 14-15 |
| 14-6 | For a simple pendulum undergoing small oscillations ($\theta < 30°$), the period $T$ is directly related to the length of the pendulum | $T = 2\pi\sqrt{\dfrac{I}{g}}$ | 14-16 |

| Section | Topic | Equation | Equation No. |
|---|---|---|---|
| 14-7 | The restoring torque in a torsion pendulum based on the torsion constant $k'$ and the angular displacement $\theta$ | $\tau = -k'\theta$ | 14-17 |
| 14-7 | Period of a torsion pendulum based on the torsion constant $k'$ and the moment of inertia $I$ | $T = 2\pi\sqrt{\dfrac{I}{k'}}$ | 14-18 |

## True-False Questions

T   F   **1.** When an object is vibrating with simple harmonic motion, its acceleration is a minimum when it passes through its equilibrium position.

T   F   **2.** Simple harmonic motion is realized only in the absence of friction.

T   F   **3.** In simple harmonic motion, the velocity is greatest when the oscillating body reaches its amplitude.

T   F   **4.** The greater the period of vibration, the greater the maximum acceleration of a body vibrating with simple harmonic motion.

T   F   **5.** The maximum velocity in simple harmonic motion occurs when the angle on the reference circle is 90 or 270°.

T   F   **6.** In simple harmonic motion, the acceleration is quadrupled when the frequency is increased by a factor of 2.

T   F   **7.** Since the acceleration due to gravity is less at higher elevations, the length of a pendulum in a pendulum clock should be shortened.

T   F   **8.** For a torsion pendulum, increasing the moment of inertia of the vibrating disk will increase the frequency of vibration.

T   F   **9.** The acceleration of a harmonic oscillator is a function of displacement but independent of amplitude.

T   F   **10.** The velocity of a harmonic oscillator depends on the frequency of vibration but is independent of amplitude.

## Multiple-Choice Questions

**1.** If the frequency does not change in simple harmonic motion, the acceleration of a mass is directly proportional to its
(a) velocity    (b) displacement
(c) mass    (d) amplitude

**2.** In simple harmonic motion, the velocity at any instant is not a direct function of the
(a) period    (b) amplitude
(c) time    (d) frequency

**3.** In simple harmonic motion, the radius of the reference circle corresponds most closely with the actual

(a) displacement         (b) velocity

(c) amplitude           (d) period

**4.** The period of a pendulum is determined by its

(a) maximum speed      (b) length

(c) amplitude            (d) mass

**5.** A body vibrating with simple harmonic motion experiences its maximum restoring force when it is at its

(a) equilibrium position    (b) amplitude

(c) greatest speed         (d) lowest acceleration

**6.** A 2-kg mass $m$ moves in simple harmonic motion with a frequency $f$. What mass will cause the system to vibrate with twice the frequency?

(a) 0.5 kg            (b) 4 kg

(c) 8 kg              (d) 16 kg

**7.** At the instant a harmonic oscillator has a displacement of $-8$ cm, its acceleration is 2 cm/s². The period is

(a) $8\pi$ s            (b) $4\pi$ s

(c) $2\pi$ s            (d) $\pi$ s

**8.** A harmonic oscillator vibrates with a frequency of 4 Hz and an amplitude of 2 cm. Its maximum velocity is

(a) $2\pi$ cm/s        (b) $4\pi$ cm/s

(c) $8\pi$ cm/s        (d) $16\pi$ cm/s

**9.** In Question 8, the maximum acceleration is

(a) $256\pi^2$ cm/s²      (b) $128\pi^2$ cm/s²

(c) $64\pi^2$ cm/s²       (d) $32\pi^2$ cm/s²

**10.** A 2-kg steel ball is attached to the end of a flat strip of metal that is clamped at its base. If the spring constant is 8 N/m, the frequency of vibration will be approximately

(a) 0.08 Hz          (b) 0.16 Hz

(c) 0.32 Hz          (d) 0.64 Hz

---

**Completion Questions**

**1.** The _____ is used to compare the motion of an object moving in a circle with its horizontal projection.

**2.** In simple harmonic motion, when the displacement is a maximum, the _____ is zero and the _____ is a maximum.

**3.** The product of the amplitude and the cosine of the reference angle is the _____ of a body vibrating with simple harmonic motion.

**4.** The _____ and therefore the _____ of a vibrating object are zero at the center of oscillation.

**5.** In simple harmonic motion, the _____ and the _____ are always opposite in sign.

**6.** To calculate the period for a torsion pendulum, we must know the _____ of the disk and the _____.

7. The period can be computed if the acceleration is known at a particular

   _____.

8. In simple harmonic motion, the restoring force is directly proportional to the

   _____ and _____ in direction.

9. If the frequency $f$ is to be calculated from the known spring constant $k$, we must

   know the _____ of the vibrating body.

10. For the vibration of a pendulum to approximate simple harmonic motion, the

    _____ must be small.

**Chapter**

# 15

# Fluids

**DEFINITIONS OF KEY TERMS**

**Weight density**   The ratio of an object's weight to its volume. It depends on the acceleration due to gravity; hence, it will vary with its location.

**Mass density**   The ratio of an object's mass to its volume. It is a universal constant for a particular material.

**Pressure**   The perpendicular force per unit area.

**Total force**   For a given surface, the sum of all the forces acting on that surface.

**Pascal's law**   An external pressure applied to an enclosed fluid is transmitted uniformly throughout the volume of the liquid.

**Absolute pressure**   The total fluid pressure, including atmospheric pressure.

**Gauge pressure**   The difference between absolute pressure and atmospheric pressure. Most devices that measure pressure actually indicate gauge pressures.

**Manometer**   An open, U-shaped tube partially filled with a liquid (usually mercury). When one end of the tube is connected to a pressurized chamber and the other end is left open, the difference in the levels of the liquid is an indication of gauge pressure.

**Archimedes' principle**   An object that is completely or partly submerged in a fluid experiences an upward force equal to the weight of the fluid displaced.

**Buoyant force**   The upward force in Archimedes' principle that is equal to the weight of the displaced fluid.

**Turbulent flow**   Fluid motion in which streamlines break down as obstacles are encountered, setting up swirls and eddies.

**Venturi effect**   The principle by which a drop in pressure occurs as a result of the increased velocity of a fluid flowing through a constriction.

**Bernoulli's principle**   A statement of the law of conservation of energy for the steady flow of an incompressible fluid. For every unit of volume, the sum of pressure, kinetic energy, and potential energy is a constant.

**Torricelli's theorem**   The emergent velocity of a fluid at a depth $h$ is equal to the square root of twice the product of $h$ and the acceleration due to gravity.

**Rate of flow**   A measure of how much fluid flows through a cross-sectional area per unit time, given in units of liters per second.

## Fluid Pressure

**Fluid Pressure**  Fluid pressure is the normal force per unit area applied to a fluid. The following principles are important for problems that deal with fluids at rest:

1. Forces exerted by a fluid on the walls of its container are always perpendicular.

2. The fluid pressure is directly proportional to the depth of the fluid and to its density.

3. At any particular depth, the fluid pressure is the same in all directions.

4. Fluid pressure is independent of the shape or area of its container.

5. External pressure applied to an enclosed fluid is transmitted uniformly throughout the volume of the liquid (Pascal's law).

6. An object that is completely or partly submerged in a fluid experiences an upward buoyant force equal to the weight of the displaced fluid (Archimedes' principle).

---

**EXAMPLE 15-1**

A cylindrical container has a diameter of 20 cm and a height of 40 cm. It is filled with kerosene whose density is known to be 820 kg/m³. What is the absolute pressure at the bottom and what is the total force on the bottom?

*Solution*

The absolute pressure at the bottom depends only on the height of the column of kerosene and atmospheric pressure at the top surface. First, we calculate the gauge pressure at the bottom:

$$P = \rho g h = (820 \text{ kg/m}^3)(9.8 \text{ m/s}^2)(0.4 \text{ m}) = 3214 \text{ N/m}^2$$

Thus, the gauge pressure at the bottom is 3.21 kPa. The absolute pressure is obtained by adding atmospheric pressure (101.3 kPa) to this gauge pressure. Therefore, the absolute pressure at the bottom is 104.5 kPa.

   Now the total force on the bottom depends on the gauge pressure and the area of the bottom, and since $P = F/A$, we may write

$$F = PA = P\pi R^2$$
$$= (3214 \text{ N/m}^2)\pi(0.1 \text{ m})^2 = \boxed{\textbf{101 N}}$$

The total force depends on the gauge pressure and not the absolute pressure because atmospheric pressure acts on the bottom of the container as well as on the top of the fluid. It's important to distinguish total force from pressure.

---

**Buoyant Force**  Problems involving buoyant forces that liquids exert on floating or submerged objects usually involve applications of Archimedes' principle. Since we know that the magnitude of the buoyant force is determined by the weight of the displaced fluid, it can often be determined from known densities and pressures.

---

**EXAMPLE 15-2**

A 6-kg block of steel is completely submerged in a container filled with water. In the process 0.770 L of water is displaced. What is the buoyant force?

*Solution*

The buoyant force is equal to the weight of the displaced water. Recall that 1 L = 1000 cm³ = 0.001 m³. The volume of water displaced is, therefore,

$$V = (0.770 \text{ L})(0.001 \text{ m}^3/\text{L}) = 0.000770 \text{ m}^3$$

Now, the density of water is 1000 kg/m³. Thus, the mass is

$$m = \rho V = (1000 \text{ kg/m}^3)(0.000770 \text{ m}^3) = 0.770 \text{ kg}$$

The weight of this mass of water is the buoyant force:

$$F_B = mg = (0.770 \text{ kg})(9.8 \text{ m/s}^2) = \boxed{7.54 \text{ N}}$$

The apparent weight of the steel in water is 7.54 N less than its weight in air. The volume of the steel is, of course, the same as that of the displaced water (0.000770 m³).

**Rate of Flow**   The rate of flow of a fluid is the volume passing a given point per unit of time (cubic meters per second, liters per second, gallons per minute, etc.). Its magnitude is equal to the product of velocity and area ($R = vA$). If the fluid is incompressible and we ignore friction, the rate of flow will be constant. Thus, a variation in cross-section of a pipe results in a change in speed of the fluid.

**EXAMPLE 15-3**   Gasoline flows from a pump at a velocity of 8.5 cm/s through a hose 5 cm in diameter. What is the rate of flow? What is the emergent velocity if the gasoline passes through a nozzle 2 cm in diameter?

*Solution*

The rate of flow must be the same through each nozzle. Since $vA$ is constant and $A$ is proportional to the square of the diameter $d$, we may write

$$v_1 A_1 = v_2 A_2 \qquad \text{or} \qquad v_1 d_1^2 = v_2 d_2^2$$

The emergent velocity is found by solving for $v_2$:

$$v_2 = v_1 \left(\frac{d_1}{d_2}\right)^2 = (8.5 \text{ cm/s})\left(\frac{5 \text{ cm}}{2 \text{ cm}}\right)^2$$

$$= 53.1 \text{ cm/s} \qquad \text{emergent velocity}$$

**Bernoulli's Equation**   The work done on a fluid is equal to the changes in kinetic and potential energy of the fluid. Bernoulli's equation expresses this fact in terms of the pressure $P$, the density $\rho$, the height $h$ of the fluid, and its velocity $v$:

$$P + \rho g h + \tfrac{1}{2}\rho g v^2 = \text{constant}$$

Consider Fig. 15-1 for a volume of fluid that changes from state 1 to state 2. In this general case, we write

$$P_1 + \rho g h_1 + \tfrac{1}{2}\rho v_1^2 = P_2 + \rho g h_2 + \tfrac{1}{2}\rho v_2^2$$

Special applications occur when one of the parameters does not change:

Stationary liquid:         $P_2 - P_1 = \rho g(h_1 - h_2)$

Constant pressure:         $v = \sqrt{2gh}$

Horizontal pipe:         $P_1 + \tfrac{1}{2}\rho v_1^2 = P_2 + \tfrac{1}{2}\rho v_2^2$

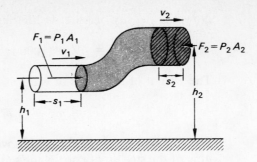

**Fig. 15-1** Applications of Bernoulli's equation.

---

**EXAMPLE 15-4**    Water flows through a horizontal pipe $A$ at a velocity of 20 cm/s. The pipe then narrows to a section $B$ where the speed increases to 250 cm/s. Determine the drop in pressure as the water enters the constriction.

**Solution**

Since the pipe is horizontal, $h_1 = h_2$ and Bernoulli's equation reduces to the third special case listed above:

$$P_1 + \tfrac{1}{2}\rho v_1^2 = P_2 + \tfrac{1}{2}\rho v_2^2$$

The drop in pressure is $P_1 - P_2$, which is given by

$$P_1 - P_2 = \tfrac{1}{2}\rho(v_2 - v_1)$$
$$= \tfrac{1}{2}(1000 \text{ kg/m}^3)[(2.5 \text{ m/s})^2 - (0.2 \text{ m/s})^2]$$
$$= (500 \text{ kg/m}^3)(6.21 \text{ m}^2/\text{s}^2) = 3105 \text{ Pa}$$

Therefore, the pressure in the constriction is 3.105 kPa less that of the pressure in Sec. $A$.

---

**Summary**

- An important physical property of matter is its density.
- Important points to remember about fluid pressure are:
  **a.** The forces exerted by a fluid on the walls of its container are always perpendicular to the walls.
  **b.** The fluid pressure is directly proportional to the depth of the fluid and to its density.
  **c.** At any particular depth, the fluid pressure is the same in all directions.
  **d.** Fluid pressure is independent of the shape or area of its container.
- Pascal's law states that an external pressure applied to an enclosed fluid is transmitted uniformly throughout the volume of the liquid.
- Archimedes' principle is expressed as: An object that is completely or partly submerged in a fluid experiences an upward force equal to the weight of the fluid displaced.
- The *rate of flow* is defined as the volume of fluid that passes a certain cross section $A$ per unit of time $t$.
- The net work done on a fluid is equal to the changes in kinetic and potential energy of the fluid.

# Key Equations by Section

| Section | Topic | Equation | Equation No. |
|---|---|---|---|
| 15-1 | The weight density $D$ of a body is its weight $W$ divided by its volume $V$ | $$D = \frac{W}{V} \quad W = DV$$ | 15-1 |
| 15-1 | The mass density $\rho$ of a body is its mass $m$ divided by its volume $V$ | $$\rho = \frac{m}{V} \quad m = \rho V$$ | 15-2 |
| 15-1 | The relationship between weight density and mass density | $$D = \frac{mg}{V} = \rho g$$ | 15-3 |
| 15-2 | Pressure, equal to force $F$ per unit area $A$ | $$P = \frac{F}{A}$$ | 15-4 |
| 15-3 | Fluid pressure at a depth $h$ | $$P = Dh = \rho g h$$ | 15-5 |
| 15-5 | Pascal's law applied to the hydraulic press | Input pressure = output pressure $$\frac{F_i}{A_i} = \frac{F_o}{A_o}$$ | 15-6 |
| 15-5 | Mechanical advantage of a hydraulic press | $$M_I = \frac{F_o}{F_i} = \frac{A_o}{A_i}$$ | 15-7 |
| 15-5 | Mechanical advantage for a hydraulic press based on force or displacement | $$M_I = \frac{F_o}{F_i} = \frac{s_i}{s_o}$$ | 15-9 |
| 15-6 | Archimedes' principle | $$F_B = V\rho g = mg$$ Buoyant force = weight of displaced fluid | 15-10 |
| 15-7 | The rate of fluid flow $R$ based on the cross-sectional area $A$ and the velocity of the fluid $v$ | $$R = \frac{Avt}{t} = vA$$ | 15-11 |
| 15-7 | Fluid flow through a pipe is equal at every point along the pipe | $$R = v_1 A_1 = v_2 A_2$$ | 15-12 |

| Section | Topic | Equation | Equation No. |
|---|---|---|---|
| 15-8 | The difference in pressure at two points is proportional to the difference in height between the points, for a stationary liquid | $P_A - P_B = \rho g h$ | 15-13 |
| 15-9 | Bernoulli's equation | $P_1 + \rho g h_1 + \frac{1}{2}\rho v_1^2 = P_2 + \rho g h_2 + \frac{1}{2}\rho v_2^2$ | 15-14 |
| 15-9 | Bernoulli's equation restated | $\boxed{P + \rho g h + \frac{1}{2}\rho v^2 = \text{constant}}$ | 15-15 |
| 15-10 | Bernoulli's equation for the case of a stationary liquid | $P_2 - P_1 = \rho g(h_1 - h_2)$ | 15-16 |
| 15-10 | Torricelli's theorem for the rate of flow out an opening at the bottom of a tank | $\boxed{v = \sqrt{2gh}}$ | 15-17 |
| 15-10 | Flow rate of fluid through an opening at the bottom of a tank | $R = vA = A\sqrt{2gh}$ | 15-18 |
| 15-10 | Bernoulli's equation for the case of constant height; this demonstrates the venturi effect | $P_1 + \frac{1}{2}\rho v_1^2 = P_2 + \frac{1}{2}\rho v_2^2$ | 15-19 |

## True-False Questions

T  F   1. The fluid pressure at the bottom of a container depends on the area of the bottom as well as the height of the fluid.

T  F   2. Fluid pressure is independent of the shape or size of the container.

T  F   3. At a particular depth in a liquid, the fluid pressure is the same in all directions except the upward direction.

T  F   4. Pascal's law states that the buoyant force will always be equal to the weight of the displaced fluid.

T  F   5. If a rock has a density of a 4 kg/m$^3$ on the earth, it will have a density of 4 kg/m$^3$ on the moon.

T  F   6. Since the weight of the overlying fluid in a container is proportional to its density, the pressure at any depth is also proportional to the density of the fluid.

T  F   7. The buoyant force is equal to the resultant force acting on a submerged object.

**T  F  8.** The buoyant force on a weather balloon does not depend directly on the density of the gas inside the balloon.

**T  F  9.** The smaller the area of the input cylinder of a hydraulic press in comparison with the area of the output cylinder, the larger the mechanical advantage.

**T  F  10.** An open-tube manometer is a device used to measure absolute pressure.

**T  F  11.** The rate of flow is defined as the speed with which a fluid passes a certain cross section.

**T  F  12.** If a fluid is considered incompressible, and if we neglect internal friction, the rate of flow through a pipe remains constant, even when the cross-sectional area changes.

**T  F  13.** Decreasing the diameter of a pipe by one-half will cause the velocity of a fluid through the pipe to be quadrupled.

**T  F  14.** An increase in fluid velocity results in an increased pressure at the constriction in a venturi meter.

**T  F  15.** The pressure $P$ in Bernoulli's equation represents the absolute pressure and not the gauge pressure.

**T  F  16.** Bernoulli's equation applies for fluids at rest as well as for fluids in motion.

**T  F  17.** If several holes are cut in the side of a container filled with water, the discharge velocity increases with depth below the surface; however, the range is a maximum at the midpoint.

**T  F  18.** If we consider a fluid to be incompressible and neglect the effects of friction, the change in velocity as a fluid goes through the constriction in a horizontal venturi tube does not depend on the density of the fluid.

**T  F  19.** Either weight density or mass density might be used in Bernoulli's equation.

**T  F  20.** Torricelli's theorem might be thought of as a special case of the more general Bernoulli's equation.

## Multiple-Choice Questions

**1.** Which of the following is independent of the density of a liquid?
(a) The total force at the bottom of the container
(b) The pressure at the surface of the liquid
(c) The pressure at the bottom of the container
(d) The pressure at the sides of the container

**2.** The hydraulic press operates primarily on
(a) Archimedes' principle  (b) Pascal's law
(c) Boyle's law  (d) Newton's laws

**3.** The ratio of an object's weight density to its mass density is
(a) absolute pressure  (b) less than 1
(c) equal to g  (d) unitless

**4.** The fluid pressure at any point in a liquid is
(a) proportional to its density
(b) of the same magnitude
(c) independent of height
(d) directed downward only

**5.** An open-tube manometer measures
  (a) atmospheric pressure   (b) absolute pressure
  (c) gauge pressure         (d) sea-level pressure

**6.** A cork float has a volume of 2 ft$^3$ and a density of 12 lb/ft$^3$. Its weight is
  (a) 6 lb      (b) 24 lb      (c) 0.17 lb      (d) 12 lb

**7.** Gasoline has a density of 680 kg/m$^3$. The pressure at the bottom of a container filled with gasoline to a height of 2 m is approximately
  (a) 1360 N/m$^2$   (b) 3400 N/m$^2$   (c) 13,328 N/m$^2$   (d) 1500 N/m$^2$

**8.** A square chunk of cork weighing 2 N floats with exactly half its volume submerged in water. The weight of the displaced water is
  (a) 4 N      (b) 2 N      (c) 1 N      (d) 0.5 N

**9.** A force of 600 lb is applied to the small piston of a hydraulic press. The diameter of the input piston is exactly one-half the diameter of the output piston. The lifting force exerted by the output piston is
  (a) 150 lb      (b) 300 lb      (c) 1200 lb      (d) 2400 lb

**10.** What must the volume of a balloon be if it is to support a total mass of 1000 kg at a point where the density of air is 1.2 kg/m$^3$? (The total mass includes that of the balloon and the helium with which it is filled, as well as the payload.)
  (a) 85 m$^3$      (b) 833 m$^3$      (c) 1200 m$^3$      (d) 8166 m$^3$

**11.** The rate of flow of a fluid out of an opening at the bottom of a container does not depend on
  (a) the depth of the hole      (b) the area of the opening
  (c) the density of the fluid   (d) the acceleration of gravity

**12.** The product of velocity and cross-sectional area for a liquid flowing through a pipe is a measure of
  (a) rate of flow        (b) fluid pressure
  (c) volume of fluid     (d) none of these

**13.** The velocity of discharge of a fluid through an orifice is associated most closely with
  (a) Bernoulli      (b) Torricelli
  (c) Venturi        (d) Archimedes

**14.** When one blows air across the top of a sheet of paper, the paper rises because of the
  (a) force of the blow
  (b) drop in the pressure above the paper
  (c) increase in pressure below the paper
  (d) increase in temperature of the air

**15.** When there is no change in pressure at the beginning and end of a flow process, Bernoulli's equation reduces to
  (a) $P = \rho gh$                      (b) $P + \frac{1}{2}\rho v^2 = $ constant
  (c) $v = \sqrt{2gh}$                   (d) $\rho gh = $ constant

**16.** Water flowing at a velocity of 20 cm/s in a 6-cm-diameter pipe encounters a constriction 3 cm in diameter. The velocity through the constriction is
  (a) 80 cm/s      (b) 40 cm/s      (c) 5 cm/s      (d) 10 cm/s

**17.** A container 4 ft high is filled to the top with liquid. The discharge velocity from a hole at the bottom of the container is
  (a) 12 ft/s      (b) 16 ft/s      (c) 20 ft/s      (d) 24 ft/s

**18.** In a horizontal venturi tube, the pressure at the inlet is 27 lb/ft$^2$, and the velocity of a liquid flowing through it is 10 ft/s. The density of the liquid is

2 slugs/ft³, and the pressure at the throat of the venturi drops to 6 lb/ft². The velocity of the liquid in the throat is approximately
(a) 11 ft/s     (b) 15 ft/s     (c) 20 ft/s     (d) 25 ft/s

**19.** The dimensions of every term in Bernoulli's equation are those of
(a) length     (b) velocity     (c) density     (d) pressure

**20.** Water rushes out the end of a pipe at the rate of 2 m³/s and with a velocity of 4 m/s. The area of the opening is
(a) 2 m²     (b) 1 m²     (c) 0.5 m²     (d) 0.25 m²

---

## Completion Questions

**1.** Gauge pressure is _____ pressure less _____ pressure.

**2.** The forces exerted by a fluid on the walls of its container are always _____.

**3.** The fluid pressure is directly proportional to the _____ of the fluid and to its _____.

**4.** Fluid pressure is independent of the _____ or _____ of its container.

**5.** The weight density of a body is equal to the ratio of its _____ to its _____. Is it a universal constant for a given material? _____.

**6.** An object that is completely or partly submerged in a fluid experiences an upward force, called the _____ force, which is equal to the _____ of the fluid displaced. This is _____ principle.

**7.** An open, U-shaped tube partially filled with mercury can be used to measure _____ pressure. Such a device is called a(n) _____.

**8.** If the weight of the displaced fluid exceeds the weight of a submerged body, the body will _____.

**9.** The _____ press operates on the principle that an output pressure is essentially equal to an input pressure. This principle was named after _____.

**10.** The ideal mechanical advantage for a hydraulic press is the ratio of the _____ of the output piston to that of the input piston.

**11.** The _____ is defined as the volume of fluid that passes a certain cross section per unit of time. Its unit is _____ in the USCS.

12. The motion of a fluid in which every particle in the fluid flows the same path past a particular point as that followed by previous particles is called

_____.

13. The _____ principle is responsible for mixing fuel with air in the throat of a carburetor as air rushes through it.

14. In our discussion of fluids, four physical quantities are derived from the fundamental quantities of mass, length, and time and play important roles. They are

_____, _____, _____,

and _____.

15. The discharge velocity increases with _____ below the surface of a liquid. The range of the discharged liquid is a maximum at the

_____.

16. The velocity of discharge from an orifice depends on the

_____ and the _____. The magnitude of the velocity is predicted by _____ theorem.

17. The product of _____ and _____ is a constant and is a measure of the rate of flow.

18. Fluid flow in which swirls and eddies increase the frictional drag is called

_____ flow.

19. In applying Bernoulli's equation, it must be remembered that $P$ represents

_____ pressure and $\rho$ represents _____

density.

20. In Bernoulli's equation, the sum of _____,

_____, and _____ must remain constant.

# Chapter 16

# Temperature and Expansion

**DEFINITIONS OF KEY TERMS**

**Thermal energy**   The total internal energy of an object: the sum of its molecular kinetic and potential energies.

**Thermal equilibrium**   Two objects are said to be in thermal equilibrium if and only if they have the same temperature.

**Temperature**   The degree of "hotness" or "coldness" of an object, as measured on a standard scale.

**Thermometer**   A device that, through marked scales, can give an indication of its own temperature.

**Ice point**   The temperature at which water and ice coexist in thermal equilibrium under a pressure of 1 atm (0°C or 32°F).

**Steam point**   The temperature at which water and steam coexist in thermal equilibrium under a pressure of 1 atm (100°C or 212°F).

**Celsius scale**   A temperature scale that assigns the number 0 to the ice point and the number 100 to the steam point.

**Fahrenheit scale**   A temperature scale on which the ice point is assigned the number 32 and the steam point is given the number 212.

**Absolute zero**   The true zero of temperature. The temperature at which the volume of an ideal gas would be zero.

**Kelvin scale**   An absolute temperature scale based on the same intervals as the Celsius scale, except that its zero point corresponds to a temperature of −273°C.

**Coefficient of linear expansion**   A property of a material that determines the change in length per unit length per degree change in temperature.

**CONCEPTS AND EXAMPLES**

**Temperature**   The SI unit of temperature is the kelvin (K), which is an absolute temperature having its zero point 273 intervals below the ice point of water. Since the Kelvin scale is based on the Celsius interval, the Celsius scale is also recognized and used internationally. Two other units, the degree Fahrenheit (°F) and the degree Rankine (°R), continue to be used in the United States. Therefore, it is necessary to be able to convert temperature intervals and specific temperatures on each of the four scales. A comparison of the scales is given in Fig. 16-1, which shows values for the steam point, the ice point, and absolute zero on each scale.

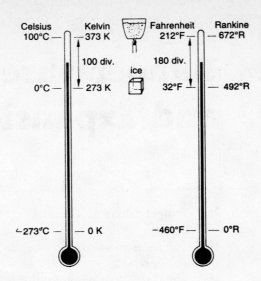

**Fig. 16-1** Comparison of four temperature scales.

For temperature intervals:

$$5 \text{ C}° = 9 \text{ F}° \qquad 1 \text{ K} = 1 \text{ C}°$$

For specific temperatures:

$$t_C = \frac{5}{9}(t_F - 32) \qquad t_F = \frac{9}{5}t_C + 32$$
$$T_K = t_C + 273$$

---

**EXAMPLE 16-1**

A piece of lead cools from 400 to 250°C after it is removed from a furnace. What is the change in temperature in Kelvin and Fahrenheit degrees?

*Solution*
We first subtract to obtain the change of temperature in Celsius degrees:

$$\Delta t = 400°C - 250°C = 150 \text{ C}°$$

Since 1 K = 1 C°, the change is also **150 K**, which is read as "150 Kelvins." To convert this change to Fahrenheit degrees, we recognize that 5 C° = 9 F° and make the conversion:

$$\Delta t = (150 \text{ C}°)\left(\frac{9 \text{ F}°}{5 \text{ C}°}\right) = \boxed{270 \text{ F}°}$$

---

For specific temperatures, you must correct the numerical differences as well as for the interval differences. It is essential that you not confuse an interval with a specific temperature on a given scale.

**EXAMPLE 16-2**    Convert 78°F to the corresponding Kelvin temperature.

*Solution*

We recognize that first we must subtract 32 to correct numerical scale differences, and then we multiply by $\frac{5}{9}$ to correct for the interval difference:

$$t_C = \frac{5(t_F - 32)}{9} = \frac{5(78 - 32)}{9} = \frac{5(46)}{9} = \boxed{25.6°C}$$

**Thermal Expansion**    Different materials expand at different rates when subjected to the same quantity of heat. In general, the change in length, area, or volume is in direct proportion to the original dimension and to the change in temperature:

| | | |
|---|---|---|
| Linear expansion: | $\Delta L = \alpha L_0 \, \Delta t$ | $\alpha = \alpha$ |
| Area expansion: | $\Delta A = \gamma A_0 \, \Delta t$ | $\gamma = 2\alpha$ |
| Volume expansion: | $\Delta V = \beta V_0 \, \Delta t$ | $\beta = 3\alpha$ |

When you find the new dimension, it is best first to determine the change in dimension and then add it to the original value. For example, $L = L_0 + \Delta L$, or $V = V_0 + \Delta V$.

**EXAMPLE 16-3**    A slab of concrete on a highway is 60 ft long. How much longer will it be at 90°F than it is at −5°F? The expansion coefficient for concrete is $0.7 \times 10^{-5}/F°$.

*Solution*

The change in length is found by substitution:

$$\Delta L = \alpha L_0 \, \Delta t = (0.7 \times 10^{-5}/F°)(60 \text{ ft})[90°F - (-5°F)]$$

$$= (42 \times 10^{-4} \text{ ft/F°})(95 \text{ F°}) = \boxed{0.399 \text{ ft}}$$

Thus, the expansion is 0.399 ft or about 5 in.

**Summary**

- It is important to be thoroughly familiar with the four temperature scales: Celsius, Kelvin, Fahrenheit, and Rankine.

- For specific temperatures, you must correct for the interval difference.

- It is necessary to also correct for the fact that different numbers are assigned for the same temperatures.

- The volume expansion of a liquid uses the same relation as for a solid except that there is no linear expansion coefficient $\alpha$ for a liquid; only $\beta$ is needed.

| **Key Equations by Section** | Section | Topic | Equation | Equation No. |
|---|---|---|---|---|
| | 16-2 | Temperature intervals | $100\ C° = 180\ F°$  or  $5\ C° = 9\ F°$ | 16-1 |
| | 16-2 | Conversion factors for differences in temperatures | $\dfrac{5\ C°}{9\ F°} = 1 = \dfrac{9\ F°}{5\ C°}$ | 16-2 |
| | 16-2 | Equivalent temperatures | $\boxed{t_C = \tfrac{5}{9}(t_F - 32)}$ | 16-3 |
| | 16-2 | Solving for $t_F$ | $\boxed{t_F = \tfrac{9}{5}(t_C + 32)}$ | 16-4 |
| | 16-4 | The absolute temperature scale | $\boxed{T_K = t_C + 273}$ | 16-5 |
| | 16-4 | Fahrenheit degree interval | $\boxed{T_R = t_F + 460}$ | 16-6 |
| | 16-4 | Equivalent intervals | $\boxed{1\ K = 1\ C°}$ | 16-7 |
| | 16-5 | Linear expansion based on $\alpha$, the coefficient of linear expansion | $\Delta L = \alpha L_0\, \Delta t$ | 16-8 |
| | 16-5 | $\alpha$, the coefficient of linear expansion | $\alpha = \dfrac{\Delta L}{L_0\, \Delta t}$ | 16-9 |
| | 16-5 | Linear expansion due to temperature change, equation for calculating the final length | $\boxed{L = L_0 + \alpha L_0\, \Delta t}$ | 16-10 |
| | 16-6 | Area expansion | $\Delta A = 2\alpha A_0\, \Delta t$ | 16-11 |
| | 16-6 | Definition of the area expansion coefficient | $\gamma = 2\alpha$ | 16-12 |
| | 16-7 | Area expansion | $\Delta A = \gamma A_0\, \Delta t$ | 16-13 |
| | | Final area | $A = A_0 + \gamma A_0\, \Delta t$ | 16-14 |
| | 16-8 | Volume expansion | $\Delta V = \beta V_0\, \Delta t$ | 16-15 |
| | | Final volume | $V = V_0 + \beta V_0\, \Delta t$ | 16-16 |
| | 16-8 | Definition of volume expansion coefficient | $\beta = 3\alpha$ | 16-17 |

## True-False Questions

**T  F  1.** Two objects that have the same temperature also have the same thermal energy.

**T  F  2.** Two objects are said to be in thermal equilibrium if and only if they are at the same temperature.

**T  F  3.** Twenty Celsius degrees represents the same temperature interval as twenty kelvins.

**T  F  4.** Water freezes at 460° on the Rankine scale.

**T  F  5.** When a temperature interval is converted from Rankine degrees to kelvins, the number of Rankine degrees should be multiplied by 5/9.

**T  F  6.** When a specific temperature in degrees Fahrenheit is converted to the corresponding temperature in degrees Celsius, the number is multiplied by 9/5 and added to 32.

**T  F  7.** For a given temperature interval, the same linear expansion coefficient may be used for the same material, regardless of the choice of units for length.

**T  F  8.** For a solid disk with a hole in its center, the diameter of the disk and the diameter of the hole will increase in length per unit length at a rate given by its linear expansion coefficient.

**T  F  9.** The volume expansion coefficient for a solid is approximately equal to 3 times the linear expansion coefficient, but this does not represent a true equality.

**T  F  10.** The temperature at the bottom of a frozen lake of fresh water is 4°C.

## Multiple-Choice Questions

**1.** Which of the following represents the steam point for water?
(a) 100°F  (b) 212°C  (c) 273 K  (d) 672°R

**2.** Which of the following represents the largest temperature interval?
(a) 40 F°  (b) 30 K  (c) 50 R°  (d) 20 C°

**3.** Two objects are in thermal equilibrium when they have the same
(a) kinetic energy  (b) temperature
(c) thermal energy  (d) potential energy

**4.** Which of the following represents the smallest specific temperature?
(a) 40°F  (b) 5°F  (c) 510°R  (d) 280 K

**5.** The coefficient of linear expansion will vary only with a change in
(a) temperature  (b) initial length
(c) thermal energy  (d) material

**6.** When a flame is held to the bulb of a mercury-in-glass thermometer, the mercury level will
(a) rise  (b) drop
(c) drop and then rise  (d) rise and then drop

**7.** The boiling point of oxygen is −183°C. This temperature is also
(a) −329.4°F  (b) 162.6°R  (c) 456 K  (d) −83.9°F

**8.** The linear expansion coefficient for silver is $2 \times 10^{-5}$/C°. A 6-in. bar of silver is heated from 0 to 100°C. The increase in length is approximately
(a) 0.06 in.  (b) 0.12 in.  (c) 0.012 in.  (d) 0.006 in.

9. The area expansion coefficient for the silver bar in Question 8 is approximately
   (a) $1 \times 10^{-5}/\text{F}°$            (b) $4 \times 10^{-5}/\text{F}°$
   (c) $7.2 \times 10^{-5}/\text{F}°$          (d) $2.2 \times 10^{-5}/\text{F}°$

10. The volume expansion coefficient for ethyl alcohol is $11 \times 10^{-4}/\text{C}°$. What change in temperature must occur to increase the volume of 16 L of the alcohol to 17 L?
    (a) 56.8 C°       (b) 1454 C°       (c) 53.5 C°       (d) 90.9 C°

---

**Completion Questions**

1. The temperature at which the volume of an ideal gas is zero is referred to as _____.

2. Temperature is a measure of the _____ per molecule, and two objects that are at the same temperature are in _____.

3. A metal bar 1 ft in length increases its length by 0.0006 ft when its temperature is increased by 1 C°. Under the same conditions, a 1-m length of the same material would increase its length by _____ m.

4. The coefficient of linear expansion may be defined as the change in _____ per unit _____ per degree change in _____.

5. A device that can give an indication of its own temperature is called a(n) _____.

6. The temperature interval on the Kelvin scale is the same as the _____ temperature interval; however, it is larger than the Fahrenheit interval by a factor of _____.

7. Two fixed points often used as standards for calibration of thermometers are the _____ and the _____.

8. One hundred eighty division on the Fahrenheit scale would correspond to _____ division on the Kelvin scale.

9. Thermal energy represents the sum of the _____ and _____ of all molecules present in a substance.

10. Water experiences its maximum _____ at 4°C.

# Chapter 17

# Quantity of Heat

**DEFINITIONS OF KEY TERMS**

**Heat**    A physical quantity that is a measure of the change in thermal energy during a given process.

**Calorie**    The quantity of heat required to change the temperature of 1 g of water through 1 C°.

**British thermal unit (Btu)**    The quantity of heat required to change the temperature of 1 lb of water through 1 F°.

**Mechanical equivalent of heat**    The quantitative relationship between thermal energy units and mechanical energy units:

$$1 \text{ Btu} = 778 \text{ ft} \cdot \text{lb} \qquad 1 \text{ cal} = 4.186 \text{ J}$$

**Specific heat capacity**    The quantity of heat required to change the temperature of a unit mass of a substance through one unit degree.

**Water equivalent**    The mass of water that would gain or lose the same quantity of heat in a given process as a particular object, in other words, the water equivalent of a thermometer.

**Fusion**    The process in which the phase of a substance changes from a solid to a liquid, also called melting. The opposite process is called *solidification.*

**Melting point**    The temperature at which fusion occurs at 1 atm of pressure.

**Latent heat of fusion**    The quantity of heat per unit of mass required to change a substance from its solid phase to its liquid phase at its melting point. For water, $L_f = 80$ cal/g.

**Vaporization**    The process in which the phase of a substance changes from a liquid to a vapor. The opposite process is called condensation.

**Boiling point**    The temperature at which vaporization occurs under 1 atm of pressure.

**Latent heat of vaporization**    The quantity of heat per unit of mass required to change a substance from its liquid phase to its vapor phase at its boiling point. For water, $L_v = 540$ cal/g.

**Sublimation**    The process in which the phase of a substance changes directly from a solid to a vapor without passing through the liquid phase.

**Heat of combustion**    The quantity of heat per unit mass or per unit volume liberated when a substance is burned.

**Calorimeter**   A gauge for measuring heat content.

**Freezing**   The phase change process of matter changing from liquid to solid. Another name for *solidification*.

**Temperature**   A measure of heat content.

**Specific Heat Capacity**   The specific heat capacity $c$ is used to determine the quantity of heat $Q$ absorbed or released by a unit mass $m$ as the temperature changes by an interval $\Delta t$.

---

**EXAMPLE 17-1**   How much heat is required to raise the temperature of 60 g of copper from 10 to 80°C?

*Solution*
The specific heat of copper is 0.094, so

$$Q = mc\,\Delta t$$
$$= (60 \text{ g})[0.094 \text{ cal}/(\text{g} \cdot \text{C}°)](80°\text{C} - 10°\text{C}) = \boxed{395 \text{ cal}}$$

---

**Calorimetry**   Calorimetry is the application of the principle of conservation of thermal energy. In any given thermodynamic process, the heat gained must equal the heat lost. By setting up an equation based on this principle, we can solve for unknown parameters such as mass, equilibrium temperatures, or specific heats.

---

**EXAMPLE 17-2**   A 70-g aluminum calorimeter cup is partially filled with 200 g of water, and the system is at an initial temperature of 20°C. How many grams of steel shot at 98°C must be added to the system if the equilibrium temperature of the mixture is 31°C? (See Fig. 17-1.)

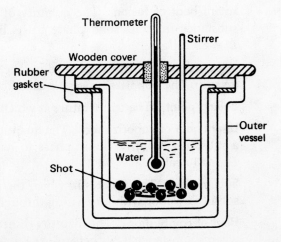

**Fig. 17-1**   A laboratory calorimeter.

*Solution*

The heat lost by the steel shot must equal the heat gained by the water $Q_w$ and by the aluminum cup $Q_c$. First, we determine the total heat gained:

$$Q_w = (200 \text{ g})[1 \text{ cal}/(\text{g} \cdot \text{C}°)](31°\text{C} - 20°\text{C}) = 2200 \text{ cal}$$

$$Q_c = (70 \text{ g})[0.22 \text{ cal}/(\text{g} \cdot \text{C}°)](11 \text{ C}°) = 169.4 \text{ cal}$$

$$\text{Heat gained} = 2200 \text{ cal} + 169.4 \text{ cal} = 2369.4 \text{ cal}$$

Now the heat lost by the shot must also equal 2369.4 cal:

$$\text{Heat lost} = mc \, \Delta t = 2369.4 \text{ cal}$$

$$m(0.114 \text{ cal}/\text{g} \cdot \text{C}°)(98°\text{C} - 31°\text{C}) = 2369.4 \text{ cal}$$

Solving for $m$, we obtain

$$m = \frac{2369.4 \text{ cal}}{(0.114 \text{ cal}/\text{g} \cdot \text{C}°)(67 \text{ C}°)} = \boxed{310 \text{ g}}$$

**Fusion and Vaporization**   When a substance undergoes a change of phase, the thermal energy causes changes in potential energy associated with the separation of molecules. In such cases, there is no accompanying change of temperature. Each substance has its own characteristic boiling point or melting point, and a quantity of latent heat is released or absorbed during a change of phase.

**EXAMPLE 17-3**

How much heat is required to change 60 g of lead at 24°C to molten lead at its melting temperature of 327°C? The heat of fusion for lead is 5.9 cal/g.

*Solution*

There are two steps in the process. First, the temperature of lead must be raised to 327°C, then each gram must be melted at 5.9 cal/g:

$$Q_T = Q_1 + Q_2 = mc \, \Delta t + mL_f$$

$$= (60 \text{ g})(0.031 \text{ cal}/\text{g} \cdot \text{C}°)(327°\text{C} - 24°\text{C}) + (60 \text{ g})(5.9 \text{ cal}/\text{g})$$

$$= 563.6 \text{ cal} + 354 \text{ cal} = 917.6 \text{ cal}$$

## Summary

- The British thermal unit (Btu) is the heat required to change the temperature of one pound-mass of water one Fahrenheit degree.
- The calorie is the heat required to raise the temperature of one gram of water by one Celsius degree.
- The specific heat capacity is used to determine the quantity of heat $Q$ absorbed or released by a unit of mass $m$ as the temperature changes by an interval $\Delta t$.
- Conservation of thermal energy requires that in any exchange of thermal energy the heat lost must equal the heat gained.
- The latent heat of fusion $L_f$ and the latent heat of vaporization $L_v$ are heat losses or gains by a unit mass $m$ during a phase change; there is no temperature change.

| Key Equations by Section | Section | Topic | Equation | Equation No. |
|---|---|---|---|---|
| | 17-2 | Units of quantity of heat | $1 \text{ Btu} = 252 \text{ cal} = 0.252 \text{ kcal}$ | 17-1 |
| | 17-3 | Heat capacity | $\text{Heat capacity} = \dfrac{Q}{\Delta t}$ | 17-2 |
| | 17-3 | Specific heat capacity | $c = \dfrac{Q}{m\Delta t} \qquad Q = mc\,\Delta t$ | 17-3 |
| | 17-4 | Conservation of thermal energy | $\textit{Heat lost} = \textit{heat gained}$ | 17-5 |
| | 17-5 | Latent heat of fusion | $L_f = \dfrac{Q}{m} \qquad Q = mL_f$ | 17-6 |
| | 17-6 | Latent heat of vaporization | $L_v = \dfrac{Q}{m} \qquad Q = mL_v$ | 17-7 |

## True-False Questions

T  F  **1.** Since the Btu is based on the pound, its magnitude depends on the acceleration due to gravity.

T  F  **2.** The Btu is a larger unit of heat than the kilocalorie.

T  F  **3.** It is proper to speak of the heat capacity of a penny and the *specific* heat capacity of copper.

T  F  **4.** The specific heat of any material is the same numerically in either the metric or British system of units.

T  F  **5.** A unit mass of water will absorb a larger quantity of heat per degree change in temperature than any other common substance.

T  F  **6.** Since temperature is a measure of the average kinetic energy per molecule, the absorption or emission of thermal energy is always accompanied by a change in temperature.

T  F  **7.** More than 5 times the thermal energy is required to vaporize 1 g of water than is required to melt 1 g of ice.

T  F  **8.** Two objects that have the same heat capacity must be made of the same material.

T  F  **9.** The heat lost is equal to the heat gained unless a change of phase occurs.

T  F  **10.** The mass of water displaced by a thermometer in a calorimetry experiment is known as the *water equivalent* of the thermometer.

## Multiple-Choice Questions

**1.** Which of the following is not a property of a material?
(a) Specific heat
(b) Heat capacity
(c) Heat of fusion
(d) Heat of combustion

**2.** Which of the following represents the largest transfer of heat?
(a) 600 cal
(b) 3 Btu
(c) 0.7 kcal
(d) 2200 ft · lb

3. Which of the following is best associated with the term *heat*?
   (a) A change of temperature      (b) A change in kinetic energy
   (c) A change in thermal energy      (d) A change in heat capacity

4. The specific heat of aluminum is 0.22 cal/(g · C°). The quantity of heat required to change the temperature of 10 lb of aluminum from 20 to 100°F is approximately
   (a) 220 Btu      (b) 200 Btu      (c) 176 Btu      (d) 88 Btu

5. When a liquid freezes, it
   (a) evolves heat      (b) absorbs heat
   (c) decreases in temperature      (d) sublimes

6. The quantity of heat required to convert 10 g of ice at 0°C to steam at 100°C is
   (a) 6300 cal      (b) 7200 cal      (c) 720 cal      (d) 1350 cal

7. If 50 g of aluminum shot ($c = 0.22$) is heated to 100°C and dropped into 200 g of water at 20°C, the equilibrium temperature (neglecting other heat transfers) is approximately
   (a) 24°C      (b) 26°C      (c) 34°C      (d) 46°C

8. A 50-g aluminum calorimeter cup is partially filled with 200 g of water at 20°C. What mass of copper ($c = 0.093$) at 100°C should be added to the system for the equilibrium temperature to be 30°C?
   (a) 32.4 g      (b) 47.5 g      (c) 324 g      (d) 476 g

9. For a given substance, which of the following processes transfers the largest amount of thermal energy per unit of mass?
   (a) Fusion      (b) Vaporization
   (c) Freezing      (d) Sublimation

10. If 5 g of steam at 100°C is mixed with 10 g of ice at 0°C, the equilibrium mixture includes
    (a) 3.33 g of steam and 11.67 g of water
    (b) 1.67 g of steam and 13.33 g of water
    (c) 1.67 g of steam and 3.33 g of water
    (d) 15 g of water

---

**Completion Questions**

1. The _____ is the quantity of heat required to change the temperature of 1 lb of water through 1 F°.

2. The calorie is the quantity of heat required to change the _____ of _____ of water through one _____.

3. The quantity of heat per unit mass of a substance required to change its temperature through one degree is called its _____.

4. The latent heat of fusion of a substance is the _____ necessary to change _____ of the substance from a(n) _____ to a(n) _____ at its _____.

**5.** The latent heat of vaporization is the heat per unit _____ of a

substance required to change it from a(n) _____ to

a(n) _____ at its _____.

**6.** The _____ of _____ is the heat per unit

mass required to burn a substance completely.

**7.** The latent heat of vaporization for water is _____ cal/g, or

_____ Btu/lb.

**8.** The process in which a substance changes directly from its solid phase to its va-

por phase is called _____.

**9.** The quantitative relationship between thermal energy units and mechanical

energy units is called _____.

**10.** The conservation of heat energy states that in any given transfer of thermal

energy the _____ must equal the _____.

## Chapter 18

# Transfer of Heat

**DEFINITIONS OF KEY TERMS**

**Conduction**   The process in which heat energy is transferred by adjacent molecular collisions throughout a material medium. The medium itself does not move.

**Radiation**   The process in which heat is transferred by electromagnetic radiation.

**Thermal conductivity**   A measure of the ability of a substance to conduct heat.

**Natural convection**   Currents produced in a fluid because of a difference in density that accompanies a change in temperature of a portion of the fluid.

**Forced convection**   Currents produced in a fluid as a result of the action of a pump, fan, or other device.

**Blackbody radiation**   Radiation emitted from an ideal radiator, which is defined as an object that absorbs *all* the radiation incident on its surface.

**Emissivity**   A measure of the ability of an object to absorb or emit radiation.

**Stefan–Boltzmann law**   The relation predicting the power per unit of area emitted from an object of known emissivity and temperature.

**Prevost law of heat exchange**   A body at the same temperature as its surroundings radiates and absorbs heat at the same rate.

**CONCEPTS AND EXAMPLES**

**Conduction**   In the transfer of heat by conduction, the quantity of heat $Q$ transferred per unit of time $\tau$ through a wall or rod of length $L$ is given by

$$H = \frac{Q}{\tau} = kA\frac{\Delta t}{L} \qquad k\text{(conductivity), kcal/m} \cdot \text{s} \cdot \text{C}°$$

where $A$ is the area and $\Delta t$ is the difference in surface temperatures. The choice of units is critical since the conductivities $k$ are determined in terms of specific units. We summarize the appropriate units:

|      | $Q$  | $L$ | $A$    | $t$ | $\tau$ |
|------|------|-----|--------|-----|--------|
| SI   | kcal | m   | $m^2$  | C°  | s      |
| USCS | Btu  | in. | $ft^2$ | F°  | h      |

Only these consistent units may be substituted into the conductivity equation.

**EXAMPLE 18-1**

A solid concrete wall is 12 ft high, 24 ft wide, and 8 in. thick. The outside surface is at 36°F and the inside surface is maintained at 81°F. How much heat is lost through the wall in 8 h?

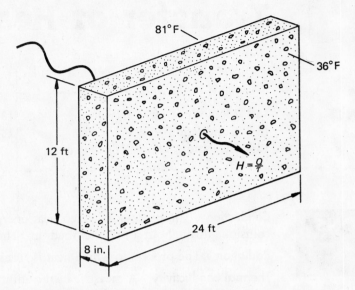

**Fig. 18-1**   Heat transfer by conduction through a concrete wall.

**Solution**

A figure is drawn as Fig. 18-1 to clarify the information. The area of the wall is 12 × 24 ft, or 288 ft². All units are checked for consistency with the constant $k$, which is 12 Btu · in./(ft² · h · F°). Solving the conductivity equation for $Q$ and substituting knowns give

$$Q = kA\tau\frac{\Delta t}{L} = (12 \text{ Btu} \cdot \text{in./ft}^2 \cdot \text{h} \cdot \text{F}°)(288 \text{ ft}^2)\frac{(81°\text{F} - 36°\text{F})}{8 \text{ in.}}$$

$$= \boxed{155,520 \text{ Btu}} \qquad \text{heat lost in 8 h}$$

It is important to substitute the units into the equation as a check on consistency. All units but the desired unit should cancel.

**Radiation**   For heat transfer by radiation, we define the rate of radiation $R$ as the energy emitted per unit area $A$ per unit time $\tau$ (or simply the power per unit area $P/A$):

$$R = \frac{E}{\tau A} = e\sigma T^4 \qquad \sigma = 5.67 \times 10^{-8} \text{ W/(m}^2 \cdot \text{K}^4)$$

The emissivity $e$ is a number from 0 (for a perfect reflector) to 1 (for a perfect absorber). It measures the ability of a surface to absorb or to emit radiation.

**EXAMPLE 18-2**    Thermal radiation is incident on a body at the rate of 200 W/m². If the body absorbs 25 percent of the incident radiation, what energy will be emitted by this body in 1 h? The surface area is 0.6 m², and its temperature is 800°C.

*Solution*

From the given information we must determine the emissivity $e$ and the absolute temperature $T$. The emissivity, by definition, is related to the percentage of absorption or emission. Therefore, $e = 0.25$ for this body, and its temperature is 1073 K (800 + 273). We solve the radiation equation for $E$ and substitute given quantities. Recall that 1 h = 3600 s.

$$E = e\sigma AtT4$$
$$= (0.25)[5.67 \times 10^{-8} \text{ W/(m}^2 \cdot \text{K}^4)](1073 \text{ K})^4$$
$$= 4.059 \times 10^7 \text{ J}$$
$$= \boxed{\textbf{40.6 MJ}} \quad \text{energy radiated in 1 h}$$

This example demonstrates a radiated power of 11.3 kW.

## Summary

- Heat is the transfer of thermal energy from one place to another.
- The rate of transfer by conduction, convection, and radiation can be predicted from experimental formulas.
- In the transfer of heat by conduction, the quantity of heat $Q$ transferred per unit of time $\tau$ through a wall or rod of length $L$ is given by the conduction equation.
- Other commonly used units are kcal/s and Btu/h.
- The $R$-value is an engineering term to measure the thermal resistance offered to the conduction of heat.
- For heat transfer by radiation, we define the rate of radiation as the energy emitted per unit area per unit time (or the power per unit area).
- Prevost's law of heat exchange states that a body at the same temperature as its surroundings radiates and absorbs heat at the same rates.

## Key Equations by Section

| Section | Topic | Equation | Equation No. |
|---|---|---|---|
| 18-2 | Conduction of heat | $$\boxed{H = \frac{Q}{\tau} = kA\frac{\Delta t}{L}}$$ | 18-1 |
| 18-2 | Thermal conductivity of a material | $$k = \frac{QL}{A\tau\,\Delta t}$$ | 18-2 |
| 18-3 | Insulation and $R$-value | $$\frac{Q}{\tau} = \frac{A\,\Delta t}{\Sigma_i(L_i/k_i)} = \frac{A\,\Delta t}{\Sigma_i/R_i}$$ | 18-3 |

| Section | Topic | Equation | Equation No. |
|---|---|---|---|
| 18-5 | Rate of radiation in terms of radiant energy, area, and time; or power and area | $R = \dfrac{E}{\tau A} = \dfrac{P}{A}$ | 18-4 |
| 18-5 | The Stefan–Boltzmann law; the rate of radiation in terms of power and area; or emissivity (0–1), Stefan's constant $\sigma$ ($5.67 \times 10^{-8}$ W/m$^2 \cdot k^4$), and absolute temperature | $\boxed{R = \dfrac{P}{A} = e\sigma T^4}$ | 18-5 |
| 18-5 | Net rate of radiation | $\boxed{R = e\sigma(T_1^4 - T_2^4)}$ | 18-6 |

## True-False Questions

T   F   **1.** The relationships that predict heat transfer are based on empirical observations and depend on ideal conditions.

T   F   **2.** The quantity of heat transferred through a slab of area 4 ft$^2$ is greater than the quantity of heat conducted through an area of 8 ft$^2$, assuming that all other parameters are constant.

T   F   **3.** When a radiator is used to heat a room, the principal method of heat transfer warming the room is convection.

T   F   **4.** Air-conditioning outlets in the ceiling are more efficient than those on the floor.

T   F   **5.** All objects emit electromagnetic radiation, regardless of their temperature or the temperature of their surroundings.

T   F   **6.** An object that absorbs a large percentage of incident radiation will be a poor emitter of radiation.

T   F   **7.** Because of a similarity in the definition of heat units, the thermal conductivities are the same numerically in the engineering system and the metric system of units.

T   F   **8.** A body at the same temperature as its surroundings radiates and absorbs heat at the same rate.

T   F   **9.** In a composite wall of two or more different materials, the same number of calories is transferred per unit area per unit time through each material after time is allowed for steady flow to be established.

T   F   **10.** The units of thermal conductivity in the metric system may be written cal $\cdot$ cm/(cm$^2 \cdot$ s $\cdot$ C°), but the numerical value would be different.

## Multiple-Choice Questions

**1.** Which of the following geometries will result in the largest convection coefficient?
(a) Vertical plate
(b) Horizontal plate, facing upward
(c) Diagonal plate
(d) Horizontal plate, facing downward

**2.** When the temperature of an object is doubled, its rate of radiation is increased by a factor of

(a) 2       (b) 4       (c) 8       (d) 16

**3.** The dead airspace between the walls of a calorimeter cup and its outside container minimizes heat loss due to

(a) conduction               (b) convection

(c) radiation                 (d) contamination

**4.** The rate at which heat flows through a solid plate of some materials does *not* depend on the

(a) temperature difference       (b) thickness

(c) specific heat              (d) area

**5.** The direction of heat flow is always from

(a) high temperature to low temperature

(b) high pressure to low pressure

(c) high density to low density

(d) a point of higher emissivity

**6.** Which of the following does *not* indicate heat flow as a quantity of heat per unit time?

(a) $kA\,\Delta t/L$       (b) $hA\,\Delta t$       (c) $H$       (d) $e\sigma T^4$

**7.** The thermal conductivity of a plate is 0.01 kcal/s · m · C°. The plate is 2 cm thick and has a cross section of 4000 cm². If one side is at 150°C and the other at 50°C, the number of kilocalories transferred every second is approximately

(a) 10 kcal       (b) 20 kcal       (c) 40 kcal       (d) 80 kcal

**8.** The convection coefficient for a vertical plate is $12.7 \times 10^{-4}$ kcal/s · m² · C° when the difference of temperature between the plate and its surroundings is 810°C. How much heat is transferred by convection from each side of the plate in 1 h if the area is 20 cm²?

(a) 0.206 cal       (b) 12.4 cal       (c) 7410 cal       (d) 8410 cal

**9.** A body having an emissivity of 0.2 and a surface area of 0.2 m² is heated to 727°C. The power radiated from the surface is approximately

(a) 634 W       (b) 1134 W       (c) 1830 W       (d) 2268 W

**10.** The units Btu/(h · ft² · F°) are appropriate for the

(a) convection coefficient       (b) thermal conductivity

(c) rate of radiation           (d) emissivity

---

**Completion Questions**

**1.** Two types of convection that apply to most heating systems are

_____ convection and _____ convection.

**2.** For the common laboratory calorimeter, heat losses resulting from

_____ are minimized by a dead airspace. The rubber ring

prevents heat losses by _____, and radiation losses are

minimized by the _____.

**3.** Heat is transferred from the sun to the earth by means of

_____.

**4.** The rate at which thermal radiation is emitted from a surface varies directly

with the _____ power of the _____.

5. The _____ of a body is a measure of its ability to absorb or emit thermal radiation, and it may vary from a value of _____ to a value of _____.

6. The convection coefficient is not a property of the solid or fluid but depends primarily on the _____ of the solid. The convection coefficients for a wall, a floor, and a ceiling are largest for the _____ and lowest for the _____.

7. The British units commonly used for thermal conductivity are the _____, and the metric units are the _____.

8. Copper has about twice the thermal conductivity of aluminum, but its specific heat is a little less than half that of aluminum. A rectangular block is made from each material, so that they have identical masses and the same surface area at their bases. Each block is heated 300°C and placed on top of a large cube of ice. The _____ block will sink deeper into the ice because it has a higher _____. The _____ block will stop sinking first because it has a higher _____.

9. The warm air over a burning fire will rise under the influence of _____ currents.

10. On a cold day, a piece of iron feels colder to the touch than a piece of wood at the same temperature because the iron has a higher _____.

# Chapter 19

# Thermal Properties of Matter

## DEFINITIONS OF KEY TERMS

**Ideal gas**   A gas whose behavior is completely unaffected by cohesive forces or molecular volumes. Such gases obey simple laws and can never be liquefied.

**Boyle's law**   Provided that the mass and temperature of a sample of gas are maintained constant, the volume of the gas is inversely proportional to its absolute pressure.

**Charles's law**   Provided that the mass and pressure of a gas sample are kept constant, the volume of the sample is directly proportional to its absolute temperature.

**Gay-Lussac's law**   Provided that the mass and volume of a gas remain constant, the pressure of the sample is directly proportional to its absolute temperature.

**Atomic mass**   The atomic mass of an element is the mass of an atom of that element compared with the mass of an atom of carbon, taken as 12 atomic mass units (u).

**Molecular mass**   The sum of the atomic masses of all the atoms make up a molecule.

**Mole**   One mole of a substance is the mass in grams equal numerically to the molecular mass of a substance.

**Avogadro's number**   The number of molecules in 1 mol of any substance.

**Critical temperature**   That temperature above which a gas will not liquefy, regardless of the amount of pressure applied to it.

**Evaporation**   Vaporization at the surface of a liquid because of the exodus of the more energetic molecules within the liquid.

**Saturated vapor pressure**   The additional pressure exerted by vapor molecules on a substance and its surroundings under a condition of saturation.

**Boiling**   Vaporization within the body of a liquid when its vapor pressure equals the pressure on its surface.

**Triple point**   The temperature and pressure for which a substance may coexist in its solid, liquid, and vapor phases under a condition of equilibrium.

**Absolute humidity**   The mass of water vapor per unit volume of air.

**Ideal gas law**   This is written as $PV = nRT$.

**Sublimation**   The phase change of matter from a solid directly to a gas.

**Dew point**   The temperature to which the air must be cooled, at constant pressure, to produce saturation.

**General Gas Laws**  Boyle's law, Charles's law, and Gay-Lussac's law are each special cases of a general law that relates the pressure, volume, mass, and temperature of a gas in one state with their values at any other state. In general,

$$\frac{P_1 V_1}{m_1 T_1} = \frac{P_2 V_2}{m_2 T_2}$$

where

$$P = \text{absolute pressure}$$
$$T = \text{absolute temperature}$$
$$m = \text{mass of gas}$$
$$V = \text{volume of gas}$$

**EXAMPLE 19-1**

An 18-g sample of a gas occupies a 12-L storage tank. When the temperature is 27°C, the gauge pressure reading is 250 kPa. After a few days, 6 g of gas escapes through a crack in the container. What is the new gauge pressure when the temperature is at 40°C?

*Solution*
The volume will not change because the walls of the tank are not flexible. Thus, $V_1 = V_2$, and volume drops out of the general equation. The absolute pressure and temperatures are calculated first:

$$P_1 = 250 \text{ kPa} + 101.3 \text{ kPa} = 351.3 \text{ kPa}$$

$$T_1 = 27 + 273 = 300 \text{ K} \qquad T_2 = 40 + 273 = 313 \text{ K}$$

Now since the volume is constant, we write

$$\frac{P_1}{m_1 T_1} = \frac{P_2}{m_2 T_2} \qquad or \qquad P_2 = \frac{P_1 m_2 T_2}{m_1 T_1}$$

so

$$P_2 = \frac{(351.3 \text{ kPa})(12 \text{ g})(313 \text{ K})}{(18 \text{ g})(300 \text{ K})} = 244.3 \text{ kPa}$$

But this is the absolute pressure. The new gauge pressure is found by subtracting atmospheric pressure, to obtain

$$P_2 = 244.3 \text{ kPa} - 101.3 \text{ kPa} = 143 \text{ kPa}$$

In this example, we did not change grams to kilograms or kilopascals to newtons per square meter. As long as we use the same units for the final and initial states, any dimensionally correct unit will work.

**Ideal Gas Law**  It is often desirable to work with a sample of gas in a single state rather than to deal with initial and final conditions. In such cases it is necessary for us to learn the concepts of molecular mass, Avogadro's number, and the mole. Remember that 1 mol is the mass in grams equal numerically to the molecular mass of a substance. The number of molecules in 1 mol of any gas is Avogadro's number $N_A$:

$$N_A = 6.023 \times 10^{23} \text{ molecules per mole}$$

The number of moles is found by dividing the mass of a gas (in grams) by its molecular mass $M$. For example, the molecular mass of $CO_2$ is 44 u. This means that each mole of this gas has a mass of 44 g. A 600-g sample would contain 600/44 mol, or 13.6 mol.

Now the ideal gas law can be stated in terms of the molar concept to arrive at a more specific equation:

$$PV = nRT \qquad R = 0.0821 \text{ L} \cdot \text{atm/mol} \cdot \text{K}$$

In applying this equation, the units of $P$, $V$, $T$, and $n$ must be restricted to those appearing in the constant $R$.

---

**EXAMPLE 19-2**

The molecular mass of oxygen is 32 u. What mass of oxygen is contained in a 6-L tank at 300 K and an absolute pressure of 40 atm?

**Solution**

We solve for the number of moles $n$, using the ideal gas law:

$$n = \frac{PV}{RT} = \frac{(40 \text{ atm})(6 \text{ L})}{[0.0821 \text{ L} \cdot \text{atm/(mol} \cdot \text{K)}](300 \text{ K})} = 9.744 \text{ mol}$$

Now 1 mol has a mass of 32 g. Therefore, the mass of our sample is

$$m = (32 \text{ g/mol})(9.744 \text{ mol}) = 312 \text{ g}$$

---

**Relative Humidity**   The relative humidity is the ratio of the actual vapor pressure in the air to the saturated vapor pressure at that temperature. Tables have been developed to give the saturated vapor pressure of water at various temperatures (Table 19-1 of the text). The *dew point* is the temperature to which air must be cooled under constant pressure to produce saturation. Given the temperature of the air and the dew point, one can use saturated vapor tables to find the relative humidity.

---

**EXAMPLE 19-3**

The air temperature on a summer day is 82°F. When iced tea is poured into a glass, moisture forms on the outside surface when its temperature cools to 72°F. What is the relative humidity?

**Solution**

The actual vapor pressure in the air at 82°F is determined from tables by looking for the saturated vapor pressure at the dew point (72°F). At 72°F the tables show a pressure of approximately 19.8 mm of mercury. If the air were saturated at 82°F, the pressure would be 28.3 mm. Therefore, the relative humidity is

$$\text{Relative humidity} = \frac{\text{actual vapor pressure}}{\text{saturated vapor pressure}}$$

$$= \frac{19.8 \text{ mm}}{28.3 \text{ mm}} = 0.70$$

The relative humidity is 70 percent.

---

## Summary

- Boyle's law: If the temperature of a sample of gas is held constant, the volume of the gas is inversely proportional to its absolute pressure.
- Charles's law: If the pressure of a gas is held constant, the volume of the gas is directly proportional to its absolute temperature.
- Gay-Lussac's law: If the volume of a gas remains constant, the absolute pressure of the gas is directly proportional to its absolute temperature.
- When applying the general gas law in any of its forms, you must remember that the pressure is *absolute pressure* and the temperature is *absolute temperature*.
- A more general form of the gas law is obtained by using the concepts of molecular mass $M$ and the number of moles $n$ for a gas.
- The relative humidity can be computed from saturated-vapor-pressure tables.
- The actual vapor pressure at a particular temperature is the same as the *saturated* vapor pressure for the dew-point temperature.

## Key Equations by Section

| Section | Topic | Equations | Equation No. |
|---------|-------|-----------|--------------|
| 19-1 | Boyle's law ($T$ constant) | $\boxed{P_1 V_1 = P_2 V_2}$ | 19-1 |
| 19-1 | Charles's law ($P$ constant) | $\boxed{\dfrac{V_1}{T_1} = \dfrac{V_2}{T_2}}$ | 19-2 |
| 19-2 | Gay-Lussac's law ($V$ constant) | $\boxed{\dfrac{P_1}{T_1} = \dfrac{P_2}{T_2}}$ | 19-3 |
| 19-3 | General gas law | $\boxed{\dfrac{P_1 V_1}{T_1} = \dfrac{P_2 V_2}{T_2}}$ | 19-4 |
| 19-3 | General gas law including changes in gas mass | $\boxed{\dfrac{P_1 V_1}{m_1 T_1} = \dfrac{P_2 V_2}{m_2 T_2}}$ | 19-5 |
| 19-4 | Avogadro's number | $N_A = \dfrac{N}{n} = 6.023 \times 10^{23}$ particles per mole | 19-6, 19-7 |
| 19-4 | Method for determining the number of moles | $\boxed{n = \dfrac{m}{M}}$ | 19-8 |
| 19-5 | General gas law | $\boxed{\dfrac{P_1 V_1}{n_1 T_1} = \dfrac{P_2 V_2}{n_2 T_2}}$ | 19-9 |
| 19-5 | Ideal gas law | $\boxed{PV = nRT}$ | 19-11 |
| 19-5 | Restatement of the ideal gas law | $PV = \dfrac{m}{M} RT$ | 19-12 |
| 19-10 | Relative humidity | $\dfrac{\textit{Actual vapor pressure}}{\textit{Saturated vapor pressure}}$ | 19-13 |

## True-False Questions

**T  F  1.** Provided that the mass and temperature of an ideal gas are maintained constant, the volume of a gas is directly proportional to its absolute pressure.

**T  F  2.** If the mass and volume of a gas remain constant, doubling the pressure will also double the temperature.

**T  F  3.** The mass of a single molecule of a substance is known as its molecular mass.

**T  F  4.** At a temperature of 273 K and a pressure of 1 atm, 1 mol of any gas will occupy a volume of 22.4 L.

**T  F  5.** Only absolute temperatures and absolute pressures can be used in applying the general gas laws.

**T  F  6.** If the relative humidities inside and outside the house are the same, the dew points must also be the same.

**T  F  7.** The saturated vapor pressure for a substance is greater at higher temperatures.

**T  F  8.** All forms of vaporization are cooling processes.

**T  F  9.** It is possible for ice to be in thermal equilibrium with boiling water.

**T  F  10.** The same mass of any ideal gas will occupy the same volume at standard temperature and pressure.

## Multiple-Choice Questions

**1.** Boyle's law states that, when other parameters are held constant,
  (a) pressure varies directly with volume
  (b) pressure varies directly with temperature
  (c) pressure varies inversely with volume
  (d) volume varies directly with temperature

**2.** If the mass and pressure of a gas are held constant while its volume doubles, the temperature is changed by a factor of
  (a) $\frac{1}{4}$        (b) $\frac{1}{2}$        (c) 2        (d) 4

**3.** The amount of water contained in the air of a given room is described most accurately by the
  (a) absolute humidity        (b) relative humidity
  (c) vapor pressure        (d) dew point

**4.** It is possible for a substance to coexist in all three of its phases in equilibrium when the substance is at its
  (a) critical pressure        (b) critical temperature
  (c) triple point        (d) dew point

**5.** At a temperature of 273 K and a pressure of 1 atm, 1 mol of any gas will occupy a volume
  (a) of 1 L        (b) of 22.4 m$^3$
  (c) equal to its molecular mass        (d) of 22.4 L

**6.** A weather balloon is filled to a volume of 400 L at 0°C. What will its volume be at 100°C if the pressure is constant?
  (a) 147 L        (b) 255 L
  (c) 293 L        (d) 547 L

7. The molecular mass of oxygen is 32 g/mol. How many molecules are present in 64 g of oxygen? Avogadro's number is $6.02 \times 10^{23}$.
   - (a) $3.012 \times 10^{23}$ molecules
   - (b) $6.02 \times 10^{23}$ molecules
   - (c) $12.04 \times 10^{23}$ molecules
   - (d) $24.092 \times 10^{23}$ molecules

8. How many grams of $CO_2$ ($M = 44$ g) will occupy a volume of 200 L at a pressure of 3 atm and an absolute temperature of 300 K?
   - (a) 10,719 g
   - (b) 512 g
   - (c) 107 g
   - (d) 244 g

9. When the air temperature is 26°C and the dew point is 10°C, the relative humidity is
   - (a) 32 percent
   - (b) 36.5 percent
   - (c) 43.2 percent
   - (d) 54.1 percent

10. A 5000-cm³ container holds 6 g of gas under a pressure of 2 atm and a temperature of 20°C. When 10 g of the same gas fills a 2500-cm³ container, the temperature rises to 30°C. The new pressure is
    - (a) 2.71 atm
    - (b) 3.3 atm
    - (c) 6.89 atm
    - (d) 9.31 atm

---

## Completion Questions

1. Provided that the _____ and _____ of a sample of gas are maintained constant, the _____ of the gas varies inversely with its absolute pressure. This is known as _____ law.

2. In the ideal gas law, the ratio of _____ to _____ is always equal to constant $R$, known as the _____ constant.

3. The temperature above which a gas will not liquefy, regardless of the pressure applied, is called the _____.

4. Three types of vaporization are _____, _____, and _____.

5. The temperature to which air must be cooled at constant pressure to produce saturation is called the _____.

6. _____ humidity represents the quantity of water vapor per unit volume. _____ humidity, however, is a percentage based on the ratio of _____ to _____.

7. A _____ is that quantity of a substance that contains the same number of particles as there are atoms in 12 g of carbon 12.

8. If the _____ and _____ of a gas are held constant, Charles's law states that the _____ of a gas is directly proportional to its _____.

9. A mole of any gas contains _____ molecules. This number is referred to as _____.

10. The units for the constant $R$ in the ideal gas law are _____, as determined from the equation.

# Chapter
# 20

# Thermodynamics

## DEFINITIONS OF KEY TERMS

**Thermodynamics**   The science that treats the transformations of thermal energy into mechanical energy and the reverse process, the conversion of work to heat.

**P-V diagram**   A graph that plots the change in volume as a function of pressure in a thermodynamic process.

**Adiabatic process**   A thermodynamic process in which there is no exchange of thermal energy between a system and its surroundings. Work is done at the expense of internal energy.

**Isochoric process**   A thermodynamic process in which the volume of the system remains constant. All absorbed thermal energy contributes to an increase in internal energy.

**Isothermal process**   A thermodynamic process in which fluid at high pressure seeps adiabatically through a porous wall or narrow opening to a region of low pressure.

**First law of thermodynamics**   In any thermodynamic process, the net heat absorbed by a system is equal to the sum of the thermal equivalent of the work done by the system and the change in internal energy of the system.

**Second law of thermodynamics**   It is impossible to construct an engine that, operating in a full cycle, produces no other effect than the extraction of heat from a reservoir and the performance of an equivalent amount of work; a 100 percent efficient engine is not possible.

**Carnot cycle**   An ideal thermodynamic cycle that represents the maximum possible efficiency. For a Carnot engine, it consists of four stages: isothermal expansion, adiabatic expansion, isothermal compression, and adiabatic compression.

**Heat engine**   A device that converts heat energy to mechanical work.

**Refrigerator**   A device that uses mechanical work to extract heat from one reservoir and exhaust it to another.

**Coefficient of performance**   A measure of the cooling efficiency of a refrigerator; the ratio of the heat extracted from a cold reservoir to the thermal equivalent of the work supplied to the refrigerator.

**Carnot efficiency**   The total efficiency of the four Carnot engine processes.

**Compressor**   A mechanical device that does work to reduce the volume of a gas.

**Condenser**   A mechanical device that extracts energy from a gas in the process of liquefaction.

**Refrigerant**   A gas that is liquefied easily and is used to transfer heat from one part of a refrigeration system to another.

## CONCEPTS AND EXAMPLES

**First Law of Thermodynamics**   The first law of thermodynamics is simply a restatement of the principle of conservation of energy. Applied to thermodynamics, the law can be stated as follows:

*In any thermodynamic process, the net heat absorbed by a system $\Delta Q$ is equal to the sum of the net work done by the system $\Delta W$ and the change in internal energy of the system $\Delta U$.*

In equation form,

$$\Delta Q = \Delta W + \Delta U \qquad \text{first law}$$

## EXAMPLE 20-1

Consider the system described by Fig. 20-1. A quantity of heat equal to 600 cal is supplied to the system, and 200 cal of heat is lost in the process. How much work is done by the system if its internal energy increases by 150 cal?

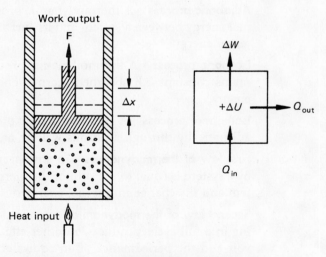

**Fig. 20-1**   The first law of thermodynamics states that the net heat input, $Q_{in} - Q_{out}$, must equal the work output plus the increase in internal energy.

### Solution
According to the first law, the work done by the system is equal to the difference between the *net* heat delivered $\Delta Q$ and the change in internal energy. First, we determine $\Delta Q$:

$$\Delta Q = Q_{in} - Q_{out} = 600 \text{ cal} - 200 \text{ cal} = 400 \text{ cal}$$

Now the work done is

$$\Delta W = \Delta Q - \Delta U = 400 \text{ cal} - 150 \text{ cal} = 250 \text{ cal}$$

The mechanical equivalent of 250 cal is

$$\Delta W = (250 \text{ cal})(4.186 \text{ J/cal}) = \boxed{1046 \text{ J}}$$

**Second Law of Thermodynamics**   The second law sets limits on the accomplishment of the first. In any real thermodynamic process, some input energy is always lost to friction or other dissipative forces and cannot be retrieved. Thus, it is impossible to construct any device that operates with 100 percent efficiency.

**Heat Engines**   A heat engine is any device that converts heat energy to mechanical work. A schematic diagram, such as the one in Fig. 20-2, is useful for describing the operation of any engine. The internal energy returns to its initial state during each cycle, making $\Delta U = 0$. Thus, the first law shows us that $\Delta W = \Delta Q$, and the work output must equal the difference between the heat input and the heat output:

$$W_{\text{out}} = Q_{\text{in}} - Q_{\text{out}}$$

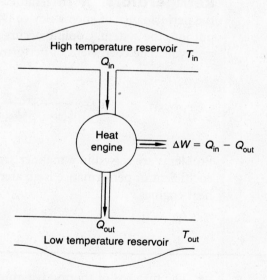

**Fig. 20-2**   Schematic diagram for a heat engine.

The *efficiency E* of an engine is the ratio of the work output to the heat input. For an ideal engine, it can be calculated from either of the following relations:

$$E = \frac{Q_{\text{in}} - Q_{\text{out}}}{Q_{\text{in}}} \qquad E = \frac{T_{\text{in}} - T_{\text{out}}}{T_{\text{in}}} \qquad \textit{efficiency}$$

**EXAMPLE 20-2**   An ideal engine has an efficiency of 30 percent. During each cycle the engine absorbs 900 cal of heat at a temperature of 800 K. How much work is done during each cycle? What is the exhaust temperature?

*Solution*
Since $e = W_{out}/Q_{in}$, we find that

$$W_{out} = eQ_{in} = 0.30(900 \text{ cal}) = 270 \text{ cal}$$

The mechanical equivalent in joules is

$$W_{out} = (270 \text{ cal})(4.186 \text{ J/cal}) = \boxed{1130 \text{ J}}$$

The exhaust temperature is $T_{out}$. Solving for $T_{out}$ from the efficiency equation gives

$$T_{out} = (1 - e)(T_{in}) = (1 - 0.3)(800 \text{ K})$$
$$= (0.7)(800 \text{ K}) = \boxed{560 \text{ K}}$$

You should also verify that the heat lost during each cycle is equal to 630 cal.

**Refrigerators**   A refrigerator is a heat engine operating in reverse. A measure of the performance of such a device is the amount of cooling you get for the work you put into the system. Cooling occurs as a result of extraction of heat $Q_{cold}$ from a cold reservoir. The coefficient of performance K is given by

$$K = \frac{Q_{cold}}{Q_{hot} - Q_{cold}} \quad \text{or} \quad K = \frac{T_{cold}}{T_{hot} - T_{cold}}$$

Problems are solved in a manner similar to that used for heat engines except that the coefficient of performance is greater than 1, whereas the efficiency was less than 1 for heat engines.

## Summary

- The first law of thermodynamics states that the net heat $\Delta Q$ put into a system is equal to the net work done by the system plus the net change in internal energy $\Delta U$ of the system.
- In thermodynamics, work $\Delta W$ is often done on a gas.
- Special cases of the first law occur when one of the quantities does not undergo a change.
- According to the second law of thermodynamics, in every process there is some loss of energy due to frictional forces or other dissipative forces.
- The efficiency $e$ of an engine is the ratio of the work output to the heat input.
- A *refrigerator* is a heat engine operated in reverse.

## Key Equations by Section

| Section | Topic | Equations | Equation No. |
|---|---|---|---|
| 20-3 | The first law of thermodynamics | $\boxed{\Delta Q = \Delta W + \Delta U}$ | 20-2 |
| 20-4 | Work done on a piston at constant pressure | $\boxed{\Delta W = P\,\Delta V}$ | 20-3 |
| 20-6 | Adiabatic process ($\Delta Q = 0$) | $\Delta W = -\Delta U$ | 20-4 |
| 20-7 | Isochoric process ($\Delta W = 0$) | $\Delta Q = \Delta U$ | 20-5 |
| 20-8 | Isothermal process ($\Delta U = 0$) | $\Delta Q = \Delta W$ | 20-6 |
| 20-9 | The second law of thermodynamics | $\boxed{W_{\text{out}} = Q_{\text{in}} - Q_{\text{out}}}$ | 20-7 |
| 20-9 | Efficiency $= \dfrac{\text{work output}}{\text{heat input}}$ | $\boxed{e = \dfrac{Q_{\text{in}} - Q_{\text{out}}}{Q_{\text{in}}}}$ | 20-8 |
| 20-11 | Efficiency of an engine | $\boxed{e = \dfrac{T_{\text{in}} - T_{\text{out}}}{T_{\text{in}}}}$ | 20-9 |
| 20-12 | Efficiency of an internal combustion engine based on compression ratio ($V_1/V_2$) and adiabatic constant ($\gamma$) | $e = 1 - \dfrac{1}{(V_1/V_2)^{\gamma-1}}$ | 20-10 |
| 20-13 | Efficiency of refrigeration | $K = \dfrac{Q_{\text{cold}}}{W} = \dfrac{Q_{\text{cold}}}{Q_{\text{hot}} - Q_{\text{cold}}}$ | 20-11 |
| 20-13 | Maximum efficiency in terms of absolute temperatures | $K = \dfrac{T_{\text{cold}}}{T_{\text{hot}} - T_{\text{cold}}}$ | 20-12 |

## True-False Questions

**T   F   1.** In the absence of friction, heat engines are 100 percent efficient.

**T   F   2.** If the first law of thermodynamics is satisfied, the second law will also be satisfied.

**T   F   3.** In every thermodynamic process, the heat absorbed by a system must equal the sum of the work done by the system and its change in internal energy.

**T   F   4.** An isochoric process is graphed as a straight line on a P-V diagram.

**T   F   5.** In an adiabatic process, the internal energy will increase when work is done *on* the system, whereas it will decrease when work is done *by* the system.

**T   F   6.** During an isothermal expansion, all the absorbed thermal energy is converted to useful work.

**T  F  7.** All Carnot engines are perfect engines and, therefore, operate at 100 percent efficiency.

**T  F  8.** The greater the difference between the input and output temperatures of a steam engine, the greater the efficiency of the engine.

**T  F  9.** A high compression ratio of an internal combustion engine means a higher operating efficiency.

**T  F  10.** The coefficient of performance for a refrigerator is a measure of cooling efficiency and is expressed as a percentage.

## Multiple-Choice Questions

**1.** The thermodynamic state of a gas refers to
(a) its pressure
(b) its volume
(c) its temperature
(d) all of these

**2.** The net work accomplished by an engine undergoing adiabatic compression is equal to
(a) $\Delta U$
(b) $-\Delta U$
(c) $\Delta Q$
(d) $-\Delta Q$

**3.** An engine that operates with 100 percent efficiency
(a) is a Carnot engine
(b) violates the first law
(c) has an Otto cycle
(d) violates the second law

**4.** If a heat engine absorbs heat at 600 K and rejects heat at 200 K, its efficiency is
(a) 33 percent
(b) 50 percent
(c) 67 percent
(d) 80 percent

**5.** In a Carnot cycle, 1600 cal is absorbed at 600 K, and 400 cal is exhausted to a cold reservoir. The temperature of the cold reservoir is
(a) 150 K
(b) 200 K
(c) 450 K
(d) 800 K

**6.** An adiabatic process is one in which
(a) the temperature is constant
(b) the pressure is constant
(c) the volume is constant
(d) no heat enters or leaves the system

**7.** If the adiabatic constant is 1.4 and the compression ratio is 6, a gasoline engine has an efficiency of
(a) 45 percent
(b) 51 percent
(c) 56 percent
(d) 64 percent

**8.** For a Carnot engine to operate with an efficiency of 100 percent, the exhaust temperature must be
(a) 0°C
(b) 0 K
(c) infinite
(d) equal to the input temperature

**9.** In a mechanical refrigerator, the low-temperature coils of the evaporator are at −23°C, and the compressed gas in the condenser has a temperature of 77°C. The coefficient of performance is
(a) 20 percent
(b) 70 percent
(c) 0.23
(d) 2.5

**10.** In a typical refrigerator, heat is extracted from the interior by the
(a) compressor
(b) evaporator
(c) condenser
(d) throttling valve

**Completion Questions**

1. The efficiency of a heat engine is the ratio of the _____ to the _____.

2. A(n) _____ process is one in which the volume remains constant, whereas in a(n) _____ process the temperature is constant.

3. The throttling process is an example of a(n) _____ process.

4. The area under the curve on a P-V diagram represents the _____ in a thermodynamic process.

5. Three coordinates used to describe the thermodynamic state of a system are _____, _____, and _____.

6. The _____ is essentially a restatement of the conservation of energy.

7. A(n) _____ is a heat engine operating in reverse. Its effectiveness is measured by the _____.

8. Four essential elements of a typical refrigerator include the _____, _____, _____, and _____.

9. The fact that all natural spontaneous processes are irreversible is a consequence of the _____.

10. A(n) _____ engine has the maximum possible efficiency for an engine that absorbs heat from one reservoir, performs work, and rejects heat to another reservoir at a lower temperature.

**Chapter 21**

# Mechanical Waves

**DEFINITIONS OF KEY TERMS**

**Wave motion**   Energy propagation by means of a disturbance in a medium.

**Mechanical wave**   A physical disturbance in an elastic medium.

**Transverse wave**   Wave motion in which the vibration of the individual particles of the medium is perpendicular to the direction of wave propagation.

**Longitudinal wave**   Wave motion in which the vibration of the individual particles is parallel to the direction of wave propagation.

**Linear density**   Mass per unit length, normally used for a vibrating string or wire.

**Wavelength**   The distance between any two closest particles along a wave train that are in phase; the distance between adjacent crests in a transverse wave.

**Amplitude**   The maximum displacement of a particle in a wave train from its equilibrium position.

**Frequency**   The number of complete vibrations or oscillations per second.

**Wave speed**   The distance traveled per unit of time by a wave pulse.

**Phase**   Two particles along a wave train are in phase if they have the same displacement and if they are moving in the same direction.

**Superposition principle**   When two or more waves exist simultaneously in the same medium, the resultant displacement at any point in time is the algebraic sum of the displacements of each wave.

**Constructive interference**   The effect of two superimposed waves that results in a wave of larger amplitude than either of the composite waves.

**Destructive interference**   The effect of two superimposed waves that results in a wave of smaller amplitude than either of the composite waves.

**Standing waves**   Wave trains for which boundary conditions and frequencies of composite waves result in one or more stationary positions (nodes) along the waves.

**Nodes**   Points along a standing wave at which the vibrating medium is not moving.

**Antinodes**   Points of maximum displacement in a standing wave.

**Characteristic frequencies**   Frequencies of vibration, for a given medium under fixed boundary conditions, which result in regular standing waves.

**Fundamental**   The simplest possible standing wave for a given situation that results in the least number of nodal positions.

**Frequency**   The number of waves per unit time. Usually expressed in hertz, or waves per second.

**Harmonics**   The series of characteristic frequencies beginning with $n = 1$ for the fundamental and continuing with integral multiples of the fundamental.

**Overtones**   Integral multiples of the fundamental, beginning with the next standing wave that occurs after the fundamental. The first overtone is the second harmonic.

**CONCEPTS AND EXAMPLES**

**Velocity of a Transverse Wave**   The velocity of a transverse wave in a vibrating string depends on the tension or force $F$ in the string, the length $l$ of the string, and the mass $m$ of the string. The mass per unit length $m/l$ is often defined as the linear density $\mu$. The relation is given by

$$v = \sqrt{\frac{F}{\mu}} \quad \text{or} \quad v = \sqrt{\frac{Fl}{m}} \quad \text{wave speed}$$

**EXAMPLE 21-1**   A 9-g string is stretched between two supports that are 600 mm apart. What must be the tension in the string if the wave speed is to be 200 m/s?

*Solution*
To solve for the tension, we must first square both sides of the wave speed equation and then solve for $F$:

$$v^2 = \frac{Fl}{m} \quad \text{or} \quad F = \frac{mv^2}{l}$$

Now, $m$, $l$, and $v$ are substituted after we convert to consistent units:

$$F = \frac{mv^2}{l} = \frac{(0.009 \text{ kg})(200 \text{ m/s})^2}{0.6 \text{ m}} = \boxed{600 \text{ N}}$$

**Energy and Power in Waves**   The energy per unit length $E/l$ and the power $P$ of a wave can be found from

$$\frac{E}{l} = 2\pi^2 f^2 A^2 \mu \qquad P = 2\pi^2 f^2 A^2 \mu v$$

**Standing Waves**   When the boundary conditions and frequency of vibration for interfering waves are such that some particles have no displacement (nodes), the composite wave is called a standing wave. The characteristic frequencies for which standing waves exist are integral multiples of the simplest such wave pattern, which is called the fundamental. The characteristic frequencies for the possible modes of vibration in a stretched string are found from

$$f_n = \frac{n}{2l} \sqrt{\frac{F}{\mu}} \qquad n = 1, 2, 3, \ldots$$

The series $f_n = nf_1$ is called the harmonics. They are integral multiples of the fundamental $f_1$. Since harmonics are mathematical values, not all may exist in a real situation with certain physical restrictions. The actual possibilities beyond the fundamental are called the overtones. Since all harmonics are present for a vibrating string, the first overtone is the second harmonic, the second overtone is the third harmonic, and so on.

**EXAMPLE 21-2**

The third overtone for a vibrating string 2 m long is 600 Hz. What is the tension in the string if the linear density is 0.004 kg/m?

*Solution*

To find the tension, we must square again both sides of the harmonic equation and solve for $F$:

$$f_n^2 = \frac{n^2}{4l^2}\left(\frac{F}{\mu}\right) \qquad \text{or} \qquad F = \frac{4f_n^2 l^2 \mu}{n^2}$$

Now the third overtone for a vibrating string is the fourth harmonic, so that $n = 4$ and $f_n = f_4 = 600$ Hz. Direct substitution of known quantities yields

$$F = \frac{4(600 \text{ Hz})^2(2 \text{ m})^2(0.004 \text{ kg/m})}{4^2} = \boxed{1440 \text{ N}}$$

## Summary

- The fundamental laws of mechanical waves are important because they also apply to many other types of waves.
- Harmonics are integral multiples of the fundamental $f_1$ that are mathematical values.
- All harmonics may not exist.
- The actual possibilities beyond the fundamental are called *overtones*.
- Since all harmonics are possible for the vibrating string, the first overtone is the second harmonic, the second overtone is the third harmonic, and so on.

## Key Equations by Section

| Section | Topic | Equation | Equation No. |
|---|---|---|---|
| 21-2 | Calculating wave speed of a wave traveling down a string based on string tension $F$ and mass per unit length of string ($\mu = m/l$) | $v = \sqrt{\dfrac{F}{\mu}} = \sqrt{\dfrac{F}{m/l}}$ | 21-1 |
| 21-4 | The relation of the wave speed to the wavelength ($\lambda$) and period (T) | $v = \dfrac{\lambda}{T}$ | 21-2 |

| Section | Topic | Equation | Equation No. |
|---|---|---|---|
| 21-4 | The relation of wave speed to frequency and wavelength | $\boxed{v = f\lambda}$ | 21-3 |
| 21-5 | The total energy of a point on a vibrating string | $E = E_p + E_k = (E_k)_{\max}$ $= \frac{1}{2}mv_{\max}^2 = \frac{1}{2}m(2\pi fA)^2$ $= 2\pi^2 f^2 A^2 m$ | 21-4 |
| 21-5 | Wave energy per unit length of a vibrating string | $\boxed{\dfrac{E}{l} = 2\pi^2 f^2 A^2 \mu}$ | 21-5 |
| 21-5 | Power of a wave | $P = \dfrac{E}{t} = \dfrac{E}{l/v} = \dfrac{E}{l}v$ | 21-6 |
| 21-5 | Rate of energy propagation | $\boxed{P = 2\pi^2 f^2 A^2 \mu v}$ | 21-7 |
| 21-8 | Characteristic frequencies | $\lambda_n = \dfrac{2l}{n} \quad n = 1, 2, 3, \ldots$ | 21-8 |
| 21-8 | Corresponding frequencies of vibration | $\boxed{f_n = \dfrac{nv}{2l} = n\dfrac{v}{2l} \quad n = 1, 2, 3, \ldots}$ | 21-9 |
| 21-8 | Characteristic frequencies | $f_n = \dfrac{n}{2l}\sqrt{\dfrac{F}{\mu}} \quad n = 1, 2, 3, \ldots$ | 21-10 |
| 21-8 | Overtones | $f_n = nf_1 \quad n = 1, 2, 3, \ldots$ | 21-11 |

**True-False Questions**

T  F  **1.** A physical medium is necessary for the transmission of all kinds of waves.

T  F  **2.** The speed of a wave in a string is a function of the linear density of the string but is really independent of the actual length of the string.

T  F  **3.** In a longitudinal wave, the wavelength is equal to the distance between adjacent condensations or between adjacent rarefactions.

T  F  **4.** Increasing the frequency of a wave results in a decrease in its wavelength if other parameters are held constant.

T  F  **5.** For a standing wave, the distance between adjacent nodes or between adjacent antinodes is equal to the wavelength.

T  F  **6.** The superposition principle applies only for transverse waves.

T  F  **7.** The third harmonic is equivalent to the second overtone when characteristic frequencies are described for a vibrating string.

T  F  **8.** Constructive interference results in a wave of greater energy than the sum of the energies of its component waves.

T  F  **9.** Standing waves are the result of constructive interference.

T  F  **10.** If the frequency of a wave is doubled and other parameters remain the same, the energy of the wave per unit of length will be quadrupled.

## Multiple-Choice Questions

1. In a longitudinal wave, the individual particles of the medium move
   (a) in circles
   (b) in elipses
   (c) parallel to wave propagation
   (d) perpendicular to wave propagation

2. For a vibrating string, the third overtone will be the same as the
   (a) second harmonic          (b) third harmonic
   (c) fourth harmonic          (d) fifth harmonic

3. Two particles along a wave train are in phase if they have the same
   (a) displacement             (b) speed
   (c) amplitude                (d) energy

4. A longitudinal wave traveling at 300 m/s has a wavelength of 2 m. Its frequency is
   (a) 100 Hz                   (b) 150 Hz
   (c) 167 Hz                   (d) 600 Hz

5. If 120 waves strike a wall in 1 min and the distance between adjacent crests is 2 m, the speed of the waves is
   (a) 2 m/s                    (b) 4 m/s
   (c) 8 m/s                    (d) 30 m/s

6. A flexible cable 20 m long weighs 16 N and is stretched between two poles with a force of 450 N. The speed of a transverse wave through this medium is
   (a) 16 m/s                   (b) 23.7 m/s
   (c) 57.3 m/s                 (d) 74.2 m/s

7. If the frequency of the fundamental for a vibrating string is 200 Hz, the second overtone has a frequency of
   (a) 200 Hz                   (b) 400 Hz
   (c) 600 Hz                   (d) 800 Hz

8. A metal string of mass 250 g and length 25 cm is under a tension of 400 N. The fundamental frequency for this string is
   (a) 40 Hz                    (b) 400 Hz
   (c) 126 Hz                   (d) 800 Hz

9. The rate at which energy is propagated down a string is *not* dependent on the
   (a) frequency                (b) amplitude
   (c) linear density           (d) length of the string

10. The ratio of the wavelength to the period is a measure of
    (a) frequency               (b) speed
    (c) period                  (d) amplitude

## Completion Questions

1. In a(n) _____ wave, the vibration of the individual particles is perpendicular to the direction of wave propagation.

2. The speed of a wave in a vibrating string is equal to the square root of the _____ divided by the _____.

3. The distance between any two particles that are in phase is known as the _____.

4. The energy transmitted per unit length of a string is proportional to the square of the _____ and to the square of the _____.

5. For characteristic frequencies of a vibrating string, the fifth harmonic is the _____ overtone.

6. When two or more waves interfere, the resultant _____ at any point in time is the algebraic sum of the _____ of each wave. This is a statement of the _____ principle.

7. For a standing wave, the points along a vibrating string that remain at rest are called _____. The points where the amplitude is a maximum are called _____.

8. The speed of any wave may be found from the product of _____ and _____.

9. For a standing wave, the wavelength of the component waves is the distance between alternate _____ or between alternate _____.

10. The characteristic frequencies consisting of the fundamental and all its overtones are known as the _____ series.

# Chapter 22

# Sound

**DEFINITIONS OF KEY TERMS**

**Sound**   A longitudinal mechanical wave that travels through an elastic medium.

**Compression**   A high-pressure region in a longitudinal wave train where particles are packed tightly together.

**Rarefaction**   A low-pressure region in a longitudinal wave train where particles experience their maximum displacements.

**Audible sound**   Sound waves in the frequency range from 20 to 20,000 Hz, that is, within the range of human hearing.

**Infrasonic**   Sound waves of frequencies lower than the audible range (below 20 Hz).

**Ultrasonic**   Sound waves of frequencies higher than the audible range (above 20,000 Hz).

**Intensity**   For a sound wave, the power transmitted per unit area perpendicular to the direction of wave propagation.

**Hearing threshold**   The standard zero of sound intensity. Its magnitude, by agreement, is $1 \times 10^{-12}$ W/m$^2$.

**Pain threshold**   The maximum intensity that the average ear can record without a sensation of feeling or pain. The accepted standard intensity for the pain threshold is 1 W/m$^2$.

**Intensity level**   A logarithmic scale of sound intensity that compares the relative intensity of a given sound with that of another sound, usually the hearing threshold. Units are the bel and the decibel.

**Loudness**   A subjective sensory effect that describes human perception of intensity.

**Frequency**   A physical property of sound that is a measure of the number of longitudinal oscillations per second.

**Pitch**   A sensory effect that describes human perception of the frequency of sound.

**Quality**   The sensory effect produced by a sound wave as determined by the number and relative intensities of the overtones that make up the sound.

**Waveform**   The complex nature of a sound wave, as might be represented electronically with the use of an oscilloscope (a physical measure of sound quality).

**Forced vibrations**   Secondary vibrations set up in an object as a result of the proximity of a primary vibration.

**Beats** Regular pulsations alternating in intensity because of the interference of two sources of sounds whose frequencies differ only slightly.

**Doppler effect** The apparent change in frequency of a source of sound when there is relative motion between the source and the listener.

**Decibels** (dB) Unit of measurement of intensity of sound. This measure follows a log scale; that is, a 2-dB sound is 10 times louder than a 1-dB sound.

**Speed of Sound** In a general sense, the speed of sound in any medium depends on the elasticity of that medium and its density. Since the density is directly affected by temperature, the speed also depends on temperature. For air, a useful approximation of sound speed as a function of temperature is

$$v = 331 \text{ m/s} + [0.6 \text{ m/(s} \cdot \text{C}^\circ)]t_c$$

You should verify this expression by showing that the speed of sound is approximately 343 m/s when the temperature is 20° C.

In other media, the speed of sound can be found from the following equations:

$$v = \sqrt{\frac{Y}{\rho}} \quad \text{rod} \qquad v = \sqrt{\frac{\gamma P}{\rho}} = \sqrt{\frac{\gamma RT}{M}} \quad \text{gas}$$

$$v = \sqrt{\frac{B}{\rho}} \quad \text{fluid} \qquad v = \sqrt{\frac{B + \frac{4}{3}S}{\rho}} \quad \text{extended solid}$$

**Characteristic Frequencies** Standing longitudinal sound waves may be set up in a vibrating air column for a pipe that is open at both ends or for one that is closed at one end only. The characteristic frequencies are

$$f_n = \frac{nv}{2l} \qquad n = 1, 2, 3, \ldots \qquad \text{open pipe of length } l$$

$$f_n = \frac{nv}{4l} \qquad n = 1, 3, 5, \ldots \qquad \text{closed pipe of length } l$$

Only the odd harmonics are possible for a closed pipe. Thus, the third harmonic is the first overtone, the fifth harmonic is the second overtone, and so on.

---

**EXAMPLE 22-1**

What must be the length of a closed pipe so that its first overtone will have a frequency of 800 Hz on a day when the speed of sound in air is 340 m/s?

*Solution*
We must first solve for $l$ in the closed-pipe equation:

$$f_n = \frac{nv}{4l} \qquad \text{or} \qquad l = \frac{nv}{4f_n}$$

Recalling that $n = 3$ for the first overtone, we substitute:

$$l = \frac{3(340 \text{ m/s})}{4(800 \text{ Hz})} = 0.319 \text{ m} \qquad \text{or} \qquad \boxed{l = 319 \text{ mm}}$$

---

**Sound Intensity**  The intensity $I$ of sound is the power per unit area $P/A$ normal to the direction of the sound wave:

$$I = \frac{P}{A} = 2\pi^2 f^2 A^2 v \qquad \text{intensity, W/m}^2$$

A more useful measure of sound intensity is the intensity level, a logarithmic comparison given in decibels (dB):

$$\beta = 10 \log \frac{I}{I_0} \qquad I_0 = 1 \times 10^{-12} \text{ W/m}^2$$

**Doppler Effect**  The Doppler effect refers to the apparent change in frequency of sound when the source and observer are in relative motion. The general equation is

$$f_0 = \frac{f_s(V + v_0)}{V - v_s} \qquad \textbf{Doppler effect}$$

In this equation, $f_0$ is the frequency heard by the listener, $f_s$ is the actual source frequency, $V$ is the constant velocity of sound, $v_0$ is the velocity of the listener, and $v_s$ is the velocity of the source of sound. Velocities are reckoned as positive for speeds of approach and negative for speeds of recession. For example, if the sound source is chasing the listener who is running away from the sound, $v_s$ is positive and $v_0$ is negative.

---

**EXAMPLE 22-2**    On a day when the speed of sound is 340 m/s, a woman is traveling north at 20 m/s. What frequency will she hear if a 600-Hz source of sound is traveling north with a speed of 60 m/s?

**Solution**
By convention $v_0 = +20$ m/s and $v_s = -60$ m/s. Direct substitution into the Doppler equation gives

$$f_0 = \frac{(600 \text{ Hz})(340 \text{ m/s} + 20 \text{ m/s})}{340 \text{ m/s} - (-60 \text{ m/s})} = \frac{216,000 \text{ m/s}}{400 \text{ m/s}}$$

$$= \boxed{540 \text{ Hz}} \qquad \text{(lower apparent frequency)}$$

The frequency was reduced in Example 22-2 because the relative velocity was such as to increase the separation.

---

**Summary**

- Under certain conditions, standing sound waves can produce characteristic frequencies that we observe as the pitch of the sound.
- Sound is a longitudinal wave traveling through an elastic medium.
- The speed of sound in air at 0°C is 331 m/s or 1087 ft/s.
- Standing and longitudinal sound waves may be set up in a vibrating air column for a pipe that is open at both ends or for one that is closed at one end.

- The intensity of a sound is the power $P$ per unit area $A$ perpendicular to the direction of propagation.
- The Doppler effect refers to the apparent change in frequency of a source of sound when there is relative motion of the source and the listener.

| Key Equations by Section | Section | Topic | Equation | Equation No. |
|---|---|---|---|---|
| | 22-2 | The speed of longitudinal sound waves in a rod, based on Young's modulus and the density of the material | $v = \sqrt{\dfrac{Y}{\rho}}$ | 22-1 |
| | 22-2 | The longitudinal wave speed in an extended solid, based on the bulk modulus of elasticity, the shear modulus, and the density | $v = \sqrt{\dfrac{B + \frac{4}{3}S}{\rho}}$ | 22-2 |
| | 22-2 | The longitudinal wave speed in a fluid, based on the bulk modulus of elasticity and the density | $v = \sqrt{\dfrac{B}{\rho}}$ | 22-3 |
| | 22-2 | The speed of longitudinal waves in a gas, based on the bulk modulus and the density, or on the adiabatic constant, the pressure, and the density | $v = \sqrt{\dfrac{B}{\rho}} = \sqrt{\dfrac{\gamma P}{\rho}}$ | 22-4 |
| | 22-2 | A restatement of the ideal gas law | $\dfrac{P}{\rho} = \dfrac{RT}{M}$ | 22-5 |
| | 22-2 | The speed of sound waves in a gas | $v = \sqrt{\dfrac{\gamma P}{\rho}} = \sqrt{\dfrac{\gamma RT}{M}}$ | 22-6 |
| | 22-2 | The speed of sound $v$ | $v = 331 \text{ m/s} + \left(0.6 \dfrac{\text{m/s}}{\text{C}^\circ}\right)t$ | 22-7 |
| | 22-3 | Possible harmonics of sound waves | $f_n = \dfrac{nv}{2l} \quad n = 1, 2, 3, \ldots$ | 22-8 |

| Section | Topic | Equation | Equation No. |
|---|---|---|---|
| 22-3 | Possible wavelengths of sound waves | $\lambda_n = \dfrac{4l}{n} \quad n = 1, 3, 5, \ldots$ | 22-9 |
| 22-3 | Possible harmonics in a closed pipe | $\boxed{f_n = \dfrac{nv}{4l} \quad n = 1, 3, 5, \ldots}$ | 22-10 |
| 22-3 | Possible wavelengths in an open pipe | $\lambda_n = \dfrac{2l}{n} \quad n = 1, 2, 3, \ldots$ | 22-11 |
| 22-3 | Possible frequencies in an open pipe | $f_n = \dfrac{nv}{2l} \quad n = 1, 2, 3, \ldots$ | 22-12 |
| 22-5 | Sound intensity | $\boxed{I = \dfrac{P}{A}}$ | 22-13 |
| 22-5 | Alternative expression for sound intensity in a gas | $I = 2\pi^2 f^2 A^2 \rho v$ | 22-14 |
| 22-5 | Hearing threshold | $I_0 = 1 \times 10^{-12} \text{ W/m}^2$ | 22-15 |
| 22-5 | Pain threshold | $I_p = 1 \text{ W/m}^2$ | 22-16 |
| 22-5 | Difference in intensity levels, in bels (B) | $B = \log \dfrac{I_1}{I_2} \text{ bels (B)}$ | 22-17 |
| 22-5 | The intensity level in decibels | $\beta = 10 \log \dfrac{I}{I_0} \quad decibels \text{ (dB)}$ | 22-18 |
| 22-5 | Intensity of a sound as related to distance from the source | $\dfrac{I_1}{I_2} = \dfrac{r_2^2}{r_1^2} \quad \text{or} \quad I_1 r_1^2 = I_2 r_2^2$ | 22-19 |
| 22-7 | Interference | Number of beats per second $= |f - f'|$ | 22-20 |
| 22-8 | Doppler effect giving the apparent wavelength $\lambda'$ based on the speed of the wave, the motion of the source, and the frequency of the wave | $\lambda' = \dfrac{V - v_s}{f_s}$ | 22-21 |

| Section | Topic | Equation | Equation No. |
|---|---|---|---|
| 22-8 | Doppler effect, giving the observed frequency based on the frequency of the sound, the traveling velocity of the sound wave, and the observer's speed | $f_o = \dfrac{f_s(V + v_o)}{V}$ | 22-23 |
| 22-8 | Doppler effect for moving sound source and moving observer | $f_o = f_s \dfrac{V + v_o}{V - v_s}$ | 22-24 |

## True-False Questions

T  F  **1.** Sound waves are longitudinal waves that require a medium for transmission.

T  F  **2.** If a tree falls in a forest, there is no sound unless an ear can pick up the vibrations.

T  F  **3.** Sound waves travel faster in air than in metals because the air is less dense.

T  F  **4.** The speed of sound is increased significantly with rising temperatures.

T  F  **5.** The quality of different sounds is demonstrated by the difference in tones when a C note is sounded on a flute, a violin, and a trumpet.

T  F  **6.** A sound of intensity level 40 dB is twice as intense as a sound of 20 dB.

T  F  **7.** Sound that fluctuates in intensity because of the simultaneous output of two sources is a consequence primarily of the Doppler effect.

T  F  **8.** In applying Doppler's equation, speeds are reckoned as positive for speeds of approach and negative for speeds of recession.

T  F  **9.** Opening the end of a closed pipe will double the frequency produced.

T  F  **10.** The speed of sound in gases is larger for the gases with higher molecular masses.

## Multiple-Choice Questions

**1.** The speed of sound is greatest when the medium is
(a) a vacuum    (b) air    (c) water    (d) metal

**2.** Which of the following is a sensory effect rather than a measurable physical quantity?
(a) Frequency    (b) Quality    (c) Intensity    (d) Waveform

**3.** For a closed pipe, the second overtone is the
(a) second harmonic          (b) third harmonic
(c) fourth harmonic          (d) fifth harmonic

**4.** The physical property that is most responsible for resonance is
(a) waveform    (b) frequency    (c) quality    (d) intensity

**5.** Which of the following sounds is loudest?
(a) 40 dB                    (b) $10^{-5}$ $\mu$W/cm$^2$
(c) $10^{-10}$ W/m$^2$        (d) 3 B

6. The speed of sound in air at 0°C is 331 m/s. Its speed at 30°C is approximately
   (a) 331 m/s      (b) 343 m/s      (c) 349 m/s    (d) 350 m/s

7. The fundamental frequency for a 20-cm closed pipe when the speed of sound is 340 m/s is
   (a) 4.25 Hz      (b) 8.25 Hz      (c) 425 Hz     (d) 825 Hz

8. The intensity level in decibels of a sound whose intensity is $2 \times 10^{-6}$ $\mu$W/cm$^2$ is approximately
   (a) 27 dB        (b) 43 dB        (c) 50 dB      (d) 103 dB

9. A car horn emits sound at a frequency of 200 Hz. The pitch heard when the car is moving at 31 m/s toward an observer in air at 0°C is approximately
   (a) 221 Hz       (b) 219 Hz       (c) 181 Hz     (d) 183 Hz

10. Two tuning forks of 340 and 343 Hz are sounded together. The resulting beats per second will be
    (a) 1            (b) 2            (c) 3          (d) 4

---

**Completion Questions**

1. The closed end of a pipe must be a displacement _____; the open end must be a displacement _____.

2. The velocity of sound waves in a liquid is equal to the square root of the ratio of its _____ to its _____.

3. The three physical properties of sound that correspond to loudness, pitch, and quality are _____, _____, and _____, respectively.

4. The _____ represents the standard zero of sound intensity. Its value is _____ W/cm$^2$.

5. Sound waves having frequencies below the range of audible sound are termed _____; sounds having frequencies above this range are termed _____.

6. For a closed pipe, only the _____ harmonics are possible. Thus, the seventh harmonic will be the _____ overtone.

7. The power transmitted by a sound wave through a unit of area is a measure of _____.

8. The _____ refers to the apparent change in frequency of a source of sound when there is relative motion between the source and the observer. The pitch heard by an observer is _____ when a sound approaches her or him and _____ when the sound leaves.

9. Beats are the product of alternating _____ and _____ interference of sound waves of slightly different frequency.

10. When the intensity $I_1$ of one sound is 10 times as great as the intensity $I_2$ of another, the difference in intensity levels is said to be 1 _____.

# Chapter 23

# The Electric Force

## DEFINITIONS OF KEY TERMS

**Electrostatics**   The science that treats electric charges at rest.

**Charging**   A process that transfers electrons in such a way that an excess or a deficiency of electrons on a body results.

**Negative charge**   Electrification resulting from an excess of electrons.

**Positive charge**   Electrification resulting from a deficiency of electrons.

**Induced charge**   Electrification from a redistribution of charge as a result of the presence of a nearby charged object.

**Conductor**   A material through which charge can be transferred easily.

**Insulator**   A material that offers resistance to the flow of charge.

**Electroscope**   A laboratory device used to detect the presence of charge.

**Coulomb's law**   The force of attraction or repulsion between two point charges is directly proportional to the product of the two charges and inversely proportional to the square of the distance between them.

**Coulomb**   A unit of charge equivalent to $6.25 \times 10^{18}$ electrons. It is the charge transferred through any cross section of a conductor in 1 s by a constant current of 1 A.

**Electron**   A subatomic particle that can be traded freely in conductors (e.g., metals) and is tightly bound in insulators (e.g., air).

**Ion**   An element or a molecule with more or fewer electrons than the neutral element or molecule of the same type.

**Microcoulomb**   One one-millionth of a Coulomb, or $10^{-6}$ C.

**Semiconductor**   A material that conducts less than a conductor and more than a resistor.

## CONCEPTS AND EXAMPLES

**First Law of Electrostatics**   An object that has an excess of electrons is labeled as negative; an object that has a deficiency of electrons is labeled as positive. Given these two kinds of charges, the first law of electrostatics states that like charges repel each other and unlike charges attract each other.

**Coulomb's Law**   The force of attraction or repulsion between two point charges is directly proportional to the product of the two charges and inversely proportional to their separation:

$$F = \frac{kqq'}{r^2} \qquad k = 9 \times 10^9 \ \text{N} \cdot \text{m}^2/\text{C}^2$$

The sign (+ or −) of the charges is used to determine the direction of the electric force, and the magnitude of the force is determined from Coulomb's law.

**EXAMPLE 23-1**   Charges $A$, $B$, and $C$ are positioned at the corners of a right triangle, as shown in Fig. 23-1. If we take the magnitudes of charges and the distance from the figure, what is the resultant force on charge $C$ due to charges $A$ and $B$?

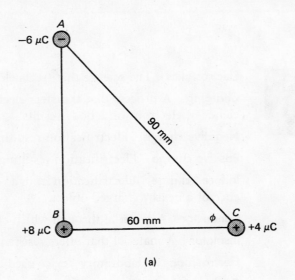

(a)

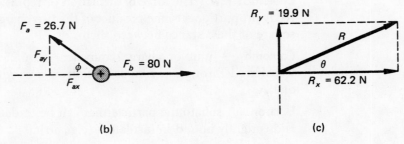

(b)                                                  (c)

**Fig. 23-1**

*Solution*
First, we draw a free-body diagram of the forces acting on the 4-$\mu$C charge. We label the force on charge $C$ due to charge $A$ as $F_a$ and the force on charge $C$ due to charge $B$ as $F_b$. Because of the nature of the charges, $F_a$ is an attractive force and $F_b$ is a repulsive force. The angle $\phi$ is found from the triangle:

$$\cos \phi = \frac{60 \text{ mm}}{90 \text{ mm}} \qquad \phi = 48.2°$$

We now find the magnitude of each force from Coulomb's law:

$$F_a = \frac{(9 \times 10^9 \text{ N} \cdot \text{m}^2/\text{C}^2)(6 \times 10^{-6} \text{ C})(4 \times 10^{-6} \text{ C})}{(0.09 \text{ m})^2} = 26.7 \text{ N}$$

$$F_b = \frac{(9 \times 10^9 \text{ N} \cdot \text{m}^2/\text{C}^2)(8 \times 10^{-6} \text{ C})(4 \times 10^{-6} \text{ C})}{(0.06 \text{ m})^2} = 80.0 \text{ N}$$

The $x$ and $y$ components of the resultant vector are found by referring to the figure and summing the components of each vector:

$$R_x = 80 \text{ N} - (26.7 \text{ N})(\cos 48.2°) = +62.2 \text{ N}$$
$$R_y = 0 + 19.9 \text{ N} = +19.9 \text{ N}$$

Since both components are positive, the resultant is in the first quadrant. The magnitude and direction are found by the methods of Chap. 3:

$$R = \sqrt{R_x^2 + R_y^2} \qquad R = 65.3 \text{ N}$$

$$\tan \theta = \frac{R_y}{R_x} \qquad \theta = 17.7°$$

The resultant force on charge $C$ is, therefore, 65.3 N in a direction of 17.7° north of east.

## Summary

- The first law of electrostatics states that *like charges* repel each other and *unlike charges* attract each other.
- Coulomb's law states that the force of attraction or repulsion between two point charges is directly proportional to the product of the two charges and inversely proportional to the separation of the two charges.
- Remember to use the sign of the charges to determine the direction of forces and Coulomb's law to determine the magnitude of forces.
- The resultant force on a particular charge is found by the methods of vector mechanics.

## Key Equations by Section

| Section | Topic | Equation | Equation No. |
|---------|-------|----------|--------------|
| 23-1 | Coulomb's law | $F = \dfrac{kqq'}{r^2}$ | 23-1 |
| 23-1 | The charge of one electron expressed in coulombs | $e^- = -1.6 \times 10^{-19}$ C | 23-2 |
| 23-1 | The microcoulomb | $1 \ \mu\text{C} = 10^{-6}$ C | 23-3 |
| 23-1 | The proportionality constant | $k = 9 \times 10^9$ N $\cdot$ m$^2$/C$^2$ | 23-4 |
| 23-1 | Coulomb's law in SI units | $F = \dfrac{(9 \times 10^9 \text{ N} \cdot \text{m}^2/\text{C}^2)qq'}{r^2}$ | 23-5 |

## True-False Questions

T  F  **1.** Rubbing a glass rod with a silk cloth leaves a negative charge on the cloth.

T  F  **2.** If two objects placed closely together repel each other electrically, we can be sure that they are *both* charged.

T  F  **3.** Bringing a negatively charged rod closer and closer to a positively charged electroscope causes the leaf to converge.

T  F  **4.** The process of charging an object by induction leaves a charge on the object that is opposite to that of the charging device.

T  F  **5.** Because of its large size, the coulomb is not a practical unit for static electricity.

T  F  **6.** According to Coulomb's law, the electric force will be doubled if the separation of two equal charges is cut in half.

T  F  **7.** If two nearby objects experience a mutual force of electric attraction, they must *both* be electrically charged.

T  F  **8.** When two or more charges are in the vicinity of another charge, the latter charge experiences an electric force equal to the algebraic sum of the forces due to each charge.

T  F  **9.** The plus and minus signs used to identify charge have significance primarily for determining direction when they are applied to Coulomb's law.

T  F  **10.** One coulomb is that quantity of charge that, when placed one meter away from an equal charge of the same sign, will experience a repulsive force of one newton.

## Multiple-Choice Questions

**1.** A negatively charged body
 (a) has a deficiency of electrons
 (b) has an excess of electrons
 (c) is produced on glass by rubbing with silk
 (d) repels a positively charged body

**2.** Which of the following represents the largest measure of charge?
 (a) 1 $\mu$C  (b) 1 nC
 (c) $10^{12}$ electrons  (d) $10^{-7}$ C

**3.** Charging a single body by induction always leaves a residual charge that is
 (a) greater than that of the charging object
 (b) the same sign as that of the charging body
 (c) opposite in sign to that of the charging object
 (d) an excess of electrons

**4.** Decreasing the separation of two identical positive charges by one-half will cause the force of repulsion to change by a factor of
 (a) 4  (b) 2  (c) $\frac{1}{2}$  (d) $\frac{1}{4}$

**5.** When two suspended objects are seen to attract each other electrically,
 (a) they are both charged  (b) one must be charged
 (c) either (a) or (b) is true  (d) neither (a) nor (b) is true

**6.** Two balls each having a charge of +12 $\mu$C are 8 cm apart. The electric force is approximately
 (a) 0.02 N  (b) 40 N  (c) 202 N  (d) 404 N

**7.** If a repulsive force of 2.0 N is observed between two identical 9-$\mu$C charges, their separation must be approximately
(a) 6 cm     (b) 3.6 cm   (c) 60 cm   (d) 36 cm

**8.** Three charges of +4, +8, and −2 nC are at the corners of an equilateral triangle 6 cm on a side. The magnitude of the force on the 8-nC charge is approximately
(a) $6.93 \times 10^{-5}$ N          (b) $3.47 \times 10^{-5}$ N
(c) $6 \times 10^{-5}$ N          (d) $2.7 \times 10^{-5}$ N

**9.** A charge of 6 $\mu$C is 10 cm to the right of a $-4$-$\mu$C charge. The resultant force on a 2-nC charge placed 4 cm to the right of the 6-$\mu$C charge is
(a) $4.75 \times 10^{-2}$ N          (b) $8.75 \times 10^{-2}$ N
(c) $6.75 \times 10^{-2}$ N          (d) $2 \times 10^{-2}$ N

**10.** As a positively charged rod is brought closer and closer to a positively charged electroscope, the gold leaf
(a) diverges          (b) converges
(c) is neutralized          (d) is unaffected

---

## Completion Questions

**1.** An object that has an excess of electrons is _____ charged and will repel a(n) _____ charged body.

**2.** A(n) _____ is a material through which charge may be transferred easily, whereas a(n) _____ resists the flow of charge.

**3.** A charge of 1 $\mu$C is equivalent to a charge of _____ C.

**4.** The first law of electrostatics states that like charges _____ and unlike charges _____.

**5.** Rubbing a wool cloth against a rubber rod transfers _____ from the _____ to the _____.

**6.** The process of charging without the necessity of direct contact with a charged body is called _____.

**7.** According to _____ law, the electric force is inversely proportional to the square of the _____ between two charges.

**8.** A charge of 1 C is equivalent to that charge represented by _____ electrons.

**9.** The smallest unit of charge is the _____, which has a charge of _____ C.

**10.** The _____ is a laboratory device used to detect the presence of charge.

# Chapter 24

# Electric Field

## DEFINITIONS OF KEY TERMS

**Electric field**   A region of space in which an electric charge will experience an electric force.

**Electric field intensity**   At a given point in space, the electric field intensity is the force per unit positive charge placed at that point.

**Electric field lines**   Imaginary lines describing the electric field in a region of space. They are drawn so that their direction at any point is the same as the direction of the electric field at that point.

**Permittivity**   A property of the medium surrounding a charge that is a measure of the medium's ability to support an electric field.

**Gauss's law**   The net number of electric field lines crossing any closed surface is numerically equal to the net total charge within that surface.

**Gaussian surface**   An imaginary surface of simple geometrical form that is used in the application of Gauss's law.

**Charge density**   Represented by the greek letter sigma ($\sigma$), the charge density is the charge per unit area of surface.

**Faraday's ice pail experiment**   An experiment in which a charged glass ball was lowered into a conducting metal ice pail. When the ball touched the metal, the charge was neutralized.

## CONCEPTS AND EXAMPLES

**Electric Field Intensity**   An electric field is said to exist in a region of space where a test charge will experience an electric force. The magnitude is the force per unit charge $F/q$, and the direction is the same as that in which a positive charge would move at that point.

$$E = \frac{F}{q} \qquad E = \frac{kQ}{r^2} \qquad k = 9 \times 10^9 \, \text{N} \cdot \text{m}^2/\text{C}^2$$

In the second equation, $r$ is the distance from the charge $Q$ to the point in question.

**EXAMPLE 24-1**   A $-6$-$\mu$C charge is placed 6 mm to the left of a 4-$\mu$C charge. What is the electric field intensity at a point 2 mm to the right of the 4-$\mu$C charge?

*Solution*

We construct a diagram as in Fig. 24-1 that shows the given information. The direction of the field due to the positive charge is to the right, and the direction due to the negative charge is to the left. Thus,

$$E_4 = \frac{kQ_4}{r^2} = \frac{(9 \times 10^9 \text{ N} \cdot \text{m}^2/\text{C}^2)(4 \times 10^{-6} \text{ C})}{(2 \times 10^{-3} \text{ m})^2}$$

$$= 90 \times 10^8 \text{ N/C} \quad \text{to the right}$$

$$E_6 = \frac{kQ_6}{r^2} = \frac{(9 \times 10^9 \text{ N} \cdot \text{m}^2/\text{C}^2)(6 \times 10^{-6} \text{ C})}{(8 \times 10^{-3} \text{ m})^2}$$

$$= 8.44 \times 10^8 \text{ N/C} \quad \text{to the left}$$

Now, the resultant field is the *vector* sum of $E_4$ and $E_6$. Since the vectors are along the same line and opposite in direction, the vector sum is the algebraic difference:

$$E = 90 \times 10^8 \text{ N/C} - 8.44 \times 10^8 \text{ N/C} = \boxed{81.6 \text{ N/C}}$$

Therefore, the resultant field is directed to the right and is equal to $8.16 \times 10^9$ N/C. Remember that the electric field is a vector quantity and that the directions (angels) must be considered as well as the magnitudes of the electric field intensities.

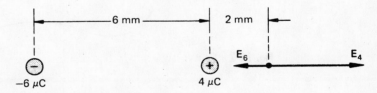

**Fig. 24-1**   The electric field intensity at a point *A* due to two charges.

## Electric Field Lines and Gauss's Law

According to Gauss's law, the net numbers of electric lines crossing any closed surface in an outward direction is numerically equal to the net total charge within that surface:

$$N = \Sigma \epsilon_0 E_n A = \Sigma q \qquad \textbf{Gauss's law}$$

The symbol $\epsilon_0$ represents the permittivity of free space:

$$\epsilon_0 = 8.85 \times 10^{-12} \text{ C}^2/(\text{N} \cdot \text{m}^2)$$

In applying Gauss's law it is often useful to speak in terms of the charge density $\sigma$, which is defined by

$$\sigma = \frac{q}{A} \qquad q = \sigma A \qquad \textbf{charge density}$$

**EXAMPLE 24-2** The electric field intensity just outside a solid conductor is 40 N/C. Use Gauss's law to determine the charge density on the surface of this conductor.

**Solution**
We construct a diagram with an imaginary gaussian surface such as that shown in Fig. 24-2. The cylinder has an area $A$ at the surface of the conductor and at the top of the cylinder. According to Gauss's law,

$$\Sigma \epsilon_0 EA = \Sigma q \qquad \text{and} \qquad q = \sigma A$$

From which we may write

$$\epsilon_0 EA = \sigma A \qquad \text{or} \qquad \sigma = \epsilon_0 E$$

Finally we substitute to find the charge density:

$$\sigma = \epsilon_0 E = [8.85 \times 10^{-12} \text{ C}^2/(\text{N} \cdot \text{m}^2)](40 \text{ N/C})$$
$$= \mathbf{3.54 \times 10^{-10} \text{ C/m}^2}$$

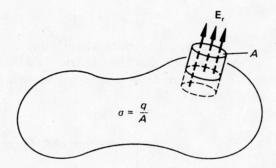

**Fig. 24-2** Gauss's law.

## Summary

- An electric field occurs wherever an electric charge will experience an electric force.
- The resultant field intensity at a point in the vicinity of a number of charges is the vector sum of the contributions from each charge.
- Gauss's law states that the net number of electric field lines crossing any closed surface in an outward direction is numerically equal to the net total charge within that surface.

## Key Equations by Section

| Section | Topic | Equation | Equation No. |
|---|---|---|---|
| | Coulomb's law for electrostatic forces | $F_e = k\dfrac{q_1 q_2}{r^2}$ | 24-2 |
| 24-1 | The magnitude of the electric field intensity based on the force $F$ experienced by a charge $q$ | $\boxed{E = \dfrac{F}{q}}$ | 24-4 |

| Section | Topic | Equation | Equation No. |
|---------|-------|----------|--------------|
| 24-2 | The force on a charge, based on Coulomb's law | $F = \dfrac{kQq}{r^2}$ | 24-6 |
| 24-2 | The field intensity on a charge | $\boxed{E = \dfrac{kQ}{r^2}}$ | 24-7 |
| 24-2 | The vector sum of all field intensities on all charges | $\boxed{E = \Sigma \dfrac{kQ}{r^2}}$ | 24-8 |
| 24-4 | The electric field intensity on the surface of a sphere a radius $r$ from the point charge | $E = \dfrac{kq}{r^2}$ | 24-9 |
| 24-4 | Permittivity of free space | $\boxed{\begin{array}{l} \epsilon_0 = \dfrac{1}{4\pi k} = \\ 8.85 \times 10^{-12} \ C^2/N \cdot m^2 \end{array}}$ | 24-11 |
| 24-4 | The number of electric lines of force, given by $\epsilon_0$ (the permittivity of free space), $E_n$ (the field intensity at a particular line of force), and the area of the surface | $N = \epsilon_0 E_n A$ | 24-13 |
| 24-4 | $E_n$ defined for a sphere | $E_n = \dfrac{1}{4\pi\epsilon_0} \dfrac{q}{r^2}$ | 24-14 |
| 24-4 | Gauss's law | $\boxed{N = \Sigma \ \epsilon_0 E_n A = \Sigma q}$ | 24-15 |
| 24-5 | The charge density $\sigma$ is equal to the charge per unit area | $\boxed{\sigma = \dfrac{q}{A} \quad q = \sigma A}$ | 24-16 |
| 24-5 | The field intensity of an infinite sheet, based on the charge density and the permittivity of free space | $E = \dfrac{\sigma}{2\epsilon_0}$ | 24-17 |
| 24-5 | The field intensity between the two plates of a capacitor | $E = \dfrac{\sigma}{\epsilon_0}$ | 24-18 |

## True-False Questions

T  F  **1.** It is necessary that a charge be placed at a point to have an electric field at that point.

T  F  **2.** Electric field lines never intersect.

T  F  **3.** Gauss's law represents a statement of equality, but it will not withstand unit analysis.

T  F  **4.** The direction of the electric field at a given point A in the vicinity of a positive charge depends on the sign of a charge placed at point A.

T  F  **5.** The field in the vicinity of a number of charges is equal to the algebraic sum of the fields due to the individual charges.

T  F  **6.** The spacing of electric field lines is such that they are close together when the field is strong and far apart when the field is weak.

T  F  **7.** Gauss's law demonstrates that all the charge lies on the surface of a conductor.

T  F  **8.** Because of the way in which an electric field is defined, the direction of the electric field and the force on a test positive charge will always be the same.

T  F  **9.** The electric field intensity at the midpoint of a line joining identical charges will always be zero.

T  F  **10.** At a point twice as far away from a certain charge, the field intensity will be reduced by one-fourth.

## Multiple-Choice Questions

**1.** The electric field intensity is zero
(a) midway between two equal charges of like sign
(b) midway between two charges of unlike sign
(c) at any point equal distances from two identical charges
(d) between two equal but oppositely charged plates

**2.** The direction of the electric field intensity is
(a) away from all negative charges
(b) toward all negative charges
(c) the same as the direction of an electric force
(d) dependent on the nature of a charge placed at the point in question

**3.** The magnitude of the electric field does *not* depend on the
(a) distance from charged objects
(b) sign of the charges causing the field
(c) magnitude of the charges causing the field
(d) force a unit positive charge will experience

**4.** The spacing of electric field lines between two identical point charges of opposite sign is
(a) not dependent on the magnitude of the charges
(b) an indication of the field direction
(c) an indication of the field strength
(d) large when the charges are large

**5.** According to Gauss's law, the number of electric field lines crossing any closed surface is
(a) numerically equal to the enclosed charge
(b) equal to the enclosed positive charge

(c) equal to the electric field inside the surface
(d) equal to the charge density on the surface

6. The permittivity of a medium
   (a) is a measure of its density
   (b) is equal to unity for air or a vacuum
   (c) is dependent on the charge density of the medium
   (d) determines the magnitude of an electric field that can be established by the medium

7. The electric field intensity at a distance of 4 m from a 6-$\mu$C charge is
   (a) $1.69 \times 10^3$ N/C          (b) $3.38 \times 10^3$ N/C
   (c) $1.35 \times 10^4$ N/C          (d) $2 \times 10^4$ N/C

8. The electric field intensity between two oppositely charged plates is $4 \times 10^5$ N/C in a downward direction. The force on a $-2$-nC charge passing between the plates is
   (a) $2 \times 10^4$ N upward          (b) $2 \times 10^4$ downward
   (c) $8 \times 10^{-4}$ N upward          (d) $8 \times 10^{-4}$ N downward

9. Two point charges of $-4$ and $-6$ $\mu$C are 10 cm apart in air. The magnitude of the electric field midway between the two charges is approximately
   (a) $7.2 \times 10^6$ N/C          (b) $3.6 \times 10^7$ N/C
   (c) $1.8 \times 10^6$ N/C          (d) $3.6 \times 10^5$ N/C

10. An 8-$\mu$C charge is 12 cm to the right of a $-5$-$\mu$C charge. The magnitude of the electric field at a point 9 cm above the 8-$\mu$C charge is approximately
    (a) 3.72 MN/C          (b) 5 MN/C          (c) 6.4 MN/C          (d) 7.85 MN/C

---

**Completion Questions**

1. An electric field is said to exist in a region of space in which a(n) _____ will experience a(n) _____.

2. The direction of the _____ at a point in space is the same as the direction in which a(n) _____ would move if it were placed at that point.

3. The electric field intensity near a known charge is directly proportional to the _____ and inversely proportional to the _____.

4. When more than one charge contributes to a field, the resultant field at a point is the _____ of the fields due to each charge.

5. The spacing of electric field lines must be such that they are _____ where the field is weak and _____ where the field is strong.

6. _____ are imaginary lines drawn so that their direction at any point is the same as the direction of the _____ at that point.

7. The direction of an electric field is _____ a positive charge and _____ is a negative charge.

8. The total number of lines passing normally through a surface is numerically equal to the _____ contained within the surface. This is known as _____.

9. It can be shown from _____ that all charge resides on the _____ of a conductor.

10. The units of the proportionality constant $k$ used in calculating the electric field intensity are _____.

**Chapter 25**

# Electric Potential

## DEFINITIONS OF KEY TERMS

**Electric work**   The scalar product of an electric force and the parallel distance through which it moves a charged particle.

**Electric potential energy**   The work done against electric forces in moving a charge $+q$ from infinity to a given point. This represents the energy that the system has available, and it would be entirely expended if the charge $+q$ were pushed to an infinite location by electric forces.

**Electric potential**   The potential energy per unit charge P.E./$q$. It is the work per unit charge done against electric forces in moving a charge from infinity to the point in question.

**Volt**   A unit of potential that represents a work per unit charge equivalent to 1 joule per coulomb (1 V = 1 J/C).

**Potential difference**   The work per unit positive charge done by electric forces in moving a test charge from the point of higher potential to the point of lower potential. It is measured in volts and is sometimes referred to as voltage.

**Potential gradient**   The electric field intensity $E$ expressed as a change in potential with distance. It is expressed in volts per meter (V/m), which is equivalent to newtons per coulomb (N/C).

**Equipotential lines**   Imaginary lines in an electrified space that connect points of equal potential. They are perpendicular to electric field lines at every point in space.

**Electronvolt**   A unit of energy equivalent to the energy acquired by an electron that is accelerated through a potential difference of 1 V.

## CONCEPTS AND EXAMPLES

**Electric Potential Energy**   When a charge is moved against electric forces for a distance $d$, the potential energy of the system increases by

$$\text{P.E.} = qEd$$

where $E$ is the electric field intensity and $q$ is the magnitude of the charge. If the charge is then released, it will acquire a final kinetic energy K.E. that is equal to the initial potential energy:

$$\text{K.E.} = \tfrac{1}{2}mv^2 = qEd$$

It is important to recognize the existence of two kinds of electric charge (positive and negative). The electric field and electric potential energy are defined in terms of *positive* charges. Thus, when a positive charge is moved *against* the electric field, the potential energy increases (the work is done by an outside agent). However, if a *negative* charge moves against electric forces, the potential energy decreases: the field does the work.

**EXAMPLE 25-1**

A positively charged horizontal plate is placed 30 mm above another plate of equal negative charge. The electric field intensity is a constant 40,000 N/C between the plates. How much work is done *by* the electric field in moving a $-5$-$\mu$C charge from the positive plate to the negative plate? What is the potential energy of the system when this charge reaches the negative plate? What will be the final kinetic energy if the charge is released?

**Solution**
The direction of the electric field is downward (positive to negative). If a *negative* charge is moved in the same direction as the electric field, an outside agent must overcome electric forces. The work done *by* the electric field is, therefore, negative:

$$\text{Work} = -|qEd| = -(5 \times 10^{-6} \text{ C})(40,000 \text{ N/C})(0.03 \text{ m})$$
$$= -6 \times 10^{-3} \text{ J} \qquad \text{work } by \text{ electric field}$$

Now, the work done *against* electric forces is done by an outside agent, so that the potential energy of the system increases by the same amount. Thus,

$$\text{P.E.} = +6 \times 10^{-3} \text{ J} \qquad \text{work } against \text{ electric field}$$

Finally, if the charge is released, electric field forces take over, and eventually the charge acquires a kinetic energy equal to $6 \times 10^{-3}$ J. If the mass of the charge were known, the final speed could be found from K.E. $= \frac{1}{2}mv^2$.

The electric potential energy of a charge $q$ placed a distance $r$ from another charge $Q$ is equal to the work done against electric forces in moving that charge from infinity. It can be shown that

$$\text{P.E.} = \frac{kQq}{r} \qquad \text{(P.E. of } q \text{ a distance } r \text{ from } Q)$$

## Electric Potential and Potential Difference

The electric potential $V$ at a point a distance $r$ from a charge $Q$ is equal to the potential energy per unit charge at that point. In other words, the electric potential represents the potential work that can be done by the field in removing a unit charge to infinity. It is found from

$$V_A = \frac{kQ}{r} \qquad \text{electric potential at point } A$$

Because electric fields are defined in terms of positive charges, it is convenient that the potential is positive in the vicinity of a positive charge and negative in the vicinity of a negative charge. Thus, the potential at a point $A$ that is near several charges is the algebraic sum of the potentials due to each charge:

$$V_A = \sum \frac{kQ}{r} = \frac{kQ_1}{r_1} + \frac{kQ_2}{r_2} + \frac{kQ_3}{r_3} + \cdots$$

The potential difference between two points $A$ and $B$ is the algebraic difference between the potentials at those points:

$$V_{AB} = V_A - V_B \qquad \text{potential difference } (1 \text{ V} = 1 \text{ J/C})$$

**EXAMPLE 25-2**  Determine the potential difference between points $A$ and $B$ as defined in Fig. 25-1.

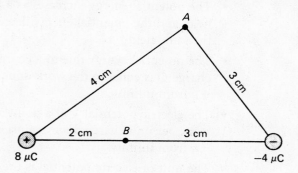

**Fig. 25-1**  The potential difference between $A$ and $B$ is $V_A - V_B$.

*Solution*

Taking the values from the figure, we determine the potentials at points $A$ and $B$:

$$V_A = \sum \frac{kQ}{r} = \frac{(9 \times 10^9)(-4 \times 10^{-6} \text{ C})}{0.03 \text{ m}} + \frac{(9 \times 10^9)(8 \times 10^{-6} \text{ C})}{0.04 \text{ m}}$$

$$= -1.2 \times 10^6 \text{ V} + 1.8 \times 10^6 \text{ V} = 0.6 \times 10^6 \text{ V}$$

$$V_B = \frac{(9 \times 10^9)(-4 \times 10^{-6} \text{ C})}{0.03 \text{ m}} + \frac{(9 \times 10^9)(8 \times 10^{-6})}{0.02 \text{ m}}$$

$$= -1.2 \times 10^6 \text{ V} + 3.6 \times 10^6 \text{ V} = 2.4 \times 10^6 \text{ V}$$

Now, the potential difference between point $A$ and point $B$ is

$$V_{AB} = 0.6 \times 10^6 \text{ V} - 2.4 \times 10^6 \text{ V} = -1.8 \times 10^6 \text{ V}$$

The negative sign in Example 25-2 means that an external agent must perform $1.8 \times 10^6$ J of work on each coulomb of positive charge it moves from point $A$ to point $B$. If $V_{AB}$ were positive, the electric field would do this work on each coulomb of positive charge. It is important not to confuse the work done *by* the electric field with the oppositely signed value of work done *against* the field. The work done by electric forces in moving a charge $+q$ from point $A$ to $B$ is

$$\text{Work}_{AB} = q(V_A - V_B) \qquad \textbf{work and potential difference}$$

As an additional example, you should show that the work done by the electric field in Example 25-2 in moving a 3-nC charge from $A$ to $B$ is equal to $-5.4 \times 10^{-3}$ J.

**Potential Gradient**  The potential difference between two oppositely charged plates is equal to the product of the constant field intensity $E$ and the plate separation $d$:

$$V = Ed \qquad E = \frac{V}{d}$$

The electric field intensity resulting from this equation would be expressed in volts per meter (V/m). When the electric field is described in this fashion, it is called the *potential gradient*.

## Summary

- The potential energy increases as a positive charge is moved against the electric field, and the potential energy decreases as a negative charge is moved against the same field.
- The potential energy due to a charge $q$ placed at a distance $r$ from another charge $Q$ is equal to the work done against electric forces in moving the charge $+q$ from infinity.
- The electric potential $V$ at a point a distance $r$ from a charge $Q$ is equal to the work per unit charge done against electric forces in bringing a positive charge $+q$ from infinity.
- The unit of electric potential is the joule per coulomb (J/C), which is renamed the volt (V).
- The potential difference between two points $A$ and $B$ is the difference in the potentials at those points.
- The potential difference between two oppositely charged plates is equal to the product of the field intensity and the plate separation.
- The electron, the basic quantity of charge, was determined to be $1.6065 \times 10^{-19}$ C by Millikan's oil-drop experiment.

## Key Equations by Section

| Section | Topic | Equation | Equation No. |
|---|---|---|---|
| 25-1 | Electric potential energy | P.E. $= qEd$ | 25-1 |
| 25-2 | The force required to move a charge from point $A$ to point $B$ | $F = \dfrac{kQq}{r_A r_B}$ | 25-2 |
| 25-2 | The work done to move a charge from point $A$ to point $B$ | $\text{Work}_{A \to B} = kQq\left(\dfrac{1}{r_B} - \dfrac{1}{r_A}\right)$ | 25-3 |
| 25-2 | The work required to move a charge $q$ from infinity to a point a distance $r$ from charge $Q$ | $\text{Work}_{\infty \to r} = \dfrac{kQq}{r}$ | 25-4 |

| Section | Topic | Equation | Equation No. |
|---|---|---|---|
| 25-2 | The potential energy of a charge | $\text{P.E.} = \dfrac{kQq}{r}$ | 25-5 |
| 25-3 | The potential energy of a charge | $\text{P.E.} = qV_A$ | 25-7 |
| 25-3 | The voltage potential | $V_A = \dfrac{kQ}{r}$ | 25-8 |
| 25-3 | The total voltage potential is equal to the sum of all potentials | $V = \sum \dfrac{kQ}{r}$ | 25-9 |
| 25-4 | Potential difference (the work done by moving a charge from $A$ to $B$ is equal to the charge times the potential difference) | $\text{Work}_{A \to B} = q(V_A - V_B)$ | 25-10 |
| 25-4 | The potential between two capacitor plates equals the field intensity times the distance | $V = Ed$ | 25-11 |
| 25-5 | Millikan's oil-drop experiment for $q$ (the net charge of oil drop), $m$ (the mass of oil drop), and $g$ (the acceleration of gravity) | $qE = mg$ | 25-12 |
| 25-5 | Millikan's oil-drop experiment | $q = \dfrac{mgd}{V}$ | 25-13 |

## True-False Questions

T F **1.** When the electric field does negative work in moving a charge from infinity to point $B$, the potential energy of the charge at $B$ will also be negative.

T F **2.** The electric potential energy is positive in the vicinity of a positive charge and negative in the vicinity of a negative charge.

T F **3.** Electric potential at a point is a property of the space, whereas electric potential energy cannot exist unless a charge is placed at that point.

T F **4.** Whenever a negative charge is moved from a point of high potential to a point of low potential, its potential energy is increased.

T F **5.** The electric potential in the vicinity of a number of charges is equal to the algebraic sum of the potentials due to each charge.

T F **6.** A negative potential means that the electric field will hold on to positive charge, and work must be done by an external agent to remove it.

T F **7.** If the potential is zero at a point, the electric field also must be zero at that point.

T F **8.** The electric field between two oppositely charged plates is equal to the product of the voltage and the plate separation.

T F **9.** The electronvolt is a unit of potential difference.

T F **10.** The surface of any conductor is an equipotential surface.

---

## Multiple-Choice Questions

1. When a negative charge is moved from a point of low potential to a point of high potential, its potential energy
   (a) increases
   (b) decreases
   (c) stays the same
   (d) increases and then decreases

2. The potential energy at a given point is independent of the
   (a) work required to bring a charge to that point
   (b) electric field
   (c) path taken to reach that point
   (d) magnitude of a charge at that point

3. In the vicinity of a negative charge,
   (a) the potential is always negative
   (b) the potential energy is always negative
   (c) the potential energy is always positive
   (d) the potential is always positive

4. Which of the following represents a unit of energy?
   (a) V        (b) N/C        (c) J/C        (d) eV

5. The Millikan oil-drop experiment was used primarily to determine the
   (a) mass of an electron
   (b) charge of an electron
   (c) electron charge to mass ratio
   (d) density of oil

6. The electric potential is zero
   (a) inside a conductor
   (b) halfway between $+q$ and $-q$
   (c) halfway between $+q$ and $+q$
   (d) on a line between $+q$ and $-q$

7. A 3-nC charge is located 2 m away from another charge of 40 $\mu$C. The potential energy is
   (a) $1.8 \times 10^{-4}$ J          (b) $2.7 \times 10^{-4}$ J
   (c) $5.4 \times 10^{-4}$ J          (d) $6.9 \times 10^{-4}$ J

8. A charge of $+4$ $\mu$C is 10 cm to the right of a $-12$-$\mu$C charge. The electric potential at a point midway between the two charges is approximately
   (a) 1.44 $\mu$V     (b) $-1.44$ $\mu$V   (c) 72 $\mu$V        (d) $-2.16$ $\mu$V

9. Points $A$ and $B$ are located 6 and 10 cm away, respectively, from a $-24$-$\mu$C charge. The potential difference $V_A - V_B$ is approximately
   (a) $-1.44$ $\mu$V   (b) $1.44$ $\mu$V   (c) $-5.04$ $\mu$V   (d) $5.04$ $\mu$V

10. A 16-$\mu$C charge is located 8 cm to the right of a $-8$-$\mu$C charge. How much work will be done by the electric field in moving a 2-nC charge from a point midway between the two charges to a point 4 cm to the left of the $-8$-$\mu$C charge?
    (a) 2.4 MJ      (b) 4.8 MJ      (c) $-2.4$ MJ     (d) $-4.8$ MJ

---

**Completion Questions**

1. The _____ at a point is equal to the negative of the work per unit charge done by electric forces in bringing a positive charge from infinity.

2. The potential in the vicinity of a positive charge is _____, and the potential in the vicinity of a negative charge is _____.

3. The potential in the vicinity of a number of charges is equal to the _____ of the potentials due to each charge.

4. A potential of 1 V means that a charge of _____ will have a potential energy of _____ when placed at that point.

5. The potential difference between two oppositely charged plates is equal to the product of the _____ and the _____.

6. The _____ is a unit of energy equivalent to the energy acquired by an electron that accelerated through a potential difference of one volt.

7. The work done by an electric field in moving a charge from a point of potential $V_A$ to a point of potential $V_B$ is equal to the product of _____ and _____.

8. The volt per meter is a unit of _____ and is equivalent to the unit _____.

9. A(n) _____ potential energy means that work must be done _____ the electric field in removing a charge from the field.

10. Whenever a positive charge is moved against the electric field, its potential energy _____; whenever a negative charge moves against an electric field, its potential energy _____.

# Chapter 26

# Capacitance

## DEFINITIONS OF KEY TERMS

**Dielectric**   An insulator, or a material containing few charges that are free to move.

**Dielectric strength**   For a given material, the electric field intensity for which that material ceases to be an insulator and becomes a conductor.

**Capacitor**   Two closely spaced conductors carrying equal and opposite charges.

**Capacitance**   For a single conductor, the ratio of the charge on the conductor to the potential produced; for a capacitor, the ratio of the charge on either plate to the resulting potential difference between the plates.

**Farad**   A measure of capacitance. A capacitor has capacitance of 1 F if an increase in charge of 1 C results in an increase in potential difference of 1 V.

**Dielectric constant**   A property of a dielectric equal to the ratio of the capacitance of a capacitor with the dielectric to its capacitance for a vacuum.

**Permittivity**   A property of a material that is a measure of its ability to establish an electric field.

**Series connection**   A group of capacitors connected along a single path.

**Parallel connection**   A group of capacitors connected directly to the same source of potential difference so that the same voltage is applied to each capacitor.

**Corona discharge**   A slow leakage of charge to ionized air from any pointed conductor.

**Variable capacitor**   A capacitor in which the area of opposing plates can be varied, changing the value of the capacitor.

## CONCEPTS AND EXAMPLES

**Computing Capacitance**   Capacitance is defined as the ratio of charge $Q$ to potential $V$ for a given conductor. In the case of a parallel-plate capacitor, the capacitance is the ratio of the charge on either plate to the potential difference placed across the plates:

$$C = \frac{Q}{V} \qquad 1 \text{ farad (F)} = \frac{1 \text{ coulomb (C)}}{1 \text{ volt (V)}}$$

**EXAMPLE 26-1**   What voltage must be placed across a 6-$\mu$F capacitor to store 2400 $\mu$C of charge?

*Solution*

We solve for $V$ and substitute:

$$V = \frac{Q}{C} = \frac{2400 \ \mu C}{6 \ \mu F} = \boxed{\textbf{400 V}}$$

Note that the prefix $\mu$ divides out, and it is not necessary to convert to the base unit in this instance.

**Dielectrics**   The material surrounding a charged conductor or between the plates of a capacitor is called a *dielectric*. For a given material, the dielectric strength is the electric field intensity $E$ for which that material ceases to be an insulator and allows charge to leak off nearby conductors. The dielectric strength is $3 \times 10^6$ N/C for air.

The insertion of a dielectric material between the plates of a capacitor has a direct effect on the electric field and the potential difference between the plates. A greater amount of charge can be stored for the same applied voltage. Thus, the capacitance is increased. The ratio of the capacitance with the dielectric to the capacitance in a vacuum is known as the dielectric constant $K$. If we use the subscript zero to refer to conditions for a vacuum, we can calculate the dielectric constant from any of the following ratios:

$$K = \frac{C}{C_0} \qquad K = \frac{E_0}{E} \qquad K = \frac{V_0}{V}$$

**EXAMPLE 26-2**   A 4-$\mu$F capacitor holds 80 $\mu$C of charge when it is connected to a battery. How much charge can be stored if mica ($K = 5$) is placed between the capacitor plates and the same battery is attached?

*Solution*

The battery voltage is found from $Q_0 = C_0 V_0$:

$$V_0 = \frac{Q_0}{C_0} = \frac{80 \ \mu C}{4 \ \mu F} = 20 \ V$$

When the dielectric is inserted, the new capacitance is

$$C = (5)(4 \ \mu F) = 20 \ \mu F$$

Now, if we apply the same 20 V, we increase the charge that can be stored:

$$Q = CV = (20 \ \mu F)(20 \ V) = 400 \ \mu C$$

We could have found this value directly from $Q = KQ_0$.

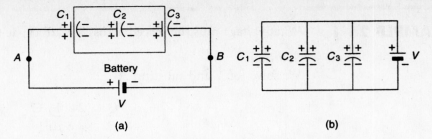

**Fig. 26-1** (a) A series connection of capacitors, (b) a parallel connection of capacitors.

The capacitance can be determined also from the physical conditions of the capacitor and knowledge of the dielectric constant. The general equation is

$$C = \epsilon \frac{A}{d} \qquad C = K\epsilon_0 \frac{A}{d} \qquad \epsilon_0 = 8.85 \times 10^{-12} \text{ C}^2/(\text{N} \cdot \text{m}^2)$$

The permittivity $\epsilon$ of the dielectric is greater than the permittivity $\epsilon_0$ of a vacuum by a factor equal to the dielectric constant $K$ ($K = 1$ for a vacuum). Refer to Example 26-4 in the text.

### Capacitors in Circuits

Capacitors may be connected in series as shown in Figure 26-1a or in parallel as shown in Figure 26-1b. For **series connections,** the charge on each capacitor is the same as the total charge, the potential difference across the battery is equal to the sum of the potential drops across each capacitor, and the net capacitance is given by

$$Q_T = Q_1 = Q_2 = Q_3$$
$$V_T = V_1 + V_2 + V_3 \qquad \text{series connections}$$
$$\frac{1}{C_e} = \frac{1}{C_1} + \frac{1}{C_2} + \frac{1}{C_3}$$

For two capacitors in series, a simpler form is

$$C_e = \frac{C_1 C_2}{C_1 + C_2} \qquad \text{two capacitors in series}$$

For *parallel connections*, the total charge is equal to the sum of the charges across each capacitor, the voltage drop across each capacitor is the same as the drop across the battery, and the effective (net) capacitance is the sum of the individual capacitances:

$$Q_T = Q_1 + Q_2 + Q_3$$
$$V_B = V_1 = V_2 = V_3 \qquad \text{parallel connections}$$
$$C_e = C_1 + C_2 + C_3$$

**EXAMPLE 26-3**    A 12-V battery is connected to the circuit shown in Fig. 26-2. What is the equivalent capacitance of the circuit? What is the charge on each capacitor? What is the voltage drop across the 3-$\mu$F capacitor?

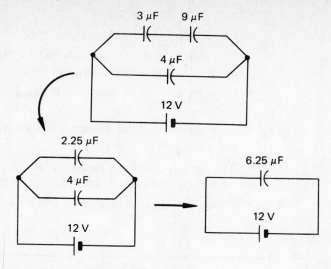

**Fig. 26-2** Calculating the equivalent capacitance of a circuit.

### Solution

The effective capacitance is determined in two steps. First, we determine the effective capacitance of $C_9$ and $C_3$:

$$C_{3,9} = \frac{C_3 C_9}{C_3 + C_9} = \frac{(3\ \mu\text{F})(9\ \mu\text{F})}{(3\ \mu\text{F} + 9\ \mu\text{F})} = \boxed{2.25\ \mu\text{F}}$$

Now $C_{3,9}$ and $C_4$ are in parallel, so $C_e = C_{3,9} + C_4$, or

$$C_e = 2.24\ \mu\text{F} + 4\ \mu\text{F} = \boxed{6.25\ \mu\text{F}}$$

Now the battery voltage of 12 V is across both $C_{3,9}$ and $C_4$. The charge $Q_4$ is equal to the product $C_4 V_4$:

$$Q_4 = (4\ \mu\text{F})(12\ \text{V}) = \boxed{48\ \mu\text{C}}$$

The total charge $Q_T$ is equal to $C_e V$:

$$Q_T = (6.25\ \mu\text{F})(12\ \text{V}) = \boxed{75\ \mu\text{C}}$$

Since $C_3$ and $C_9$ are in series, they will each have the same charge, which is the difference between $Q_T$ and $Q_4$. Thus,

$$Q_3 = Q_9 = 75\ \mu\text{C} - 48\ \mu\text{C} = \boxed{27\ \mu\text{C}}$$

Finally, the voltage drop across $C_3$ is found from the known charge $Q_3$:

$$V_3 = \frac{27\ \mu\text{C}}{3\ \mu\text{F}} = \boxed{9\ \text{V}}$$

The remaining 3 V drops across the 9-$\mu$F capacitor.

# Summary

- Capacitance is the ratio of charge $Q$ to the potential $V$ for a given conductor.
- The dielectric strength is that value for $E$ for which a given material ceases to be an insulator and becomes a conductor.
- For a parallel-plate capacitor, the material between the plates is called the *dielectric*.
- The insertion of the dielectric has an effect on the electric field and the potential between the plates.
- The capacitance for a parallel-plate capacitor depends on the surface area $A$ of each plate, the plate separation $d$, and the permittivity or dielectric constant.
- Capacitors may be connected in series or in parallel.

## Key Equations by Section

| Section | Topic | Equation | Equation No. |
|---------|-------|----------|--------------|
| 26-1 | The capacitance $C$ of a conductor | $C = \dfrac{Q}{V}$ | 26-1, 26-2 |
| 26-2 | Definition of farad (F); the basic unit of capacitance | $1\,F = \dfrac{1\,C}{1\,V}$ | 26-2 |
| 26-3 | The electric field intensity of a charged capacitor | $E = \dfrac{V}{d}$ | 26-3 |
| 26-3 | Alternative expression for the electric field intensity of a charged capacitor | $E = \dfrac{\sigma}{\epsilon_0} = \dfrac{Q}{A\epsilon_0}$ | 26-4 |
| 26-3 | Capacitance of a charged capacitor, in a vacuum | $C_0 = \dfrac{Q}{V} = \epsilon_0 \dfrac{A}{d}$ | 26-5 |
| 26-4 | The electric field intensity of a capacitor is the value of the field in a vacuum minus the dielectric field intensity | $E = E_0 - E_D$ | 26-6 |
| 26-4 | Dielectric constant $K$ | $K = \dfrac{C}{C_0}$ | 26-7 |
| 26-4 | Dielectric constant $K$, alternative definitions | $K = \dfrac{V_0}{V} = \dfrac{E_0}{E}$ | 26-8 |

| Section | Topic | Equation | Equation No. |
|---------|-------|----------|--------------|
| 26-4 | Capacitance calculation | $$C = K\epsilon_0 \frac{A}{d}$$ | 26-9 |
| 26-4 | Definition of permittivity | $\epsilon = K\epsilon_0$ | 26-10 |
| 26-4 | Capacitance in a vacuum or, approximately, in air | $C = \epsilon \frac{A}{d}$ | 26-11 |
| 26-5 | Total voltage for capacitors in series | $V = V_1 + V_2 + V_3$ | 26-12 |
| 26-5 | Total capacitance for capacitors in series | $$\frac{1}{C_e} = \frac{1}{C_1} + \frac{1}{C_2} + \frac{1}{C_3}$$ | 26-13 |
| 26-5 | Total capacitance for two capacitors in series | $C_e = \dfrac{C_1 C_2}{C_1 + C_2}$ | 26-14 |
| 26-5 | Total charge for capacitors in series | $Q = Q_1 + Q_2 + Q_3$ | 26-15 |
| 26-5 | Total voltage for capacitors in parallel | $$C = C_1 + C_2 + C_3$$ | 26-17 |
| 26-6 | Expressions for potential energy of a charged capacitor | $$\begin{aligned} \text{P.E.} &= \tfrac{1}{2}QV \\ &= \tfrac{1}{2}CV^2 \\ &= \frac{Q^2}{2C} \end{aligned}$$ | 26-18 |

---

**True-False Questions**

T   F    **1.** The total capacitance of two capacitors connected in series is less than that of either capacitor alone.

T   F    **2.** The capacitance is dependent on the potential difference placed across its plates.

T   F    **3.** If a conductor has a capacitance of one farad, a transfer of one coulomb of charge to the conductor will increase its potential by one volt.

T   F    **4.** The capacitance of a given capacitor will be higher if the separation of the plates is reduced without changing the dielectric.

T   F    **5.** Insertion of a dielectric between the plates of a capacitor decreases the voltage across it and hence reduces the capacitance.

**T F 6.** If the charge on a capacitor is doubled, the potential energy will be quadrupled.

**T F 7.** For capacitors connected in parallel, the voltage across each capacitor is the same as that across the source.

---

## Multiple-Choice Questions

1. The amount of charge that can be placed on a conductor does not depend on
   (a) the dielectric strength
   (b) its capacitance
   (c) its potential
   (d) its size or shape

2. The capacitance of a capacitor increases with a decrease in
   (a) dielectric constant
   (b) permittivity
   (c) plate area
   (d) plate separation

3. Which of the following is not a representation of the potential energy of a conductor?
   (a) $\frac{1}{2}QV^2$    (b) $Q^2/(2\,C)$    (c) $\frac{1}{2}QV$    (d) $\frac{1}{2}CV^2$

4. Which of the following is true for capacitors in series?
   (a) The total capacitance is the sum of the individual capacitances.
   (b) The total charge is the sum of the charges on each capacitor.
   (c) The total voltage is the sum of the voltages across each capacitor.
   (d) The available charge is shared between two or more capacitors.

5. The plates of a 2-pF capacitor have an extra of 20 cm². If air is the dielectric, the plate separation must be approximately
   (a) 0.885 mm    (b) 8.85 mm    (c) 88.5 mm    (d) 885 mm

6. What potential difference is required to store 24 μC of charge on a 6-μF capacitor?
   (a) 4 V    (b) 0.25 V    (c) 40 V    (d) 144 V

7. The voltage across the 3-μF capacitor in Fig. 26-3 is
   (a) 4 V    (b) 6 V    (c) 8 V    (d) 12 V

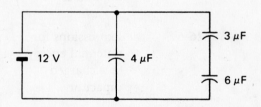

**Fig. 26-3**

8. A certain capacitor has a capacitance of 12 μF when the dielectric is air. The capacitor is charged to 400 V and disconnected from the power source. If a dielectric ($K = 4.0$) is inserted, the new voltage will be
   (a) 100 V    (b) 400 V    (c) 800 V    (d) 1600 V

9. An 8-μF capacitor is connected to a potential difference of 12 V. The potential energy is
   (a) $4.8 \times 10^{-4}$ J
   (b) $5.76 \times 10^{-4}$ J
   (c) 576 J
   (d) 480 J

**Completion Questions**

1. The capacitance of a given capacitor will be directly proportional to the _____ of the plates and inversely proportional to _____.

2. The _____ for a given material is that electric field intensity for which the material ceases to be an insulator and becomes a conductor.

3. Three advantages for the use of dielectrics with capacitors are _____, _____, and _____.

4. Three different physical ratios that can be used to calculate the dielectric constant are _____, _____, and _____.

5. The dielectric constant may also be referred to as the relative _____ of a material.

6. The total capacitance of a number of capacitors connected in series is _____ (less than, greater than, or the same as) the capacitance of any capacitor taken individually.

7. The _____ for a material is the ratio of the capacitance with that material between the plates to the capacitance for a vacuum between the plates.

8. For a given capacitor of known charge $Q$, voltage $V$, and capacitance $C$, list three expressions for calculating its potential energy: _____, _____, and _____.

9. A group of capacitors connected directly to the same source of potential difference so that the available charge is shared are said to be connected in _____.

# Chapter 27

# Current and Resistance

**DEFINITIONS OF KEY TERMS**

**Electric current**   The rate of flow of charge $Q$ past a given point $P$ on an electric conductor. The units are coulombs per second (C/s).

**Ampere**   A unit of electric current representing a flow of charge at the rate of 1 C/s.

**Source of electromotive force**   A device that converts chemical, mechanical, or other forms of energy to the electric energy necessary to maintain a continuous flow of charge.

**Electromotive force (emf)**   The work per unit charge (J/C) required to maintain a continuous flow of charge. The unit for emf is the volt (V).

**Ohm's law**   The current produced in a given conductor is directly proportional to the difference of potential between its end points.

**Electric resistance**   The opposition that a material offers to the flow of charge. A resistance of 1 Ω will support a current of 1 A for a difference of potential of 1 V.

**Electric power**   The rate at which charge gains or loses energy.

**Resistivity**   The property of a material that determines its electric resistance at a given temperature.

**Temperature coefficient of resistance**   The change in resistance per unit resistance per degree change in temperature.

**Circular mil**   The cross-sectional area of a wire whose diameter is 1 mil (0.001 in.).

**Ammeter**   A gauge to measure current.

**Ohm**   The unit of resistance Ω equal to 1 V/1 A.

**Rheostat**   Another name for a variable resistor.

**Transient current**   A spike of high current.

**Voltmeter**   A gauge to measure voltage.

**CONCEPTS AND EXAMPLES**

**Ohm's Law**   The current produced in a given conductor depends on the difference in potential of the end points:

$$R = \frac{V}{I} \qquad V = IR \qquad \text{Ohm's law} \; (1\ \Omega = 1\ \text{V/A})$$

**EXAMPLE 27-1**   A small motor is rated at 9 A for a voltage of 120 V. What is the electric resistance? If the motor is connected by mistake to a 220-V source, what is the current?

*Solution*
Applying Ohm's law yields

$$R = \frac{120 \text{ V}}{9 \text{ A}} = \boxed{13.3 \text{ } \Omega}$$

Now, the amperage under a voltage of 220 V is

$$I = \frac{V}{R} = \frac{220 \text{ V}}{13.3 \text{ } \Omega} = \boxed{16.5 \text{ A}}$$

Note that the resistance is a constant that depends on the material and its physical characteristics. For a given resistance, a change in voltage produces a proportional change in current.

**Electric Power**   Electric power is the rate at which electric work is accomplished. Since a volt is defined as a joule of work for each coulomb of charge displaced and since an ampere is a coulomb per second, electric power is given by the product of voltage $V$ and current $I$:

$$\boxed{P = VI} \quad (1 \text{ J/C})(1 \text{ C/s}) = 1 \text{ J/s} = 1 \text{ W}$$

Applying Ohm's law gives two additional expressions for power:

$$\boxed{P = I^2 R} \quad \boxed{P = \frac{V^2}{R}}$$

**EXAMPLE 27-2**   The resistance of a 20-A fuse is known to be 6 $\Omega$. What is the maximum power that can be applied?

*Solution*
Since $P = I^2 R$,

$$P = (20 \text{ A})^2 (6 \text{ } \Omega) = 2400 \text{ W}$$

**Resistivity**   The electric resistance of a given conductor is determined by four factors:

1. The *temperature* ($R \propto T$)
2. The *cross-sectional area* ($R \propto 1/A$)
3. The *length* ($R \propto l$)
4. The type of *material* (its resistivity $\rho$)

At a given temperature, the resistance of a conductor can be found by defining the resistivity $\rho$. We may write

$$R = \rho \frac{l}{A} \qquad \rho = \frac{RA}{l} \qquad \text{SI unit for } \rho \text{ is } \Omega \cdot \text{m}$$

---

**EXAMPLE 27-3**

What length of constantan wire 1 mm in diameter is needed to construct a coil with a resistance of 2000 $\Omega$ at 20°C? The resistivity of constantan is $49 \times 10^{-8}\ \Omega \cdot \text{m}$.

**Solution**
First, we must find the cross section of the wire:

$$A = \frac{\pi \, (1 \text{ mm})^2}{4} = 0.785 \text{ mm}^2 = 7.85 \times 10^{-7} \text{ m}^2$$

Now, we solve the resistivity equation for $l$ and substitute

$$l = \frac{RA}{\rho} = \frac{(2000\ \Omega)(7.85 \times 10^{-7} \text{ m}^2)}{49 \times 10^{-8}\ \Omega \cdot \text{m}} = 3206 \text{ m}$$

---

**Temperature Coefficient of Resistance** Experimental results have shown that the resistance of most materials increases with temperature. This change in resistance $\Delta R$ is proportional to the initial resistance $R_0$ and to the change in temperature $\Delta t$. By defining a property of a material called its temperature coefficient of resistance $\alpha$, we may write

$$\alpha = \frac{\Delta R}{R_0 \Delta t} \qquad \Delta R = \alpha R_0 \Delta t$$

The unit for the coefficient of resistance is reciprocal degrees, for example, 1/C°.

---

**Summary**

- Electric current $I$ is the rate of flow of charge $Q$ past a given point on a conductor.
- By convention, the direction of electric current is the same as the direction in which positive charges would move, even if the actual current consisted of a flow of negatively charged electrons.
- Ohm's law states that the current produced in a given conductor is *directly proportional to the difference of potential between its end points.*
- The resistance of a wire depends on four factors: (a) the kind of material, (b) the length, (c) the cross-sectional area, and (d) the temperature.
- The temperature coefficient of resistance $\alpha$ is the change in resistance per unit resistance per degree change in temperature.

## Key Equations by Section

| Section | Topic | Equation | Equation No. |
|---------|-------|----------|--------------|
| 27-1 | Electric current $I$, the flow of charge $Q$ per time $t$. | $$I = \frac{Q}{t}$$ | 27-1 |
| 27-4 | Ohm's law | $$R = \frac{V}{I} \qquad V = IR$$ | 27-2 |
| 27-5 | Definition of work | $\text{Work} = VIt$ | 27-4 |
| 27-5 | Definition of power | $$P = VI = I^2R = \frac{V^2}{R}$$ | 27-6, 27-7 |
| 27-6 | Resistance based on resistivity $\rho$ and cross-sectional area | $$R = \rho\frac{l}{A}$$ | 27-8 |
| 27-6 | Resistivity | $$\rho = \frac{RA}{l}$$ | 27-9 |
| 27-6 | The cross-sectional area of a wire in square mils | $A = \dfrac{\pi D^2}{4}$ (square mils) | 27-10 |
| 27-6 | Area in circular mils | $A_{\text{cmils}} = (D_{\text{mils}})^2$ | 27-12 |
| 27-6 | Resistivity in terms of circular mils | $\rho = \dfrac{RA}{I} \rightarrow \dfrac{\Omega \cdot \text{cmil}}{ft}$ | 27-12 |
| 27-7 | Definition of the temperature coefficient of resistance | $$\alpha = \frac{\Delta R}{R_0 \Delta t}$$ | 27-14 |

## True-False Questions

**T    F    1.** The electromotive force is a force exerted on an electric charge to keep it moving.

**T    F    2.** The direction of conventional current for a conductor is opposite to the direction of electron flow.

**T    F    3.** Electric current is also a measure of the average speed with which electrons move in a conductor.

**T    F    4.** According to Ohm's law, the electric current is inversely proportional to the applied voltage.

**T    F    5.** Electric resistance increases with an increase in the cross-sectional area of a conductor.

**T    F    6.** The resistivity of a wire is independent of the length of the wire.

**T    F    7.** A wire with a diameter of 0.002 in. has a cross-sectional area of 4 cmils.

**T    F    8.** The temperature coefficient of resistance is equal to the change in resistance per degree change in temperature.

**T    F    9.** The power loss in a wire is quadrupled if the current is doubled.

**T    F    10.** A rheostat is a meter that indicates the electric resistance in a circuit.

## Multiple-Choice Questions

1. A unit of electromotive force is the
   (a) joule      (b) newton      (c) volt      (d) watt

2. If one were to use a water analogy to study electric current, voltage would be most similar to
   (a) force      (b) pressure
   (c) rate of flow      (d) density

3. The resistance of a wire is *not* dependent on its
   (a) temperature      (b) length
   (c) area      (d) current

4. Which of the following is *not* a measure of electric power?
   (a) $VR^2$      (b) $VI$      (c) $I^2R$      (d) $V^2/R$

5. Which of the following is *not* a unit resistivity?
   (a) $\Omega \cdot cm$      (b) $\Omega \cdot cmil/ft$      (c) $\Omega/ft$      (d) $\Omega \cdot m$

6. The potential difference between the terminals of a small heater is 60 V. If the resistance of the heater is 30 $\Omega$, the current is
   (a) 0.5 A      (b) 2 A      (c) 1800 A      (d) 3 A

7. An emf of 12 V will move $6.25 \times 10^{18}$ electrons past a given point in 2 s. The resistance is
   (a) 24 $\Omega$      (b) 121 $\Omega$      (c) 6 $\Omega$      (d) 3.84 $\Omega$

8. A 120-V heater has a resistance of 600 $\Omega$. The heat energy generated in 1 min is
   (a) 24 J      (b) 120 J      (c) 1200 J      (d) 1440 J

9. An aluminum wire has a resistivity of 17 $\Omega$·cmils/ft and a cross-sectional diameter of 0.2 in. What length of this wire is needed to construct a 1700-$\Omega$ resistor?
   (a) 200 ft      (b) 144.5 ft      (c) $4 \times 10^6$ ft      (d) $2 \times 10^4$ ft

10. The temperature coefficient of resistance for copper is 0.004/C°. If the resistance of a copper wire is 12 $\Omega$ at 20°C, its resistance at 100°C will be
   (a) 3.8 $\Omega$      (b) 13 $\Omega$      (c) 15.84 $\Omega$      (d) 50.4 $\Omega$

## Completion Questions

1. A source of emf of one _____ will perform one _____ of _____ on each coulomb of charge that passes through it.

2. The current in a resistor is directly proportional to the _____ and inversely proportional to the _____. This is a statement of _____.

3. Four factors that affect the resistance of a wire are _____, _____, _____, and _____.

4. The temperature coefficient of resistance is the change in _____ per unit _____ per unit change in _____.

5. The rate of heat loss in a wire can be found from the product of the _____ and the square of the _____.

    

6. The direction of conventional current is always the same as the direction in which _____ would move.

7. The area of a wire in _____ equals the square of the _____ in mils.

8. The resistance of a conductor at a given temperature is directly proportional to its _____, inversely proportional to its _____, and dependent on a material constant called its _____.

9. Three laboratory devices used to study resistance, current, and voltage are _____, _____, and _____, respectively.

10. A source of electromotive force can convert _____ energy to _____ energy.

# Chapter 28

# Direct-Current Circuits

## DEFINITIONS OF KEY TERMS

**Series connection**   Two or more elements are said to be in series if they have only one point in common that is not connected to some third element. For resistors, a series connection provides only one path for current.

**Parallel connection**   A connection of resistors in which the current may be divided between two or more elements.

**DC circuit**   An electric circuit in which there is a continuous flow of charge in only one direction.

**Terminal potential difference**   The actual potential difference representing the external voltage drop across the terminals of a source of emf.

**Internal resistance**   The inherent resistance within every source of emf. The drop in voltage due to this resistance is what causes the terminal potential difference to be less than the open-circuit potential difference.

**Kirchhoff's first law**   The sum of the currents entering a junction is equal to the sum of the currents leaving that junction.

**Kirchhoff's second law**   The sum of the emf's around any closed current loop is equal to the sum of all the $IR$ drops around that loop.

**Wheatstone bridge**   A laboratory apparatus that measures an unknown resistance by varying known resistances until the system is balanced.

**Anode**   The electrodes that attracts anions. The anode has a positive charge.

**Cathode**   The electrode that attracts cations. The cathode has a negative charge.

**Electrolysis**   The process of separating molecules by use of an electric current.

**Electrolyte**   A substance that conducts electric current when it is dissolved in water or melted.

**Ionization**   The process of adding or removing charges from molecules (such as the molecules in air) to produce ions.

**Oxidation**   A process in which particles lose electrons.

**Reduction**   A process in which particles gain electrons.

**Series and Parallel Connections**   For **series** connections, the current in all parts of the circuit must be the same since there is only one path for the flow of charge. The total voltage drop in the circuit (the terminal potential difference) is equal to the sum of the drops across each resistor. The effective resistance of the entire external circuit is the sum of the individual resistances:

$$I_T = I_1 = I_2 = I_3$$
$$V_T = V_1 + V_2 + V_3 \qquad \text{series connections}$$
$$R_e = R_1 + R_2 + R_3$$

For *parallel* connections, the total current is the sum of the individual currents, the voltage drops are all equal, and the effective resistance is found from a reciprocal relationship:

$$I_T = I_1 + I_2 + I_3$$
$$V_T = V_1 = V_2 = V_3 \qquad \text{parallel connections}$$
$$\frac{1}{R_e} = \frac{1}{R_1} + \frac{1}{R_2} + \frac{1}{R_3}$$

For two resistors connected in parallel, a simpler form is

$$R_e = \frac{R_1 R_2}{R_1 + R_2} \qquad \text{two parallel resistors}$$

**EXAMPLE 28-1**   The terminal potential difference of the battery in Fig. 28-1 is 24 V. The resistors are given as follows: $R_1 = 6\ \Omega$, $R_2 = 8\ \Omega$, $R_3 = 4\ \Omega$. Determine the equivalent resistance of the circuit and the current through each resistor.

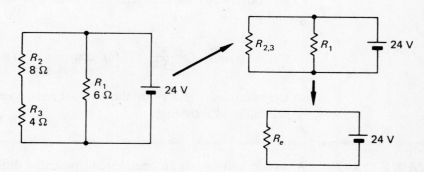

**Fig. 28-1**

**Solution**
It is better to reduce the circuit in steps, as shown in the figure. First, we find the combined resistance of the 8- and 4-$\Omega$ resistors (in series):

$$R_{2,3} = 8\ \Omega + 4\ \Omega = 12\ \Omega$$

Now, the effective resistance $R_e$ is found by combining the parallel resistors $R_{2,3}$ and $R_1$:

$$R_e = \frac{R_{2,3}R_1}{R_{2,3} + R_1} = \frac{(12 \ \Omega)(6 \ \Omega)}{12 \ \Omega + 6 \ \Omega} = \boxed{4 \ \Omega}$$

The total current to the circuit is found from Ohm's law:

$$I_T = \frac{V_T}{R_e} = \frac{24 \ V}{4 \ \Omega} = 6 \ A$$

Now, the current splits at junction $A$, but the voltage drop from $A$ to $B$ is the same through either path (24 V). The current through the 6-$\Omega$ resistor is

$$I_6 = \frac{24 \ V}{6 \ \Omega} = \boxed{4 \ A}$$

The remainder of the current (6 A − 4 A) must pass through each of the two other resistors. Thus,

$$\boxed{I_8 = 2 \ A \quad \text{and} \quad I_4 = 2 \ A}$$

As an additional exercise, you should show that the voltage drops across the 8- and 4-$\Omega$ resistors add to 24 V.

## Internal Resistance and Terminal Potential Difference (TPD)

When an electric current flows through a circuit, there is a loss of energy through the external load, but there is also a loss of energy as a result of the internal resistance in the source of emf. Thus, the actual terminal voltage $V_T$ is less than the open-circuit emf $\mathscr{E}$ by the loss through the internal resistance r.

$$V_T = \mathscr{E} - Ir \qquad \text{terminal potential difference}$$

In general, the current delivered to a circuit containing one or more sources of emf is given by

$$I = \frac{\Sigma \mathscr{E}}{\Sigma R} \qquad \text{for example} \qquad I = \frac{\mathscr{E}_1 - \mathscr{E}_2}{r_1 + r_2 + R_L}$$

In the example, $r_1$ and $r_2$ are the internal resistances of the sources of emf, and $R_L$ is the circuit load resistance.

**EXAMPLE 28-2**

A certain battery has an open-circuit potential difference of 12 V and an internal resistance of 0.6 $\Omega$. If the battery is connected to a search light with a resistance of 8 $\Omega$, what is the current delivered to the light and what is the terminal potential difference of the battery?

*Solution*
The current is equal to the emf divided by the total resistance—including the internal resistance:

$$I = \frac{12 \ V}{8 \ \Omega + 0.6 \ \Omega} = \boxed{1.40 \ A}$$

Now, the terminal potential difference is the emf less the drop through the load resistance:

$$V_T = \mathcal{E} - Ir = 12\text{ V} - (1.40\text{ A})(0.6\ \Omega) = \boxed{11.2\text{ V}}$$

## Kirchhoff's Laws

According to Kirchhoff's laws, the current entering a junction must equal the current leaving that junction and the net emf around any closed loop must equal the sum of the $IR$ drops. Symbolically,

$$\Sigma I_{\text{entering}} = \Sigma I_{\text{leaving}}$$
$$\Sigma \mathcal{E} = \Sigma IR$$

In applying Kirchhoff's laws to a complex circuit, it is important to follow a careful sequence of steps. (Refer to Fig. 28-2.)

**Step 1**  Indicate by a small arrow the direction of normal positive output of each source of emf. The arrows will point away from the positive sides.

**Step 2**  Assume a direction of current for each segment of the network and label it. Think of the emf's as pumping water through a network of pipes. The flow must be consistent and logical, even if the direction is wrong.

**Step 3**  Apply Kirchhoff's first law to write a current equation for each junction point ($\Sigma I_{\text{in}} = \Sigma I_{\text{out}}$).

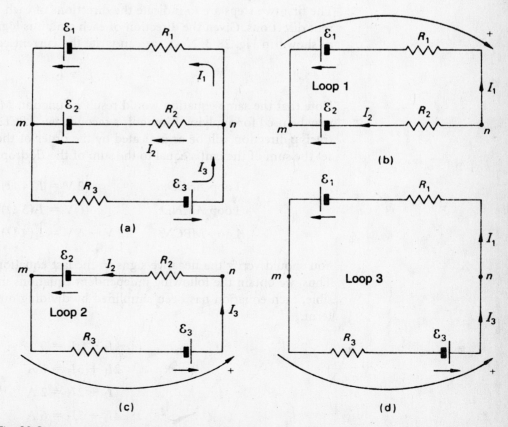

**Fig. 28-2**

**Step 4** Apply Kirchhoff's second law ($\Sigma\mathcal{E} = \Sigma IR$) to write a voltage equation for each closed loop. As you trace all the way around each loop, consider an emf positive when its direction is with the tracing direction and consider an $IR$ drop as positive if the assumed current is with the tracing direction. Otherwise, the values are considered negative.

**Step 5** Solve the equations simultaneously to determine the unknown quantities.

---

**EXAMPLE 28-3**  Consider the complex circuit described by Fig. 28-3. What are the currents $I_1$, $I_2$, and $I_3$?

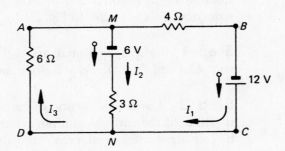

**Fig. 28-3**

**Solution**
The first two steps are to indicate the directions of each emf and the assumed current directions. Given the direction of each emf, it is logical to choose the currents as shown in Fig. 28-3. Next, we can write the current equation at the junction $N$:

$$I_1 + I_2 = I_3$$

Note that the same equation would result if junction $M$ were chosen. Three loops may be used for applying Kirchhoff's second law: $ABCD$, $AMND$, and $MBCN$. The tracing direction will be as indicated by the order of the letters. For each loop we set the sum of the emf's equal to the sum of the $IR$ drops according to convention:

| | |
|---|---|
| Loop $ABCD$: | $12 \text{ V} = I_1(4 \text{ }\Omega) + I_3(6 \text{ }\Omega)$ |
| Loop $AMND$: | $6 \text{ V} = I_2(3 \text{ }\Omega) + I_3(6 \text{ }\Omega)$ |
| Loop $MBCN$: | $12 \text{ V} - 6 \text{ V} = I_1(4 \text{ }\Omega) - I_2(3 \text{ }\Omega)$ |

You should verify the negative signs in the last equation by looking at the figure. Thus, we obtain the following independent equations in $I_1$, $I_2$, and $I_3$. Where possible, each equation has been simplified by dividing out factors common to each term.

$$I_1 + I_2 - I_3 = 0$$
$$2I_1 + 3I_3 = 6 \text{ A}$$
$$I_2 + 2I_3 = 2 \text{ A}$$
$$4I_1 - 3I_3 = 6 \text{ A}$$

The simultaneous solution of any three of these four equations will yield the following currents:

$$I_1 = 1.33 \text{ A} \qquad I_2 = -0.222 \text{ A} \qquad I_3 = 1.11 \text{ A}$$

The fourth equation can be used to check results. Notice that the solution for $I_2$ is negative. This means that the actual direction of the current through the middle segment is upward, not downward as assumed. The magnitude is still correct. Thus, current is being forced through the 6-V source in a direction opposite to its normal output. This happens when a battery is being charged.

## Summary

- In dc circuits, resistors may be connected in a series or in parallel.
  - **a.** For *series connections*, the current in all parts of the circuit is the same, and the total voltage drop is the sum of the individual drops across each resistor.
  - **b.** The effective resistance is equal to the sum of the individual resistances.
  - **c.** For *parallel connections*, the total current is the sum of the individual currents.
  - **d.** The voltage drops are all equal.
- The current supplied to an electric circuit is equal to the *net* emf divided by the total resistance of the circuit, including internal resistances.
- According to Kirchhoff's laws, the current entering a junction must equal the current leaving the junction, and the net emf around any loop must equal the sum of the *IR* drops.
- A Wheatstone bridge is a device that allows one to determine an unknown resistance $R_x$ by balancing the voltage drops in the circuit.

## Key Equations by Section

| Section | Topic | Equation | Equation No. |
|---|---|---|---|
| 28-1 | Ohm's law in terms of emf | $\mathcal{E} = IR$ | 28-1 |
| 28-1 | Ohm's law | $R = \dfrac{V}{I}$ | 28-2 |
| 28-1 | Total current over resistors in series | $I = I_1 = I_2 = I_3$ | 28-3 |
| 28-1 | Total resistance for resistors in series | $\boxed{R = R_1 + R_2 + R_3}$ | 28-5 |
| 28-2 | Voltage drop for resistors in parallel | $V = V_1 = V_2 = V_3$ | 28-7 |
| 28-2 | Total current for resistors in parallel | $I = I_1 + I_2 + I_3$ | 28-8 |

| Section | Topic | Equation | Equation No. |
|---------|-------|----------|--------------|
| 28-2 | Total resistance for resistors in parallel | $\dfrac{1}{R} = \dfrac{1}{R_1} + \dfrac{1}{R_2} + \dfrac{1}{R_3}$ | 28-9 |
| 28-2 | Total resistance for the special case of two resistors in parallel | $R = \dfrac{R_1 R_2}{R_1 + R_2}$ | 28-10 |
| 28-3 | Total voltage in terms of emf and internal resistance | $V_T = IR_L = \mathcal{E} - Ir$ | 28-12 |
| 28-3 | Current in terms of emf and total resistance | $I = \dfrac{\mathcal{E}}{R_L + r}$ | 28-13 |
| 28-5 | Total voltage when reversing the current through a source of emf such as a battery | $V_2 = \mathcal{E}_2 + Ir_2$ | 28-14 |
| 28-6 | Kirchhoff's first law | $\Sigma I_{\text{entering}} = \Sigma I_{\text{leaving}}$ | 28-16 |
| 28-6 | Kirchhoff's second law | $\Sigma \mathcal{E} = \Sigma IR$ | 28-17 |
| 28-7 | Solving for the unknown resistance in the Wheatstone bridge | $R_x = R_3 \dfrac{R_2}{R_1}$ | 28-28 |
| 28-7 | Alternative solution for the unknown resistance in the Wheatstone bridge | $R_x = R_3 \dfrac{I_2}{I_1}$ | 28-29 |
| 28-10 | Capacity rating of a battery | $\text{Life (hours)} = \dfrac{\text{ampere-hour rating}}{\text{amperes delivered}}$ | 28-35 |

**True-False Questions**

T  F  1. The rules for computing equivalent resistance are the same as those for computing equivalent capacitance.

T  F  2. The current is the same in all parts of a parallel circuit.

T  F  3. The equivalent resistance of two resistors in parallel is equal to their product divided by their sum.

T  F  4. The emf is essentially equal to the open-circuit potential difference.

T  F  5. The current supplied to an electric circuit is equal to the net emf divided by the total resistance of the circuit if we neglect internal resistance.

T  F  6. Kirchhoff's second law applies for each current loop in a complex circuit and not just for the total circuit.

T  F  **7.** In applying Kirchhoff's laws, the tracing direction must be the same as the current direction.

T  F  **8.** When the Wheatstone bridge is balanced, the voltage between the galvanometer and either terminal of the source of emf will be the same.

T  F  **9.** Kirchhoff's laws apply only for current loops that contain at least one source of emf.

T  F  **10.** When two identical resistors are connected in parallel, the voltage drop across each is one-half of the terminal potential difference at the source of emf.

---

## Multiple-Choice Questions

**1.** The variance of terminal potential difference, as compared with emf, is due to
(a) the circuit load          (b) the internal resistance
(c) Kirchhoff's law          (d) current delivered

**2.** For a parallel circuit, which of the following is *not* true?
(a) The current through each resistance is the same.
(b) The voltage across each resistance is the same.
(c) The total current is equal to the sum of the currents through each resistance.
(d) The reciprocal of the equivalent resistance is equal to the sum of the reciprocals of the individual resistances.

**3.** Which of the following is *not* affected by internal resistance?
(a) Terminal potential difference
(b) Source emf
(c) Current delivered to external circuit
(d) Power output

**4.** If a circuit contains three loops, how many *independent* equations can be obtained with Kirchhoff's two laws?
(a) Three     (b) Four     (c) Five     (d) Six

**5.** When using the Wheatstone bridge, the quantity that is balanced is
(a) voltage   (b) resistance   (c) current   (d) emf

**6.** For the circuit in Fig. 28-4, the equivalent external resistance is approximately
(a) 1.8 Ω     (b) 4 Ω     (c) 6 Ω     (d) 20 Ω

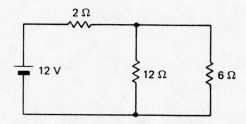

**Fig. 28-4**

**7.** In Fig. 28-4, if we neglect internal resistance, the current through the 6-Ω resistance is
(a) 1.0 A     (b) 1.33 A     (c) 1.67 A     (d) 2 A

**8.** The terminal voltage for the source of emf in Fig. 28-5 is
   (a) 24 V     (b) 22 V      (c) 21.8 V    (d) 20 V

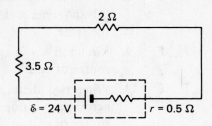

**Fig. 28-5**

**9.** The current through the 4-$\Omega$ resistance in Fig. 28-6, as found from Kirchhoff's laws is
   (a) 1 A     (b) 1.5 A      (c) 2 A     (d) 2.5 A

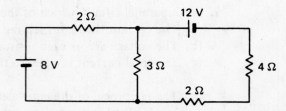

**Fig. 28-6**

**10.** A Wheatstone bridge is used to measure the unknown resistance $R_x$ of a coil of wire. The resistance box is adjusted for 8 $\Omega$, and the galvanometer indicates zero current when the contact key is positioned at the 40-cm mark. The unknown resistance is
   (a) 5.33 $\Omega$
   (b) 12 $\Omega$
   (c) neither (a) nor (b)
   (d) either (a) or (b) depending on hookup

# Chapter 29

# Magnetism and the Magnetic Field

## DEFINITIONS OF KEY TERMS

**Magnetism**   The physical phenomenon associated with the attraction of certain metals and with nonelectric forces on moving charges.

**Law of magnetic force**   Like poles repel each other; unlike poles attract each other.

**Coulomb's law for magnetic forces**   The force of attraction or repulsion between two magnetic poles is directly proportional to the product of the pole strengths and inversely proportional to the square of their separation distance.

**Magnetic domains**   A theoretical microscopic region consisting of a group of atoms in a magnetic material.

**Magnetic induction**   Magnetization occurring as a result of the influence of a magnetizing force.

**Retentivity**   The ability of some magnetic materials to retain magnetism after the magnetizing force has been removed.

**Magnetic saturation**   The maximum magnetization that can occur for a given material. (All domains are aligned.)

**Weber (Wb)**   A unit of magnetic flux equivalent to 100,000,000 magnetic field lines ($1 \text{ Wb} = 1 \times 10^8$ lines).

**Tesla (T)**   A unit of flux density equivalent to 1 weber per square meter ($1 \text{ T} = 1 \text{ Wb/m}^2$).

**Permeability**   A measure of a material's ability to support magnetic flux lines.

**Relative permeability**   The ratio of the permeability of a material to that for a vacuum.

**Diamagnetic**   Materials with a relative permeability slightly less than unity. Such materials are feebly repelled by a strong magnet.

**Paramagnetic**   Materials with permeabilities slightly greater than that of a vacuum. Such materials are feebly attracted by a strong magnet.

**Ferromagnetic**   Magnetic materials having extremely high permeabilities. Such materials are strongly attracted by a magnet.

**Hysteresis**   The lagging of the magnetization behind the magnetizing force. The area inside a hysteresis loop is a measure of the energy lost during a magnetization cycle.

**Magnet**   A material or device that exerts a magnetic field.

**Magnetic field**   An intensity field of magnetic potential. Every magnet is surrounded by a magnetic field.

**Magnetic flux lines**   Imaginary lines useful for visualizing a magnetic field.

**Magnetic poles**   The north and south ends of a magnet.

**Solenoid**   An electromechanical device with a magnetic core. A solenoid can open or close another electric circuit when energized.

**CONCEPTS
AND
EXAMPLES**

**Magnetic Flux Density**   The magnetic flux density $B$ in a region of a magnetic field is the number of flux lines that pass through a unit of area perpendicular to the flux:

$$B = \frac{\phi}{A_\perp} = \frac{\phi}{A \sin \theta} \qquad \textbf{magnetic flux density (T)}$$

where

$$\phi = \text{flux (Wb)}$$
$$A = \text{unit area (m}^2)$$
$$\theta = \text{angle area makes with flux}$$
$$B = \text{magnetic flux density (T)}$$

---

**EXAMPLE 29-1**

A circular loop of wire 30 cm in diameter makes an angle of 40° with a $B$ field of 0.6 T. What is the magnetic flux linking the loop?

**Solution**
The area of the loop is found from the diameter to be 0.0707 m². Solving for $\phi$ and substituting, we obtain

$$\phi = BA \sin \theta = (0.6 \text{ T})(0.0707 \text{ m}^2)(\sin 40°)$$
$$= \boxed{\textbf{0.0273 Wb}}$$

When a material is placed in a magnetic field, it becomes magnetized. We might think of the resulting flux density $B$ in the material as the "magnetization" and the magnetic field intensity $H$ as the "magnetizing force." In such cases, the magnetization $B$ is proportional to the magnetizing force $H$. The constant of proportionality is the permeability $\mu$ of the medium in which the field exists.

$$B = \mu H$$

For a vacuum, $\mu = \mu_0 = 4\pi \times 10^{-7}$ T · m/A. The relative permeability $\mu_r$ is the ratio $\mu/\mu_0$. Thus, in terms of the relative permeability, $B = \mu_0\mu_rH$.

---

## Magnetic Force on a Moving Charge
A magnetic field of 1 T will cause a force of 1 N on a charge of 1 C moving perpendicular to the field with a velocity of 1 m/s. The general case is shown in Fig. 29-1, in which the charge moves at an angle $\theta$ with the field.

$$F = qvB \sin \theta \qquad \textbf{magnetic force on moving charge}$$

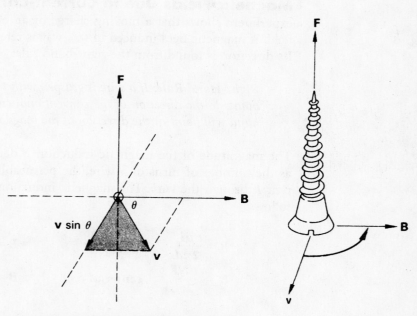

**Fig. 29-1**   Determining the direction of the magnetic force.

The direction of the magnetic force is given by the right-hand-screw rule, as illustrated in Fig. 29-1.

---

**EXAMPLE 29-2**

A proton ($q = +1.6 \times 10^{-19}$ C) moves from right to left into a perpendicular magnetic field with a velocity of $3 \times 10^6$ m/s. If the proton experiences a magnetic force of $2.4 \times 10^{-13}$ N directed out of the page, what are the magnitude and direction of the magnetic field?

**Solution**

Let us determine first the magnitude of $B$:

$$B = \frac{F}{qv \sin \theta} = \frac{2.4 \times 10^{-13} \text{ N}}{(1.6 \times 10^{-19} \text{ C})(3 \times 10^6 \text{ m/s})(1)}$$

$$= \textbf{0.50 T} \qquad \textbf{directed upward}$$

The field is directed upward because the vector **v** turned toward **B** (upward) will produce a vector **F** out of the page.

---

## Force on a Current-Carrying Conductor

The magnetic force on a wire of length $l$ carrying a current $I$ at an angle $\theta$ with a flux density $B$ is given by

$$F = BIl \sin \theta \qquad \textbf{magnetic force on a conductor}$$

The right-hand-screw rule applies for a conductor in a manner similar to that for a moving positive charge.

## Magnetic Fields Due to Current-Carrying Conductors

Oersted's experiment shows that a moving charge or an electric current produces a magnetic field. A magnetic field induced in this way is referred to as the *magnetic induction* **B**. Its direction is found from the right-hand rule:

> *Right-Hand Rule: If a wire is grasped with the right hand so that the thumb points to the direction of the conventional current, the curled fingers of that hand will point in the direction of the magnetic induction* **B**.

The magnitude of the magnetic induction $B$ depends on a number of factors, such as the number of turns of a wire, the permeability $\mu$ of the medium, and the current $I$ through the wire. The magnetic induction for many common situations is as follows:

$$B = \frac{\mu I}{2\pi d} \qquad \text{long wire} \qquad\qquad B = \frac{\mu I}{2r} \qquad \text{center of loop}$$

$$B = \frac{\mu NI}{2r} \qquad \text{center of coil} \qquad\qquad B = \frac{\mu NI}{L} \qquad \text{solenoid}$$

---

**EXAMPLE 29-3**    What current in a long wire is necessary to produce a magnetic induction of $3 \times 10^{-5}$ T at a distance of 5 cm from the wire?

*Solution*
We solve the equation for a wire to find $I$:

$$B = \frac{\mu I}{2\pi d} \qquad \text{or} \qquad I = \frac{2\pi d B}{\mu}$$

We assume that the wire is in air and that $\mu = \mu_0$. Thus,

$$I = \frac{2\pi(0.05 \text{ m})(3 \times 10^{-5} \text{ T})}{(4\pi \times 10^{-7} \text{ T} \cdot \text{m/A})} = \boxed{\textbf{7.50 A}}$$

If the medium is other than a vacuum of air, the value of $\mu$ will be determined by the relative permeability of the medium in accordance with

$$\mu = \mu_r \mu_0 \qquad \text{where} \qquad \mu_0 = 4\pi \times 10^{-7} \text{ T} \cdot \text{m/A}$$

This is extremely important in the case of solenoids or toroids, where the core often has a relative permeability in excess of 12,000.

---

# Summary

- The magnetic flux density $B$ in a region of a magnetic field is the number of flux lines that pass through a unit of area perpendicular to the flux.
- The magnetic flux density $B$ is proportional to the magnetic field intensity $H$.
- The constant of proportionality is the permeability of the medium in which the field exists.
- The relative permeability $\mu_r$ is the ratio of $\mu/\mu_0$.
- A magnetic field of flux density equal to 1 T will exert a force of 1 N on a charge of 1 C moving perpendicular to the field with a velocity of 1 m/s.

## Key Equations by Section

| Section | Topic | Equation | Equation No. |
|---|---|---|---|
| 29-4 | Magnetic flux density $B$ | $B = \dfrac{\phi \text{ (magnetic flux)}}{A_\perp \text{ (perpendicular area)}}$ | 29-2 |
| 29-4 | Definition of the tesla (T), the SI unit of flux density | $1\text{ T} = 1\text{ Wb/m}^2 = 10^4\text{ G}$ | 29-3 |
| 29-4 | Relationship between magnetic flux density and magnetic field intensity | $B = \mu H$ | 29-4 |
| 29-4 | Same as Eq. (29-4) but in a vacuum | $B = \mu_0 H$ | 29-5 |
| 29-4 | Definition of relative permeability | $\mu_r = \dfrac{\mu}{\mu_0}$ | 29-6 |
| 29-6 | Proportional relationship for the force on a moving charge in a magnetic field | $F \propto qv \sin \theta$ | 29-7 |
| 29-6 | The magnetic flux density | $\boxed{B = \dfrac{F}{qv \sin \theta}}$ | 29-8 |
| 29-6 | Definition of 1 tesla (T) | $1\text{ T} = 1\text{ N/(C} \cdot \text{m/s)} = 1\text{ N/A} \cdot \text{m}$ | 29-9 |
| 29-7 | Magnetic force on a current-carrying wire | $\boxed{F = BIl \sin \theta}$ | 29-10 |
| 29-8 | Magnetic field of a long, straight wire perpendicular to the lines of magnetic flux | $\boxed{B = \dfrac{\mu I}{2\pi d}}$ | 29-11 |

| Section | Topic | Equation | Equation No. |
|---|---|---|---|
| 29-8 | Value of $\mu_0$ for vacuum, air, and other nonmagnetic media | $\mu_0 = 4\pi \times 10^{-7} \text{ T} \cdot \text{m/A}$ | 29-12 |
| 29-9 | Magnetic induction at the center of a circular loop | $\boxed{B = \dfrac{\mu I}{2r}}$ | 29-13 |
| 29-9 | Magnetic induction at the center of a coil with $N$ turns | $\boxed{B = \dfrac{\mu NI}{2r}}$ | 29-14 |
| 29-9 | Magnetic induction in the interior of a solenoid | $\boxed{B = \dfrac{\mu NI}{L}}$ | 29-15 |
| 29-10 | Magnetic field intensity for a solenoid | $H = \dfrac{NI}{L}$ | 29-16 |

## True-False Questions

T  F  **1.** Like magnetic poles repel each other, whereas unlike magnetic poles attract each other.

T  F  **2.** Magnetic flux lines are drawn in such a way that the direction at any point is the same as the direction of the force exerted on a unit south pole placed at that point.

T  F  **3.** The weber is a unit of magnetic flux density.

T  F  **4.** An electron projected from right to left through a magnetic field directed into the page will be deflected downward.

T  F  **5.** The area of a hysteresis loop is a measure of flux density.

T  F  **6.** The right-hand rule can be used to determine the direction of a magnetic field surrounding a current-carrying conductor.

T  F  **7.** Two current-carrying conductors placed near each other will experience a force of attraction if their currents are oppositely directed.

T  F  **8.** The magnetic field lines for a solenoid are the same shape as those for a bar magnet.

T  F  **9.** A large hysteresis loop means a more efficient electromagnetic device.

T  F  **10.** Magnetic materials having a high permeability will generally have a low retentivity.

## Multiple-Choice Questions

**1.** Which of the following is *not* a unit of magnetic induction of flux density?
(a) Weber per square meter  (b) Gauss
(c) Tesla  (d) Weber

**2.** Magnetic fields have no effect on
(a) electric charges at rest
(b) electric charges in motion
(c) permanent magnets at rest
(d) permanent magnets in motion

**3.** The magnetic flux density at a distance $d$ from a long, current-carrying, straight wire is proportional to

(a) $d$       (b) $1/d$       (c) $d^2$       (d) $1/d^2$

**4.** Relative permeability is

(a) the ratio of flux density in a material to that for a vacuum
(b) large for paramagnetic materials
(c) small for ferromagnetic materials
(d) equal to $4\pi \times 10^{-7}$ T · m/A for a vacuum

**5.** The current through a wire is directed into the page. If a magnetic field is directed from right to left, the force on the wire will be

(a) to the right            (b) to the left
(c) upward                 (d) downward

**6.** An electron is projected from left to right into a flux density of 0.3 T directed into the paper. If the speed of the electron is $2 \times 10^6$ m/s, the magnetic force will be

(a) $9.6 \times 10^{-14}$ N          (b) $6.9 \times 10^{-14}$ N
(c) $5 \times 10^{-13}$ N            (d) $4.8 \times 10^{-14}$ N

**7.** A proton ($+1.6 \times 10^{-19}$ C) is projected from left to right at a velocity of $2 \times 10^6$ m/s. If an upward force of $1 \times 10^{-13}$ N is observed, the magnetic flux density perpendicular to the velocity is approximately

(a) 1.25 T out of the paper     (b) 1.25 T into the paper
(c) 0.31 T out of the paper     (d) 0.31 T into the paper

**8.** A rectangular loop of wire 20 cm wide and 30 cm long makes an angle of 40° with a magnetic flux density of 0.3 T. The flux penetrating the loop is approximately

(a) 0.01 Wb    (b) 0.02 Wb    (c) 2 Wb    (d) 115 Wb

**9.** The magnetic induction at a distance of 6 cm in air from a long, current-carrying conductor is $12 \times 10^{-6}$ T. The current in the wire is

(a) 1.8 A      (b) 3.6 A      (c) 4.8 A      (d) 6.4 A

**10.** A solenoid has 60 turns of wire and a length of 16 cm and supports a current of 10 A. The relative permeability of the core is 1200. The magnetic induction at the center is approximately

(a) 1.4 T      (b) 2.83 T      (c) 5.65 T      (d) 11 T

---

**Completion Questions**

**1.** Hysteresis is the lagging of the _____ behind the

_____.

**2.** According to the right-hand rule, if a current-carrying wire is grasped with the right hand so that the thumb points in the direction of the

_____, the curled fingers of that hand point in the direction

of _____.

**3.** A circular coil in the plane of the paper supports a clockwise electric current. The direction of the magnetic flux near the center is _____.

**4.** The relative permeability of a material is the ratio of its

_____ to that for a(n) _____.

**5.** _____ refers to the condition under which all the magnetic domains in a material are aligned.

6. The unit of flux density in the metric system of units is the
_____, and the unit of magnetic flux is the
_____.

7. In applying the right-hand screw rule for magnetic effects on a conducting wire,
the _____ is turned into the _____. Then
the direction of the _____ will be the same as the direction of
advance of a right-hand screw.

8. A magnetic field having a flux density of one tesla will exert a force of one
_____ on a charge of one _____ moving
perpendicular to the field with a velocity of one _____.

9. Materials with a relative permeability slightly less than unit are said to be
_____; _____ materials have permeabilities
slightly greater than unity.

10. The ability of some materials to retain magnetism after the magnetizing force
has been removed is referred to as _____.

# Chapter 30

# Forces and Torques in a Magnetic Field

## DEFINITIONS OF KEY TERMS

**Magnetic torque**   The torque experienced by a current element in a magnetic field.

**Galvanometer**   An electromagnetic device utilized for the detection of an electric current.

**Voltmeter**   A galvanometer that is calibrated to measure voltage. The range is determined by the **multiplier resistance** in series with the galvanometer.

**Ammeter**   A galvanometer that is calibrated to measure electric current. The range is determined by a **shunt resistance** connected in parallel with the galvanometer.

**Motor**   An electromagnetic device that converts electric energy to mechanical, rotational energy in the form of magnetic torque.

**Armature**   The central element of a motor, consisting of the core and a number of coils of wire.

**Commutator**   The metal half-rings that achieve an automatic switching of the magnetic lines of force in a dc motor by the turning action of force of the rotor.

**Full-scale deflection**   The movement of a galvanometer needle all the way across the dial.

**Multiplier**   The name for the shunt resistor used in turning a galvanometer into a voltmeter.

**Sensitivity, $I_g$**   A measure of how sensitive a galvanometer is. The current required for full-scale deflection.

**Shunt resistance**   A large multiplier resistance resistor placed in series with a galvanometer to allow the measurement of the voltage of a circuit without appreciably changing the current through the circuit.

## CONCEPTS AND EXAMPLES

**Magnetic Torque**   The magnetic torque $\tau$ on a current-carrying coil of wire is given by

$$\tau = NBIA \cos \alpha \qquad \text{torque on a coil}$$

where $N$ is the number of turns of wire, $B$ is the flux density (T), $I$ is the current (A), $A$ is the area of the coils ($m^2$), and $\alpha$ is the angle that the plane of the coil makes with the field.

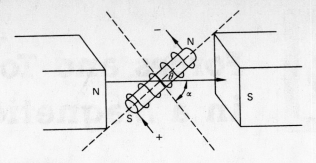

**Fig. 30-1** The magnetic torque on a solenoid.

The same equation applies for a solenoid (see Fig. 30-1), except that the angle $\alpha$ is generally replaced with $\theta$, the angle that the solenoid axis makes with the field. One angle is the complement of the other, or $\cos \alpha = \sin \theta$. For a solenoid,

$$\tau = NBIA \sin \theta \qquad \text{torque on a solenoid}$$

---

**EXAMPLE 30-1**   What current is required to produce a magnetic torque of 2.3 N · m on a solenoid when its axis makes an angle of 40° with a 0.2-T $B$ field? The coil has 400 turns each of area 0.03 m².

**Solution**
We solve first for $I$ and then substitute:

$$I = \frac{\tau}{NBA \sin \theta} = \frac{2.3 \text{ N} \cdot \text{m}}{(400)(0.2 \text{ T})(0.03 \text{ m}^2)(\sin 40°)}$$

$$= \mathbf{1.49 \text{ A}}$$

---

**Voltmeters and Ammeters**   A galvanometer is an electromagnetic device that gives a measure of electric current. It can be calibrated to read voltage or current. The dc voltmeter and the dc ammeter are described by Figs. 30-2 and 30-3.

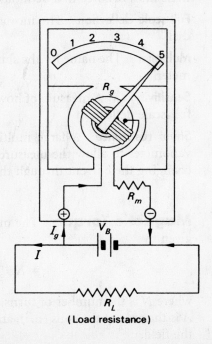

**Fig. 30-2** The dc voltmeter.

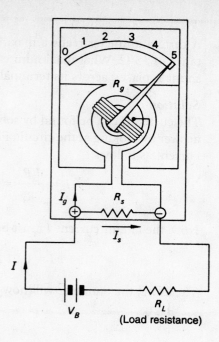

**Fig. 30-3** The dc ammeter.

The multiplier resistance $R_m$ that must be placed in series with a voltmeter to give full-scale deflection for the voltage $V_B$ is found from

$$R_m = \frac{V_B - I_g R_g}{I_g} \qquad \text{multiplier resistance}$$

**EXAMPLE 30-2**  An 80-$\Omega$ galvanometer deflects full scale for a voltage of 30 mV. What multiplier resistance must be added in series with this galvanometer to construct a voltmeter that reads 5 V full scale?

*Solution*
When it deflects full scale, the galvanometer current is

$$I_g = \frac{V_g}{R_g} = \frac{0.030\ \text{V}}{80\ \Omega} = 3.75 \times 10^{-4}\ \text{A}$$

Recognizing that $I_g R_g = 0.030$ V and $I_g = 3.75 \times 10^{-4}$ A, we have

$$R = \frac{5\ \text{V} - 0.030\ \text{V}}{3.75 \times 10^{-4}\ \text{A}} \qquad \text{and} \qquad \boxed{R_m = 13.253}$$

A galvanometer can be converted to an ammeter of a desired range by inserting a shunt resistance $R_s$ in parallel with the galvanometer. The shunt resistance required to give full-scale deflection for a current $I$ is found from

$$R_s = \frac{I_g R_g}{I - I_g} \qquad \text{shunt resistance}$$

**EXAMPLE 30-3**  A certain ammeter reads a maximum current of 2 mA and has an internal resistance of 15 $\Omega$. What maximum current can be read by this galvanometer if a 3-$\Omega$ shunt is placed across its terminals? (Refer to Fig. 30-3.)

**Solution**

The current $I$ can be found by solving the above equation for $I$, but let's deduce the answer by looking at the circuit. First, since $I_sR_s = I_gR_g$, we can determine the shunt current as follows:

$$I_s = \frac{I_gR_g}{R_s} = \frac{(0.002 \text{ A})(15 \text{ }\Omega)}{3 \text{ }\Omega} = 0.01 \text{ A}$$

Now, the circuit current $I$ must be equal to $I_s + I_g$:

$$I = 0.01 \text{ A} + 0.002 \text{ A} = \boxed{0.012 \text{ A}}$$

Therefore, the ammeter will now read 12 mA full scale.

---

## Summary

- The operation of generators, motors, ammeters, voltmeters, and many industrial instruments are directly affected by magnetic forces and torques.
- The same equation for a magnetic torque applies for a solenoid, except that the angle $\alpha$ is generally replaced with $\theta$, the angle the solenoid axis makes with the field.
- The multiplier resistance $R_m$ must be placed in series with a voltmeter to give full-scale deflection for $V_B$.
- The shunt resistance $R_s$ must be placed in parallel with an ammeter to give full-scale deflection for a current $I$.

---

## Key Equations by Section

| Section | Topic | Equation | Equation No. |
|---|---|---|---|
| 30-1 | Force on a current-carrying loop | $F = BIl \sin \theta = BI_\perp l$ | 30-1 |
| 30-1 | Force on a current-carrying loop | $F = BIa$ | 30-2 |
| 30-1 | Resultant torque on a current-carrying loop | $\tau = BI(a \times b) \cos \alpha$ | 30-3 |
| 30-1 | Torque on a current-carrying loop | $\tau = BIA \cos \alpha$ | 30-4 |
| 30-1 | Torque on a current-carrying coil with $N$ turns | $\boxed{\tau = NBIA \cos \alpha}$ | 30-5 |
| 30-2 | Torque on a solenoid | $\boxed{\tau = NBIA \sin \theta}$ | 30-6 |

| Section | Topic | Equation | Equation No. |
|---|---|---|---|
| 30-4 | The maximum voltage a galvanometer can read | $V_g = I_g R_g$ | 30-7 |
| 30-4 | The multiplier resistance to allow $I_g$ current to pass through the galvanometer for accurate voltage readings | $R_m = \dfrac{V_B}{I_g} - R_g$ | 30-8 |
| 30-5 | The value of a shunt resistor for accurate reading of current $I$ | $R_s = \dfrac{I_g R_g}{I - I_g}$ | 30-10 |

## True-False Questions

**T  F  1.** The sensitivity of a galvanometer is determined entirely by its electric resistance.

**T  F  2.** A horizontal current loop that is parallel with a magnetic field directed from right to left will experience a counterclockwise torque if the loop current is counterclockwise when viewed from the top.

**T  F  3.** The magnetic torque on a current loop is a maximum when the angle between the plane of the loop and the magnetic field is 90°.

**T  F  4.** The equation for computing the magnetic torque on a current loop of $N$ turns can also be used for a solenoid of $N$ turns.

**T  F  5.** The radial magnetic field for galvanometer helps to ensure that the pointer deflection will be directly proportional to the current in the coil.

**T  F  6.** Placing a low-resistance wire across the terminals of an ammeter will decrease the range of currents that can be measured.

**T  F  7.** Increasing the multiplier resistance for a voltmeter will increase the range of the voltmeter.

**T  F  8.** Voltmeters must be connected in parallel because of the low multiplier resistance that would short-circuit the circuit if it were placed in series.

**T  F  9.** The proper insertion of an ammeter or a voltmeter into a circuit will alter the current in that circuit slightly, introducing some error.

**T  F  10.** The torque output by a simple dc motor is not uniform.

## Multiple-Choice Questions

**1.** The torque on a solenoid in a magnetic field is *not* a function of its
(a) loop area
(b) number of loops
(c) length
(d) current

**2.** Which of the following must be a high-resistance instrument?
(a) Voltmeter    (b) Ammeter    (c) Motor    (d) Galvanometer

**3.** The minimum range of a given ammeter is determined by the
(a) value of the shunt resistance
(b) value of the multiplier resistance
(c) load resistance
(d) resistance and spring tension in the galvanometer element

4. The torque on a current-carrying loop is a maximum when the plane of the loop
   (a) is parallel with the magnetic field
   (b) is perpendicular to the magnetic field
   (c) is at an angle of 45° with the magnetic field
   (d) none of the above

5. The range of a dc voltmeter can be increased by
   (a) increasing the circuit load resistance
   (b) increasing the multiplier resistance
   (c) decreasing the multiplier resistance
   (d) placing a shunt resistance across the voltmeter terminals

6. A galvanometer has an internal resistance of 0.2 $\Omega$ and gives a full-scale deflection for a current of 3 mA. What multiplier resistance is required to convert this instrument to a voltmeter whose maximum range is 200 V?
   (a) 66 $\Omega$              (b) 33 $\Omega$
   (c) $3.3 \times 10^4$ $\Omega$    (d) $6.67 \times 10^4$ $\Omega$

7. If the galvanometer in Question 6 is used to construct an ammeter whose maximum range is 20 A, what shunt resistance must be added?
   (a) $1 \times 10^{-5}$ $\Omega$    (b) $3 \times 10^{-5}$ $\Omega$
   (c) $5 \times 10^{-5}$ $\Omega$    (d) $7 \times 10^{-5}$ $\Omega$

8. A rectangular coil of wire has a width of 12 cm and a length of 20 cm. The coil is mounted in a uniform magnetic field of flux density $4 \times 10^{-3}$ T, and a current of 20 A is sent through the windings. If the coil makes an angle of 30° with the field, how many turns of wire will be required to produce an output torque of 0.5 N · m?
   (a) 100 turns     (b) 200 turns     (c) 300 turns     (d) 400 turns

9. A commercial 5-V voltmeter requires a current of 10 mA to produce full-scale deflection. It can be converted to an instrument with a range of 50 V by adding a multiplier resistance of
   (a) 9000 $\Omega$     (b) 4500 $\Omega$     (c) 2250 $\Omega$     (d) 1125 $\Omega$

10. A laboratory ammeter has a resistance of 0.1 $\Omega$ and reads 3 A full scale. The shunt resistance that must be added to increase the range of the ammeter tenfold is
    (a) 0.0111 $\Omega$     (b) 0.0011 $\Omega$     (c) 0.022 $\Omega$     (d) 0.0022 $\Omega$

---

**Completion Questions**

1. The magnetic torque on a current-carrying loop of wire is directly proportional to the _____ of the magnetic field, the _____ of the loop, the _____ in the loop, and the _____ of the angle between the plane of the loop and the magnetic field.

2. When the plane of the loop is perpendicular to the field, the resultant torque on the loop when it supports a current is _____.

3. Any device used for the detection of an electric current is called a(n) _____.

Copyright © by Glencoe/McGraw-Hill.

**4.** A voltmeter can be constructed by placing a(n) _____ resistance in series with a(n) _____.

**5.** An ammeter is designed by placing a(n) _____ resistance in _____ with a(n) _____.

**6.** The current reversals required for continuous rotation of the armature in a dc motor are accomplished by using a(n) _____.

**7.** A galvanometer can be used to measure both _____ and _____.

**8.** An ideal ammeter has a(n) _____ resistance, whereas the ideal voltmeter has a(n) _____ resistance.

**9.** Three essential parts of a galvanometer are the _____, _____, and _____.

**10.** When you insert a voltmeter into a circuit, it must be connected in _____. An ammeter is connected in _____.

Chapter

# 31

# Electromagnetic Induction

**DEFINITIONS OF KEY TERMS**

**Lenz's law**   Whenever an emf is induced, the induced current must be in such a direction as will oppose the change by which the current is induced.

**Back emf**   The opposing or counter emf induced in a motor as a result of the rotational motion of the armature windings relative to the magnetic field.

**Series-wound motor**   A motor in which the field circuit and the armature circuit are connected in series; sometimes referred to as a *universal* motor.

**Shunt-wound motor**   A dc motor in which the field circuit and the armature circuit are connected in parallel. The speed can be controlled by varying the voltage applied to either the armature or the field.

**Compound-wound motor**   A dc motor having two separate field windings. One, usually the predominant field, is connected in parallel with the armature circuit, and the other is connected in series.

**AC generator**   A generator of ac power.

**Armature**   A loop of wire or coil of wire rotating through a magnetic field that induces a current to generate electricity.

**Commutator**   Electrically insulated fused half-rings that are used in a dc generator to transmit dc power.

**DC generator**   A generator of dc power that uses alternate windings of opposite direction to generate nearly constant power.

**Electromagnetic induction**   When a wire is moved through a magnetic field, an electric current is induced in the wire.

**Field magnet**   A permanent magnet or electromagnet that sets up a magnetic field in a generator.

**Slip rings**   A set of two rotating rings electrically connected to each end of the armature. The brushes rub against the rotating slip rings to complete the electric circuit.

**Step-up transformer**   An electromagnetic device that increases the voltage in an ac circuit.

**Step-down transformer**   An electromagnetic device that decreases the voltage in an ac circuit.

**Transformer efficiency**   The ratio of power output to power input in a transformer.

**Induced EMF**  A magnetic flux changing at the rate of 1 Wb/s will induce an emf of 1 V for each turn of a conductor:

$$\mathcal{E} = -N\frac{\Delta\phi}{\Delta t} \qquad \text{induced emf}$$

Two principle ways in which the flux changes are as follows:

$$\Delta\phi = \Delta B\,A \qquad \Delta\phi = B\,\Delta A$$

The induced emf due to a wire of length $l$ moving with a velocity $v$ at an angle $\theta$ with a field $B$ is given by

$$\mathcal{E} = Blv\sin\theta \qquad \text{emf due to moving wire}$$

**EXAMPLE 31-1**

A 30-cm length of wire moves at a constant velocity in a direction that is 37° with respect to a magnetic field of flux density 0.6 T. What must be the velocity to induce an emf of 200 mV?

**Solution**
We solve the emf question for the velocity $v$:

$$v = \frac{\mathcal{E}}{Bl\sin\theta} = \frac{0.200\ \text{V}}{(0.6\ \text{T})(0.3\ \text{m})(\sin 37°)}$$

$$= 1.85\ \text{m/s}$$

The direction of an induced current is reckoned from either Lenz's law or Flemming's rule. Flemming's rule is illustrated in Fig. 31-1. The thumb is in the direction of motion of the conductor, the index finger is in the direction of the magnetic flux, and the middle finger points in the direction of conventional current (motion-flux-current).

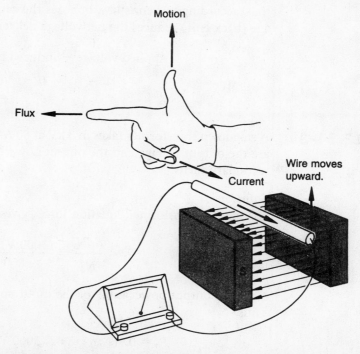

**Fig. 31-1**  The right-hand rule for determining the direction of induced current.

**Instantaneous EMF** The instantaneous emf generated by a coil of $N$ turns moving at an angular velocity $\omega$ or at a frequency of rotation $f$ is found from

$$\mathcal{E}_{inst} = NBA\omega \sin \omega t \qquad \mathcal{E}_{inst} = 2\pi f NBA \sin 2\pi ft$$

The maximum emf occurs when the argument of the sine function is equal to zero. In this case,

$$\mathcal{E}_{max} = NBA\omega \qquad \mathcal{E}_{max} = 2\pi f NBA$$

Thus, the induced emf and the induced current vary as follows:

$$\mathcal{E}_{inst} = \mathcal{E}_{max} \sin 2\pi ft \qquad i_{inst} = i_{max} \sin 2\pi ft$$

---

**EXAMPLE 31-2**

The armature of a simple ac generator consists of 200 turns of wire, each having an area of 0.23 m². How many revolutions per minute of this armature will produce a maximum emf of 12 V if it turns in a 0.8-mT field?

**Solution**
We can find the frequency in revolutions per second by solving the relation for the maximum emf for $f$:

$$f = \frac{\mathcal{E}}{2\pi NBA} = \frac{12.0 \text{ V}}{(2\pi)(200)(0.0008 \text{ T})(0.23 \text{ m}^2)} = 51.9 \text{ rev/s}$$

Now, we multiply by 60 s/min to obtain

$$f = (51.9 \text{ rev/s})(60 \text{ s/min}) = \boxed{3114 \text{ rpm}}$$

---

**Back EMF in a Motor** Every motor is also a generator that produces an emf that, according to Lenz's law, opposes the cause that produced it. This induced voltage (back emf) reduces the net voltage delivered to a circuit:

Applied voltage − induced back emf = net voltage

$$V - \mathcal{E}_b = IR \qquad \mathcal{E}_b = V - IR$$

---

**EXAMPLE 31-3**

A stepdown transformer takes in 3 kV at 80 A. Assuming 90 percent efficiency, what are the output voltage and the output power? There are 800 primary turns and 40 secondary turns.

**Solution**
Solving the transformer equation for $\mathcal{E}_s$ gives

$$\mathcal{E}_s = \frac{\mathcal{E}_p N_s}{N_p} = \frac{(3000 \text{ V})(40)}{800} = \boxed{150 \text{ V}}$$

Now, the output power is 0.9 of the input power since the efficiency is 90 percent. Thus,

$$P_0 = 0.9(\mathcal{E}_p i_p) = 0.9(3000 \text{ V})(80 \text{ A}) = \boxed{216 \text{ kW}}$$

## Summary

- The basic operating principle behind many electrical devices is that electromagnetic induction allows for the production of an electric current in a conducting wire.

- A magnetic flux changing at the rate of 1 Wb/s will induce an emf of 1 V for each turn of a conductor.

- According to Lenz's law, the induced current must be in such a direction that it produces a magnetic force that opposes the force causing the motion.

- According to Flemming's rule, if the thumb, forefinger, and middle finger of the right hand are held at right angles to each other, with the thumb pointing in the direction in which the wire is moving and the forefinger pointing in the field direction (N to S), the middle finger will point in the direction of induced conventional current (motion–flux–current).

- The back emf in a motor is the induced voltage that causes a reduction in the net voltage delivered to a circuit.

- A transformer can increase or decrease voltage by the ratio of primary windings to secondary windings.

## Key Equations by Section

| Section | Topic | Equation | Equation No. |
|---------|-------|----------|--------------|
| 31-1 | The induced emf, given by the number of turns of coil, the change in magnetic flux and the time interval | $$\boxed{\mathscr{E} = -N\frac{\Delta\phi}{\Delta t}}$$ | 31-1 |
| 31-1 | The flux based on the magnetic flux density and the area | $\phi = BA$ | 31-2 |
| 31-1 | The change in flux based on the change in magnetic flux density and constant area | $\Delta\phi = (\Delta B)A$ | 31-3 |
| 31-1 | The change in flux based on a constant flux density and changing area | $\Delta\phi = B(\Delta A)$ | 31-4 |
| 31-2 | Emf induced by a moving wire | $$\mathscr{E} = \frac{\text{work}}{q} = \frac{Fl}{q} = \frac{qvBl}{q}$$ $$= Blv$$ | 31-5 |
| 31-2 | Emf induced by a moving wire, based on the angle $\theta$ with the **B** field | $$\boxed{\mathscr{E} = Blv \sin\theta}$$ | 31-6 |

| Section | Topic | Equation | Equation No. |
|---|---|---|---|
| 31-4 | The instantaneous emf based on angular velocity | $\mathcal{E} = Bl\omega r \sin \theta$ | 31-7 |
| 31-4 | The total instantaneous emf | $\mathcal{E}_{inst} = 2Bl\omega r \sin \theta$ | 31-8 |
| 31-4 | Instantaneous emf of a coil of $N$ turns having an area of $A = l \times 2r$ | $\boxed{\mathcal{E}_{inst} = NBA\omega \sin \theta}$ | 31-9 |
| 31-4 | The maximum emf, given at $\theta = 90°$ in an ac generator | $\mathcal{E}_{max} = NBA\omega$ | 31-10 |
| 31-4 | The instantaneous emf in terms of the maximum emf, the frequency, and the time | $\boxed{\mathcal{E}_{inst} = \mathcal{E}_{max} \sin 2\pi ft}$ | 31-12 |
| 31-4 | Instantaneous current | $\boxed{i_{inst} = i_{max} \sin 2\pi ft}$ | 31-13 |
| 31-6 | Net voltage given by the applied voltage reduced by the back emf | $\boxed{V - \mathcal{E}_b = IR}$ | 31-14 |
| 31-8 | For a transformer, the emf induced by the primary coil based on the number of turns, the change in flux, and the time elapsed | $\mathcal{E}_p = -N_p\dfrac{\Delta\phi}{\Delta t}$ | 31-15 |
| 31-8 | The emf induced by the secondary coil of a transformer | $\mathcal{E}_s = -N_s\dfrac{\Delta\phi}{\Delta t}$ | 31-16 |
| 31-8 | Relation between primary and secondary emf and primary and secondary number of turns | $\boxed{\dfrac{\mathcal{E}_p}{\mathcal{E}_s} = \dfrac{N_p}{N_s}}$  $\dfrac{Primary\ voltage}{Secondary\ Voltage} = \dfrac{primary\ turns}{secondary\ turns}$ | 31-17 |
| 31-8 | Efficiency of a transformer | $E = \dfrac{power\ output}{power\ input} = \dfrac{\mathcal{E}_s i_s}{\mathcal{E}_p i_p}$ | 31-18 |
| 31-8 | Ideal transformer relation | $\dfrac{i_p}{i_s} = \dfrac{\mathcal{E}_s}{\mathcal{E}_p}$ | 31-19 |

**True-False Questions**

T  F  **1.** Lenz's law is a consequence of the fact that energy must be conserved.

T  F  **2.** The magnitude of an induced emf is directly proportional to the rate at which magnetic flux lines are cut by a conductor.

T  F  **3.** A generator is a device that converts magnetic energy to electric energy.

T  F  **4.** A transformer can be used for alternating current only.

T  F  **5.** The back emf in the armature of an electric motor is zero when the motor is started and rises as the motor's speed increases.

T  F  **6.** If the armature of a generator is rotating with a constant angular velocity in a constant magnetic field, the induced emf will be constant.

T  F  **7.** For an ac generator, the maximum current occurs when the induced emf is a minimum.

T  F  **8.** Every motor is also a generator.

T  F  **9.** In a shunt-wound motor, the output speed can be controlled by varying the input current to either the field windings or the armature windings.

T  F  **10.** A series-wound motor is preferable if a large starting torque is desirable.

**Multiple-Choice Questions**

**1.** When a wire moves perpendicular to a magnetic field, the induced emf does *not* depend on the
(a) velocity of the wire
(b) resistance of the wire
(c) flux density of the magnetic field
(d) orientation of the wire

**2.** A wire parallel to the paper moves downward into a flux density directed into the paper. The induced current will be
(a) directed to the left
(b) zero because the motion is perpendicular to the field
(c) directed to the right
(d) opposite to the induced emf, according to Lenz's law

**3.** Which of the following is *not* a component of a simple ac generator?
(a) Field magnet          (b) Split-ring commutator
(c) Armature              (d) Brushes

**4.** The instantaneous emf produced by a simple ac generator is a maximum when the angular displacement of the loop is an
(a) integral multiple of $\pi/2$
(b) odd multiple of $\pi/2$
(c) even multiple of $\pi$
(d) odd multiple of $\pi$

**5.** A coil of wire of area 0.3 m$^2$ has 90 turns of wire and is suspended with its plane perpendicular to a uniform magnetic field. An emf of $-2$ V is induced when the coil is flipped parallel to the field in 0.4 s. The flux density of the field is approximately
(a) 0.01 T      (b) 0.02 T      (c) 0.03 T      (d) 0.04 T

6. An armature in an ac generator consists of 600 turns, each having an area of 0.4 m². The coil rotates in a field of flux density $2 \times 10^{-3}$ T. What must the rotational frequency of the armature be to induce a maximum emf of 12 V?
   (a) 200 rpm   (b) 226 rpm   (c) 239 rpm   (d) 478 rpm

7. A wire 0.2 m long moves at a constant speed of 5 m/s in a direction that is 37° with respect to a magnetic field of 0.2 T directed perpendicular to the wire. The induced emf is approximately
   (a) 0.12 V   (b) 0.16 V   (c) 0.20 V   (d) 0.24 V

8. A 120-V shunt motor has a field resistance of 200 Ω and an armature resistance of 12 Ω. At full speed, the back emf of the motor is 90 V. The operating current is approximately
   (a) 3 A   (b) 6 A   (c) 12 A   (d) 16 A

9. A transformer has 400 secondary turns and only 50 primary turns. If an ac voltage of 120 V is connected to the primary coil, the output voltage is
   (a) 15 V   (b) 240 V   (c) 430 V   (d) 960 V

---

**Completion Questions**

1. A magnetic flux changing at the rate of one _____ will induce an emf of one _____ for each _____ of the conductor.

2. For a dc motor in operation, the net voltage delivered to the armature coils is equal to the _____ less the _____.

3. The major components of a simple ac generator are the _____, _____, and _____ with _____.

4. Three types of motors discussed in this text are the _____, _____, and _____.

5. The current drawn by an electric motor is a maximum when the motor is _____.

6. If an ideal transformer, the power output is equal to the _____.

7. In a step-down transformer, the number of turns in the secondary coil must be _____ the number of turns in the primary coil.

8. A simple transformer consists of three essential parts: _____, _____, and _____.

9. In a(n) _____, the field windings and the armature windings are connected in parallel.

# Chapter
# 32

# Alternating-Current Circuits

## DEFINITIONS OF KEY TERMS

**Inductor**   An element of an ac circuit that consists of a continuous loop or coil of wire. A self-induced emf opposes increases or decreases in the circuit current.

**Henry**   A unit of inductance that will induce an emf of 1 V when the current is changing at the rate of 1 A/s.

**Inductance**   The property of an ac circuit that opposes any change in the amount of current flowing in the circuit.

**Reactance**   The nonresistive opposition to the flow of an alternating current. If it is caused by an inductor, it is called *inductive* reactance; if it is caused by capacitance, it is called *capacitive* reactance.

**Impedance**   The combined opposition that a circuit offers to the flow of alternating current.

**Phase diagram**   A vector representation of the relationships of resistive, capacitive, and inductive parameters of an ac circuit.

**Resonant frequency**   The frequency for which the capacitive reactance in an ac circuit is exactly equal to its inductive reactance, resulting in a maximum current that is in phase.

**Power factor**   A measure of the effective power consumed by an ac circuit. It is equal to the cosine of the phase angle.

**Effective current**   That alternating current that will develop the same power as a direct current of the same value.

**Effective voltage**   The ratio of effective current to the impedance in an ac circuit.

**Capacitance**   The quality of holding a charge.

**Frequency**   The number of cycles per unit time.

**Resonance**   For an *LC* circuit, the inductance and capacitance can reinforce each other at a certain value to reinforce and amplify a particular frequency.

## CONCEPTS AND EXAMPLES

**Rise and Decay of Current in Capacitors and Inductors**   The charge on a capacitor will rise to 63 percent of its maximum value as the current delivered to the capacitor decreases to 37 percent of its initial value during a period of one time constant $\tau$. When the capacitor discharges, both the charge and the current decay to 37 percent of their initial values after discharging for one time constant:

$$\tau = RC \qquad \text{time constant for capacitor}$$

A capacitor may be considered fully charged or fully discharged after a period of five time constants.

When an alternating current passes through a coil of wire, an inductor, a self-induced emf arises to oppose the change. This emf is given by

$$\mathscr{E} = -L\frac{\Delta i}{\Delta t} \qquad L = -\frac{\mathscr{E}}{\Delta i/\Delta t}$$

The constant $L$ is called the *inductance*. An inductance of 1 H exists if an emf of 1 V is induced by a current changing at the rate of 1 A/s.

In a circuit containing an inductor, the current will rise to 63 percent of its maximum value or decay to 37 percent of its maximum value in a period of one time constant. The time constant for an inductor is given by

$$\tau = \frac{L}{R} \qquad \text{time constant for inductor}$$

For practical reasons, the rise and decay time for an inductor is considered to be five time constants ($5L/R$).

---

**EXAMPLE 32-1**

An $LR$ circuit has a time constant of 0.003 s. What is the inductance if the circuit resistance is 600 $\Omega$? How much time is required for this inductor to decay fully?

*Solution*
Since $\tau = L/R$, we solve for $L$:

$$L = R\tau = (600\ \Omega)(0.003\ \text{s}) = \boxed{1.80\ \text{H}}$$

The full decay will take five time constants, or

$$T = 5\tau = 5(0.003\ \text{s}) = 0.015\ \text{s}$$

---

**Alternating Currents**   With alternating currents the voltage and the current alternate between minimum and maximum values. Therefore, it becomes useful to speak of an **effective ampere** and an **effective volt,** which are defined in terms of their maxima:

$$i_{\text{eff}} = 0.707 i_{\text{max}} \qquad \mathscr{E}_{\text{eff}} = 0.707 \mathscr{E}_{\text{max}}$$

**Reactance in AC Circuits**   Both capacitors and inductors offer resistance to the flow of alternating current (called reactance). Such reactance is dependent on the frequency $f$ of the alternating current and is found from

$$X_L = 2\pi fL \qquad \text{inductive reactance } X_L\ (\Omega)$$

$$X_C = \frac{1}{2\pi fC} \qquad \text{capacitive reactance } X_C\ (\Omega)$$

You should verify by substitution that a 60-Hz ac circuit containing an 8-$\mu$F capacitor and a 0.4-H inductor has a capacitive reactance of 332 $\Omega$ and an inductive reactance of 151 $\Omega$.

**EXAMPLE 32-2**   A 110-V, 60-Hz ac circuit has a 200-$\Omega$ resistor, a 4-$\mu$F capacitor, and a 0.6-H inductor. What are the capacitive reactance and the inductive reactance of the circuit?

*Solution*
Direct substitution yields

$$X_C = \frac{1}{2\pi f C} = \frac{1}{(2\pi)(60 \text{ Hz})(4 \times 10^{-6} \text{ F})} = 663 \ \Omega$$

$$X_L = 2\pi f L = (2\pi)(60 \text{ Hz})(0.6 \text{ H}) = 226 \ \Omega$$

**Impedance and Phase Diagrams**   The voltage, current, and resistance in a series ac circuit can be represented by phase diagrams. Figure 32-1 illustrates the use of such a diagram for $X_C$, $X_L$, and $R$. The resultant of these vectors is the effective resistance of the entire circuit, called the **impedance** $Z$. Applying the phythagorean theorem to the figure yields

$$Z = \sqrt{R^2 + (X_L - X_C)^2} \qquad \textbf{impedance } (\Omega)$$

Ohm's law may be applied to the entire circuit or to each element of the circuit in a manner similar to that used for dc circuits. The following equations are deduced from Ohm's law:

$$V = iZ \qquad V_L = iX_L \qquad V_C = iX_C \qquad V_R = iR$$

$$\textbf{Ohm's law} \qquad V = i\sqrt{R^2 + (X_L - X_C)^2}$$

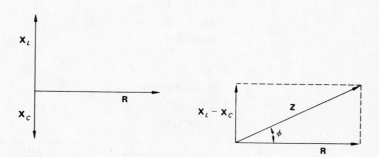

**Fig. 32-1**   Impedance diagram.

**Summary**

- The three principal elements in ac circuits: the resistor, the capacitor, and the inductor.

- A resistor is affected by ac current in the same manner as for dc circuits, and the current is determined by Ohm's law.

- The capacitor regulates and controls the flow of charge in an ac circuit; its opposition to the flow of electrons is called *capacitive reactance*.

- The inductor experiences a self-induced emf that adds *inductive reactance* to the circuit.
- The combined effect of all three elements in opposing electric current is called *impedance*.
- The charge on the capacitor will rise to 63 percent of its maximum value as the current delivered to the capacitor decreases to 37 percent of its initial value during a period of one time constant $\gamma$.
- When alternating current passes through a coil of wire, an inductor, a self-induced emf arises to oppose the change.
- In an inductive circuit, the current will rise to 63 percent of its maximum value or decay to 37 percent of its maximum in a period of one time constant.
- Since alternating currents and voltages vary continuously, an effective ampere and an effective volt are defined in terms of their maximum values.

| | Section | Topic | Equation | Equation No. |
|---|---|---|---|---|
| **Key Equations by Section** | 32-1 | The instantaneous charge of an RC circuit | $Q = CV_B(1 - e^{-t/RC})$ | 32-4 |
| | 32-1 | The instantaneous current of an RC circuit | $i = \dfrac{V_B}{R}e^{-t/RC}$ | 32-5 |
| | 32-1 | The time constant $\tau$ of the RC circuit | $\tau = RC$ | 32-6 |
| | 32-1 | The amount of current delivered to the capacitor in an RC circuit after one time constant | $i = \dfrac{V_B}{R}\dfrac{1}{e} = 0.37\dfrac{V_B}{R}$ | 32-8 |
| | 32-1 | The voltage of a discharging capacitor in an RC circuit | $-\dfrac{Q}{C} = iR$ | 32-9 |
| | 32-1 | The instantaneous charge of a discharging capacitor in an RC circuit | $Q = CV_Be^{-t/RC}$ | 32-10 |
| | 32-1 | The instantaneous current of a discharging capacitor in an RC circuit | $i = \dfrac{-V_B}{R}e^{-t/RC}$ | 32-11 |

| Section | Topic | Equation | Equation No. |
|---------|-------|----------|--------------|
| 32-2 | The emf in a coil | $\mathcal{E} = -N\dfrac{\Delta\phi}{\Delta t}$ | 32-12 |
| 32-2 | The self-induced emf resisting the change in emf, based on the inductance $L$ | $\boxed{\mathcal{E} = -L\dfrac{\Delta i}{\Delta t}}$ | 32-13 |
| 32-2 | The net emf is equal to the battery voltage minus the induced emf | $V_B - L\dfrac{\Delta i}{\Delta t} = iR$ | 32-15 |
| 32-2 | The rise in current in an $LR$ circuit | $i = \dfrac{V_B}{R}(1 - e^{-(R/L)t})$ | 32-16 |
| 32-2 | The time constant $\tau$ in an $LR$ circuit | $\boxed{\tau = \dfrac{L}{R}}$ | 32-17 |
| 32-2 | The current in a decaying $LR$ circuit | $\boxed{i = \dfrac{V_B}{R}e^{-(R/L)t}}$ | 32-18 |
| 32-3 | EMF in an ac circuit | $\mathcal{E} = \mathcal{E}_{max}\sin 2\pi ft$ | 32-19 |
| 32-3 | Current in an ac circuit | $i = i_{max}\sin 2\pi ft$ | 32-20 |
| 32-3 | The effective current of an ac circuit related to the maximum current | $i_{eff} = 0.707 i_{max}$ | 32-21 |
| 32-3 | The effective emf of an ac circuit, based on the maximum emf | $\mathcal{E}_{eff} = 0.707 \mathcal{E}_{max}$ | 32-22 |
| 32-5 | Definition of inductive reactance | $\boxed{X_L = 2\pi fL}$ | 32-23 |
| 32-5 | Voltage across an inductor based on inductive reactance | $V = iX_L$ | 32-24 |

| Section | Topic | Equation | Equation No. |
|---------|-------|----------|--------------|
| 32-5 | Definition of capacitive reactance | $X_C = \dfrac{1}{2\pi f C}$ | 32-25 |
| 32-5 | Voltage across a capacitor based on capacitive reactance | $V = iX_C$ | 32-26 |
| 32-6 | Voltage in an *LRC* series ac circuit | $V = \sqrt{V_R^2 + (V_L - V_C)^2}$ | 32-27 |
| 32-6 | Definition of the phase angle | $\tan\phi = \dfrac{V_L - V_C}{V_R}$ | 32-28 |
| 32-6 | Voltage in an *LRC* series ac circuit | $V = i\sqrt{R^2 + (X_L - X_C)^2}$ | 32-29 |
| 32-6 | Impedance *Z* of an *LRC* series ac circuit | $Z = \sqrt{R^2 + (X_L - X_C)^2}$ | 32-30 |
| 32-6 | Effective current in an ac circuit | $i = \dfrac{V}{Z}$ | 32-31 |
| 32-6 | Phase angle in terms of capacitive and inductive reactance | $\tan\phi = \dfrac{X_L - X_C}{R}$ | 32-32 |
| 32-7 | Resonance frequency of an *LRC* series ac circuit | $f_r = \dfrac{1}{2\pi\sqrt{LC}}$ | 32-33 |
| 32-8 | The power consumed by an ac circuit | $P = iV\cos\phi$ | 32-34 |
| 32-8 | The power factor | $\cos\phi = \dfrac{R}{Z} = \dfrac{R}{\sqrt{R^2 + (X_L - X_C)^2}}$ | 32-35 |

## True-False Questions

**T F 1.** In a capacitive circuit, the current delivered to a capacitor will rise to 63 percent of its maximum value after charging for a period of one time constant.

**T F 2.** In an inductive circuit, the current decays to 37 percent of its initial value in a period of one time constant.

**T F 3.** A capacitor with a breakdown voltage of 150 V would be safe to use in an ac circuit if the ac voltage were only 120 V.

**T F 4.** Capacitive reactance and inductive reactance increase with an increase in frequency of an alternating current.

**T F 5.** The effective voltage $V$ in a series ac circuit can be defined as the algebraic sum of $V_C$, $V_L$, and $V_R$.

**T F 6.** The power factor in an ac circuit is the ratio of its resistance to its impedance.

**T F 7.** Under a condition of resonance in an ac circuit, the current will be in phase with the voltage.

**T F 8.** Reactance, impedance, and resistance are all measured in the same physical unit (ohms).

**T F 9.** In a circuit containing pure capacitance, the voltage leads the current by 90°.

**T F 10.** A negative phase angle occurs when the current leads the voltage in an ac circuit.

## Multiple-Choice Questions

**1.** The voltage of a circuit containing only inductive reactance
(a) is in phase with the current
(b) leads the current by 90°
(c) lags the current by 90°
(d) leads the current by less than 90°

**2.** An alternating voltage has a maximum value of 30 V. The effective value is approximately
(a) 15 V    (b) 21.2 V    (c) 30 V    (d) 42.9 V

**3.** As the capacitance in an ac circuit increases, the resonant frequency
(a) increases              (b) decreases
(c) remains the same       (d) approaches zero

**4.** The ratio of the pure resistance in an ac circuit to the impedance is known as the
(a) power factor           (b) phase angle
(c) resonant frequency     (d) net reactance

**5.** In a circuit containing pure capacitance, during a period of one time constant, the
(a) charge decays to 63 percent of its initial value
(b) charge increases to 37 percent of its initial value
(c) current decays to 37 percent of its initial value
(d) capacitor is completely charged

**6.** In an ac circuit containing a coil with an inductance of 0.3 H, the voltage is 120 V at 60 Hz. If we neglect resistance, the effective current in the coil is
(a) 0.27 A    (b) 0.53 A    (c) 1.06 A    (d) 2.12 A

7. A 50-$\Omega$ resistor, a 0.5-H inductor, and a 12-$\mu$F capacitor are connected in series with a 120-V, 60-Hz alternating current. The effective current is approximately
   (a) 2 A      (b) 4 A      (c) 6 A      (d) 8 A

8. In Question 7, the power factor is
   (a) 0.66      (b) 0.72      (c) 0.83      (d) 0.92

9. An 800-$\Omega$ resistor, a 0.6-H inductor, and a 4-$\mu$F capacitor are connected in series with a 220-V, 50-Hz source of alternating current. The phase angle is approximately
   (a) $-31°$      (b) $-37°$      (c) $31°$      (d) $37°$

10. A circuit contains a 12-$\mu$F capacitor, a 40-$\Omega$ resistor, a switch, and a 20-V dc battery connected in series. After the switch is closed, the capacitor can be considered fully charged after
    (a) 0.00048 s      (b) 0.0012 s      (c) 0.0024 s      (d) 0.0048 s

---

## Completion Questions

1. One effective ampere is that alternating current that will develop the same _____ as _____ of direct current.

2. In a(n) _____ circuit, the current decays to 37 percent of its initial value in a period equal to one _____.

3. In a capacitive circuit, the _____ leads the _____, and the phase angle is _____.

4. For an ac circuit containing capacitance, inductance, and resistance, the current will be a maximum when the frequency of the alternating voltage is equal to _____.

5. The power factor is equal to the _____ of the circuit divided by the _____.

6. When the voltage and current reach maximum and minimum values at the same time, they are said to be _____.

7. A given inductor has an inductance of one _____ if an emf of one _____ is induced by a current changing at the rate of one _____.

8. In an inductive circuit, an increase in frequency of the applied voltage will _____ the effective current.

9. At resonant frequency, an ac circuit containing resistance, inductance, and capacitance acts as if it contained only _____.

10. A household ac circuit has a 110-V, 60-Hz source of current. The actual voltage varies from _____ to _____.

# Chapter 33

# Light and Illumination

**DEFINITIONS OF KEY TERMS**

**Light**    Electromagnetic radiation capable of affecting the sense of sight (400 to 700 nm).

**Electromagnetic wave**    A wave propagated by oscillating transverse electric and magnetic fields. Oscillations of the electric field are perpendicular to those of the magnetic field, and the collapse of one generates the other to sustain the wave without the necessity of a medium. Heat, light, and radio waves are examples.

**Quantum theory**    The theory that treats light or other electromagnetic radiation as discrete bundles of energy carried along by a wave field. Such radiation is said to be quantized because only specific energy values are allowed.

**Umbra**    The portion of a shadow that receives no light from the source.

**Penumbra**    The lighter portion of a shadow that receives light from parts of the source but not from all the source.

**Light ray**    An imaginary line drawn perpendicular to advancing wavefronts that indicates the direction of light propagation.

**Luminous flux**    That portion of the total radiant power emitted from a light source that is capable of affecting the sense of sight.

**Steradian (sr)**    The solid angle subtended at the center of a sphere by an area $A$ on its surface that is equal to the square of its radius $R$.

**Luminous intensity**    The luminous flux emitted per unit solid angle for a given light source. The unit is the candela (cd), which is 1 lumen per steradian (lm/sr).

**Isotropic source**    A source that emits light uniformly in all directions, that is, through a solid angle of $4\pi$ sr.

**Illumination**    The luminous flux per unit area. The unit is the lux (lx), which is defined as 1 lm/m$^2$.

**Nanometer**    $10^{-9}$ m.

**Photons**    The smallest quantum of light.

**Visible region**    The portion of the electromagnetic spectrum that is visible light.

**Wavelength** The wavelength $\lambda$ of electromagnetic radiation is related to its frequency $f$ by the general equation

$$c = f\lambda \qquad c = 3 \times 10^8 \text{ m/s}$$

For visible light, the wavelengths range from 400 nm for violet to 700 nm for red.

**EXAMPLE 33-1**

What is the frequency of a monochromatic beam of light whose wavelength is 570 nm?

*Solution*
Solving for $f$, we have

$$f = \frac{c}{\lambda} = \frac{3 \times 10^8 \text{ m/s}}{570 \times 10^{-9} \text{ m}} = \boxed{5.26 \times 10^{14} \text{ Hz}}$$

**Energy of Photons** The energy of light photons is proportional to the frequency:

$$E = hf \qquad E = \frac{hc}{\lambda} \qquad h = 6.626 \times 10^{-34} \text{ J} \cdot \text{s}$$

The constant $h$ is Planck's constant.

**EXAMPLE 33-2**

The energy of a monochromatic beam of light is $3.61 \times 10^{-19}$ J. What is the wavelength of this light?

*Solution*
We solve for $\lambda$ in the photon equation:

$$\lambda = \frac{hc}{E} = \frac{(6.626 \times 10^{-34} \text{ J} \cdot \text{s})(3 \times 10^8 \text{ m/s})}{3.61 \times 10^{-19} \text{ J}}$$

$$= 5.50 \times 10^{-7} \text{ m} = \boxed{550 \text{ nm}}$$

**Shadow Construction** The construction of shadows requires the application of geometry and the mathematics of ratio and proportion. The definitions of the umbra and penumbra regions are illustrated in Fig. 33-1.

**Luminous Intensity** The luminous intensity of a light source is the flux $F$ per unit solid angle $\Omega$:

$$I = \frac{F}{\Omega} \qquad \text{where} \qquad \Omega = \frac{A}{R^2}$$

For an isotropic source that emits light in all directions, the luminous flux $F$ is equal to $4\pi I$. Also 68 lm of yellow-green light (555 nm) has a radiant power of one watt (1 W).

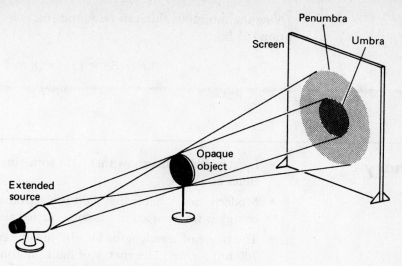

**Fig. 33-1** Shadows formed by an extended light source.

---

**EXAMPLE 33-3**    How many lumens of flux pass through a 0.23-m² opening on the surface of a sphere of radius 4 m if an 800-cd source is placed at the center of the sphere?

*Solution*
Solving for *F* in the above equation, we obtain

$$F = I\Omega \qquad \text{or} \qquad F = \frac{IA}{R^2}$$

where we have substituted the definition of solid angle. Thus,

$$F = \frac{(800 \text{ cd})(0.23 \text{ m}^2)}{(4 \text{ m})^2} = \boxed{11.5 \text{ lm}}$$

---

**Illumination**    The illumination *E* of a surface *A* is defined as the luminous flux per unit area. It may be expressed also in terms of the intensity *I* of the source and the distance from the surface *R*:

$$E = \frac{F}{A} \qquad E = \frac{I \cos \theta}{R^2} \qquad \text{illumination}$$

The SI unit for illumination is the *lux*, which is equal to a lumen per square meter.

---

**EXAMPLE 33-4**    How far must a screen 2 m × 4 m be held from a 600-cd source of light to produce an illumination of 35 lx? What is the luminous flux striking the screen?

*Solution*
We assume that the screen is held perpendicular to the incident light so that $\cos \theta = 1$ and $E = I/R^2$. We solve for *R* and then substitute:

$$R = \sqrt{\frac{I}{E}} = \sqrt{\frac{600 \text{ cd}}{35 \text{ lx}}} \qquad \text{or} \qquad \boxed{R = 4.14 \text{ m}}$$

Now the luminous flux can be found from the known area (8 m$^2$) and illumination (35 lx):

$$F = EA = (35 \text{ lx})(8 \text{ m}^2) = \boxed{280 \text{ lm}}$$

## Summary

- Light experiments show that light sometimes behaves as particles and sometimes as a wave.
- Modern theory holds that light is electromagnetic radiation and that its radiant energy is transported in photons carried along by a wave field.
- The range of wavelengths for visible light goes from 400 nm for violet to 700 nm for red. The energy of light photons is proportional to the frequency.
- *Luminous flux* is that portion of the total radiant power emitted from a light source that is capable of affecting the sense of sight.
- The luminous intensity of a light source is the luminous flux *F* per unit solid angle $\Omega$.
- *Luminous flux* is the radiant power in the visible region and is measured in *lumens*.

## Key Equations by Section

| Section | Topic | Equation | Equation No. |
|---------|-------|----------|--------------|
| 33-1 | Equation giving the speed of light as the product of the frequency and the wavelength | $\boxed{c = f\lambda}$ | 33-1 |
| 33-4 | The energy of a photon is equal to Planck's constant ($h = 6.626 \times 10^{-34}$ J · s) multiplied by the frequency of the photon | $\boxed{E = hf}$ | 33-2 |
| 33-5 | The speed of light | $c = 2.99792458 \times 10^8$ m/s (exactly) | 33-3 |
| 33-7 | An angle $\theta$ is equal to the length of arc $S$ divided by the radius $R$ | $\theta = \dfrac{S}{R}$ rad | 33-4 |
| 33-7 | A solid angle $\Omega$ is equal to the area of the part of the sphere surface included in the solid angle divided by the square of the radius | $\boxed{\Omega = \dfrac{A}{R^2}}$ sr | 33-5 |

| Section | Topic | Equation | Equation No. |
|---------|-------|----------|--------------|
| 33-8 | Luminous intensity given as the ratio of the luminous flux $F$ to the solid angle $\Omega$ | $I = \dfrac{F}{\Omega}$ | 33-6 |
| 33-8 | The total luminous flux for a light source giving illumination equally in all directions | $F = 4\pi I$ | 33-7 |
| 33-9 | Illumination $E$ given as the luminous flux $F$ per area $A$ | $E = \dfrac{F}{A}$ | 33-8 |
| 33-9 | Illumination $E$ as a function of intensity | $E = \dfrac{I \cos \theta}{R^2}$ | 33-10 |
| 33-9 | The illumination $E$ for the case of a normal surface | $E = \dfrac{I}{R^2}$ | 33-11 |

## True-False Questions

**T  F  1.** Although earlier scientists believed light was transmitted by particles, it is now established that light is a wave phenomenon.

**T  F  2.** Infrared rays are more energetic than visible or ultraviolet rays.

**T  F  3.** Interference of light waves is explained more easily on the basis of the wave theory of light.

**T  F  4.** The propagation of light through outer space is possible because of the presence of a *light-carrying ether.*

**T  F  5.** The Michelson-Morley experiment demonstrated that the velocity of light is a constant, independent of the motion of the source.

**T  F  6.** The nanometer is 10 times as large as the angstrom.

**T  F  7.** The energy of light is directly proportional to the wavelength.

**T  F  8.** The first successful terrestrial measurement of the speed of light was made by Fizeau.

**T  F  9.** The photoelectric effect occurs when light is emitted as a result of electron bombardment.

**T  F  10.** The nature of light is no different fundamentally from the nature of heat radiation.

**T  F  11.** One lumen of red light does not represent the same luminous flux as one lumen of blue light.

**T  F  12.** If a person on the earth sees a quarter moon, that person must be within the penumbra of its shadow.

**T  F  13.** If the distance between a surface and a source of light is increased by a factor of 3, the illumination will be one-ninth of its original value.

**T  F  14.** A point source of light will produce only an umbra shadow.

**T  F  15.** Since one candela is equal to one lumen per steradian, the units of intensity and illumination are the same.

**T  F  16.** The luminous intensity is not dependent on the angle a surface makes with incident flux from a source of light.

**T  F  17.** An isotropic source of 1 cd emits a luminous flux of $4\pi$ lm.

**T  F  18.** The units of flux and luminous intensity have the same physical dimensions, even though they do not represent the same physical quantity.

**T  F  19.** A lumen is the luminous flux falling on 1 m², all points of which are located 1 m from a uniform source of 1 cd.

**T  F  20.** The luminous intensity does not change as a surface is moved farther and farther from a light source.

---

## Multiple-Choice Questions

**1.** Which of the following is *not* an electromagnetic phenomenon?
(a) Heat rays    (b) Sound waves    (c) Radio waves    (d) Light

**2.** Which of the following best applies for yellow light?
(a) 640 nm    (b) 0.4 nm    (c) $5 \times 10^{14}$ Hz    (d) 580 Å

**3.** The man most responsible for explaining the photoelectric effect was
(a) Einstein    (b) Planck    (c) Huygens    (d) Maxwell

**4.** The wavelength corresponding to light with a frequency of $4 \times 10^{14}$ Hz is
(a) 1.33 m    (b) 0.075 $\mu$m    (c) 7500 nm    (d) 750 nm

**5.** A radio frequency of 780 kHz has a wavelength of approximately
(a) 38 $\mu$m    (b) 385 m    (c) 0.0026 m    (d) 26 nm

**6.** The theory that light is emitted in discrete amounts of energy rather than in a continuous fashion is known as
(a) the photoelectric effect    (b) the quantum theory
(c) Huygens's principle    (d) the electromagnetic theory

**7.** The moon is located approximately 240,000 mi from the earth. A radio signal will reach the earth from the moon in
(a) 1.3 min    (b) 0.775 s    (c) 1.29 s    (d) 0.7 min

**8.** If Planck's constant is $h = 6.626 \times 10^{-34}$ J $\cdot$ s, what energy is associated with light with a wavelength of 160 nm?
(a) $1.24 \times 10^{-18}$ J    (b) $1.88 \times 10^{15}$ J
(c) $1.24 \times 10^{-15}$ J    (d) $1.88 \times 10^{-15}$ J

**9.** Which of the following scientists measured the speed of light by using an eight-sided rotating mirror?
(a) Fizeau    (b) Roemer    (c) Galileo    (d) Michelson

**10.** Which of the following terms best describes the nature of light from the modern point of view?
(a) Photons    (b) Waves    (c) Particles    (d) Rays

**11.** Which of the following light sources of equal radiant power appears brightest?
(a) Red light    (b) Green light
(c) Blue light    (d) All are the same

**12.** During a solar eclipse, a person on earth with his or her eyes protected observes that part of the sun is darkened but never all of it. This person lies in the
(a) umbra    (b) penumbra
(c) both (a) and (b)    (d) neither (a) nor (b)

13. What is the luminous flux emitted by a 30-cd isotropic source of light?
(a) 30 lm      (b) 4 lm      (c) 2.4 lm      (d) $120\pi$ lm

14. A 200-cd lamp is 4 m directly above a surface. The illumination is approximately
(a) 12.5 lm/m$^2$   (b) 25 lm/m$^2$    (c) 50 lm/m$^2$    (d) 80 lm/m$^2$

15. What angle should a surface make with the incident flux if the illumination of a surface is to be reduced by a factor of one-half?
(a) 30°      (b) 45°      (c) 60°      (d) 90°

16. A point source of light is placed 20 cm from a pencil 8 cm high. The length of the shadow on a wall 1 m from the source is
(a) 20 cm      (b) 30 cm      (c) 40 cm      (d) 50 cm

17. The solid angle subtended at the center of a 10-cm-diameter sphere by an area of 2 cm on its surface is approximately
(a) 0.4 sr      (b) 0.04 sr      (c) 0.8 sr      (d) 0.08 sr

18. Luminous flux is associated most closely with
(a) the source
(b) the surface
(c) the space between the source and the surface
(d) none of these

19. A movie screen is 10 m from a light source in a projector. The screen makes an angle of 37° with the incident flux. What luminous intensity is required to give an illumination of 6 lx?
(a) 75.1 cd      (b) 751 cd      (c) 890 cd      (d) 997 cd

20. A 200-cd light provides an illumination of 50 lx at a distance of
(a) 1 m      (b) 2 m      (c) 3 m      (d) 4 m

---

## Completion Questions

1. Every point on an advancing wavefront can be considered as a source of secondary _____. This is known as _____.

2. The _____ is a unit of wavelength that is equal to one ten-thousandth of a micrometer.

3. Electromagnetic radiation of immediately higher energy than light is called _____ radiation.

4. Approximate values for the speed of light in a vacuum are _____ m/s and _____ mi/s.

5. The conversion factor relating light frequency to light energy is known as _____.

6. The common unit for expressing wavelengths of electromagnetic radiation is _____.

7. The process by which light is emitted when electrons strike a metallic surface is known as the _____.

8. The colors corresponding to the given wavelengths are

_____ (450 nm), _____ (480 nm),

_____ (520 nm), _____ (580 nm),

_____ (600 nm), and _____ (640 nm).

**9.** The spectrum of electromagnetic waves is divided into the following eight major regions in order of increasing energy:

(a) _____        (b) _____

(c) _____        (d) _____

(e) _____        (f) _____

(g) _____        (h) _____

**10.** Light energy is divided equally between _____ and _____ fields, which are mutually perpendicular.

**11.** One _____ is equivalent to 1/680 W of yellow-green light of wavelength _____.

**12.** The inner portion of a shadow that receives no light from the source is called the _____. The other region is known as the _____.

**13.** The luminous flux per unit solid angle is known as the _____ and is measured in _____.

**14.** Illumination $E$ may be calculated from the ratio of _____ to _____ or from the ratio of _____ to the square of the _____ between the source and the surface.

**15.** A grease-spot photometer is used to measure _____ by comparison with a standard source.

**16.** A _____ is the solid angle subtended at the center of a sphere by an area on its surface that is equal to the square of its _____.

**17.** The _____ of a surface is proportional to the luminous intensity of the light source and is inversely proportional to the square of the distance. This is sometimes called the _____ law.

**18.** When a surface makes an angle with the incident flux, the _____ is proportional to the component of the surface _____ to the flux.

**19.** A(n) _____ source is one that emits light uniformly in all directions.

**20.** The _____ is that part of the total radiant power emitted from a light source that is capable of affecting the sense of sight.

# Reflection and Mirrors

| CONTENTS | | |
|---|---|---|
| **34-1** The Laws of Reflection | | **34-5** The Mirror Equation |
| **34-2** Plane Mirrors | | **34-6** Magnification |
| **34-3** Spherical Mirrors | | **34-7** Spherical Aberration |
| **34-4** Images Formed by Spherical Mirrors | | |

**DEFINITIONS OF KEY TERMS**

**Specular reflection**  Regular reflection in which incident parallel rays of light remain parallel after reflection.

**Diffuse reflection**  Irregular reflection in which incident light is scattered as it strikes a rough surface.

**Virtual image**  An image that seems to be formed by light coming from it, although no light rays actually pass through it.

**Real image**  An image formed by actual light rays that pass through it. Real images can be projected on a screen.

**Focal length**  The distance from the vertex of a mirror to its focal point.

**Radius of curvature**  The radius of an imaginary sphere constructed by fully extending the surface of a spherical mirror.

**Magnification**  The ratio of the image size to the object size for images formed by mirrors.

**Spherical aberration**  The focusing defect of a spherical mirror in which the extreme rays of light are not sharply focused.

**Converging mirror**  A mirror whose reflections converge to a focal point.

**Diverging mirror**  A mirror whose reflections do not converge into a focal point.

**Linear aperture**  Another name for a slit.

**Mirror equation**  The relationship between the size and distance of an object and an image based on the radius of armature of the mirror.

**Parabolic mirror**  A mirror in the three-dimensional shape of a parabola.

**Plane mirror**  A flat mirror.

**Spherical mirror**  A mirror in the shape of a portion of a sphere.

**CONCEPTS AND EXAMPLES**

**Ray Tracing**  Problem solving can be assisted by careful construction of ray diagrams. The three principal rays are shown in Fig. 34-1 for converging (concave) and diverging (convex) spherical mirrors.

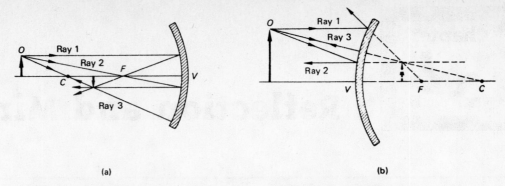

**Fig. 34-1** (a) Ray tracing for a converging mirror, (b) ray tracing for a diverging mirror.

**Ray 1** A ray parallel to the mirror axis passes through the focal point of a concave mirror or seems to come from the focal point of a convex mirror.

**Ray 2** A ray that passes through the focal point of a concave mirror or proceeds toward the focal point of a convex mirror is reflected parallel to the mirror axis.

**Ray 3** A ray that proceeds along a radius of the mirror is reflected back along its original path.

**Mirror Equations** A general set of equations can be derived that will apply to both concave and convex spherical mirrors as long as appropriate sign conventions are followed. These basic relations are

$$\frac{1}{p} + \frac{1}{q} = \frac{1}{f} \qquad f = \frac{R}{2} \qquad M = \frac{y'}{y} = \frac{-q}{p}$$

You should practice solving each of these equations for each parameter as a function of the other parameters. Remember that concave mirrors are *converging* mirrors and convex mirrors are *diverging* mirrors. When you substitute, it is essential that you apply the following sign conventions strictly:

$R$ = radius of curvature — + for converging, − for diverging
$f$ = focal length — + for converging, − for diverging
$p$ = object distance — + for real objects, − for virtual
$q$ = image distance — + for real images, − for virtual
$y$ = object size — + for erect, − if inverted
$y'$ = image size — + if erect, − if inverted
$M$ = magnification — + if both erect or both inverted

Be sure not to confuse the signs of operation that appear in formulas with the signs of the substituted quantities.

---

**EXAMPLE 34-1**

An object 5 cm tall is held 30 cm from a concave spherical mirror of radius 80 cm. What are the nature, size, and location of the image?

**Solution**
The focal length is equal to half the radius, or 40 cm. First, we find $q$ by solving the mirror equation:

$$q = \frac{pf}{p - f} = \frac{(30 \text{ cm})(40 \text{ cm})}{30 \text{ cm} - 40 \text{ cm}} = \boxed{-120 \text{ cm}}$$

Since $q$ is negative, the image is virtual and appears to be located 120 cm behind the mirror. The size of the image is found by substitution into the magnification equation. Solving for $y'$, we obtain

$$y' = \frac{-qy'}{p} = \frac{-(-120 \text{ cm})(5 \text{ cm})}{30 \text{ cm}} = \boxed{20 \text{ cm}}$$

Note that the magnification is +4 and the positive sign for $y'$ indicates an erect image.

## Summary

- The focal length and radius of curvature of converging and diverging mirrors determine the nature and size of the images they form.
- The formation of images by spherical mirrors can be visualized more easily with ray-tracing techniques.
- The three principal rays are:
  - **a. Ray 1** A ray parallel to the mirror axis passes through the focal point of a concave mirror or seems to come from the focal point of a convex mirror.
  - **b. Ray 2** A ray that passes through the focal point of a concave mirror or proceeds toward the focal point of a convex mirror is reflected parallel to the mirror axis.
  - **c. Ray 3** A ray that proceeds along a radius of the mirror is reflected back along its original path.
- The mirror equations can be applied to either converging (concave) or diverging (convex) spherical mirrors.

## Key Equations by Section

| Section | Topic | Equation | Equation No. |
|---------|-------|----------|--------------|
| 34-2 | Relation of image distance to object distance, for plane mirrors | $\boxed{p = q}$ | 34-1 |
| 34-3 | Relation between focal length and radius of curvature, for spherical mirrors | $\boxed{f = \dfrac{R}{2}}$ | 34-2 |
| 34-5 | Derivation of the mirror equation, in which $y'$ is the image size, $y$ is the object size, $R$ is the radius of curvature, $q$ is the image distance, and $p$ is the object distance | $\dfrac{-y'}{y} = \dfrac{R - q}{p - R}$ | 34-3 |

| Section | Topic | Equation | Equation No. |
|---------|-------|----------|--------------|
| 34-5 | Since the angle of incidence equals the angle of reflection, the tangents of these angles are also equal | $\tan \theta_i = \tan \theta_r$  $\dfrac{y}{p} = \dfrac{-y'}{q}$ | 34-4 |
| 34-5 | The mirror equation | $\dfrac{1}{p} + \dfrac{1}{q} = \dfrac{2}{R}$ | 34-6 |
| 34-5 | The mirror equation, in terms of focal length | $\dfrac{1}{p} + \dfrac{1}{q} = \dfrac{1}{f}$ | 34-7 |
| 34-5 | Rearrangements of the mirror equation | $p = \dfrac{qf}{q-f}$  $q = \dfrac{pf}{p-f}$  $f = \dfrac{pq}{p+q}$ | 34-8 |
| 34-6 | Magnification | Magnification $= \dfrac{\text{image size}}{\text{object size}} = \dfrac{y'}{y}$ | 34-9 |
| 34-6 | Magnification | $M = \dfrac{y'}{y} = \dfrac{-q}{p}$ | 34-10 |

**True-False Questions**

T  F  **1.** A negative magnification results whenever the image is virtual.

T  F  **2.** A virtual image cannot be formed on a screen.

T  F  **3.** Images formed by convex spherical mirrors are always virtual, erect, and enlarged.

T  F  **4.** For concave spherical mirrors, the magnification is always greater than 1 when the object is located between the center of curvature and the focal point.

T  F  **5.** A plane mirror forms real images.

T  F  **6.** Objects moving closer and closer to the vertex of a convex mirror form smaller and smaller images.

T  F  **7.** In a concave shaving mirror, greater magnification is achieved when the object is closer to the focal point.

T  F  **8.** All virtual images formed by spherical mirrors are erect and diminished.

T  F  **9.** A ray parallel to the mirror axis passes through the center of curvature after reflection from a converging mirror.

T  F  **10.** The radius of curvature is equal to twice the focal length for both concave and convex mirrors.

**Multiple-Choice Questions**

**1.** For a spherical concave mirror, virtual images are formed when the object is located

(a) between $F$ and $C$

(b) beyond $C$

(c) at $C$

(d) inside $F$

**2.** Which of the following is *not* true when an image is formed by an object located between $C$ and $F$ of a concave mirror?
(a) Negative magnification     (b) Negative image distance
(c) Inverted image             (d) Enlarged image

**3.** The focal point of a convex spherical mirror is
(a) twice the radius        (b) in front of the mirror
(c) virtual                    (d) real

**4.** Which of the following is *not* true for images formed by a plane mirror?
(a) Magnification is $+1$      (b) Image distance is negative
(c) Right and left are reversed    (d) Images are real

**5.** A source of light 12 cm high is placed 50 cm from a concave mirror of focal length 100 cm. The image distance is
(a) $-100$ cm    (b) $+100$ cm    (c) $+50$ cm     (d) $-50$ cm

**6.** An object is placed 10 cm from the vertex of a convex spherical mirror whose radius is 20 cm. The magnification is
(a) 0.667      (b) $-0.667$     (c) $+\frac{1}{2}$        (d) $-\frac{1}{2}$

**7.** A 6-ft person stands 20 ft from a plane mirror. The shortest mirror required to view the entire image is
(a) 3 ft        (b) 6 ft        (c) 9 ft        (d) 12 ft

**8.** At what distance must an object be placed to form an image on a screen 30 cm from the vertex of a mirror whose radius is 20 cm?
(a) 20 cm     (b) 15 cm     (c) 10 cm     (d) 5 cm

**9.** The magnification of a mirror is $-\frac{1}{3}$. What is the image distance when an object is placed 24 cm from this mirror?
(a) 8 cm      (b) $-8$ cm     (c) 12 cm     (d) $-12$ cm

**10.** What should be the object distance for a concave shaving mirror of radius 3.2 m to form an erect image twice as large as the object?
(a) 80 cm     (b) 1.6 m     (c) 2.4 m     (d) 3.2 m

**Completion Questions**

**1.** The focal length of a spherical mirror is equal to _____.

**2.** All images formed by convex mirrors are _____,
_____, and _____ in size.

**3.** Object and image distances must be reckoned as _____ for real objects or images and _____ for virtual ones.

**4.** A magnification less than 0 but greater than $-1$ means that the image is _____ than the object in size and is also _____.

**5.** When an object is placed inside the focus of a concave mirror, the image is _____, _____, and _____ in size.

**6.** The magnification is equal to the ratio of the _____ to the _____ and will be _____ for erect images and _____ for inverted images.

**7.** When an object is located at the focal point of a concave mirror, all reflected rays are _____.

**8.** A ray that proceeds toward the focal point of a concave mirror is reflected _____.

**9.** A(n) _____ image appears to be formed by actual rays of light, but no rays of light actually pass through it.

**10.** The linear apertures of spherical mirrors should be _____ in comparison with their focal lengths to reduce the effects of _____.

# Chapter 35

# Refraction

## DEFINITIONS OF KEY TERMS

**Refraction**  The bending of a light ray as it passes obliquely from one medium to another.

**Index of refraction**  The ratio of the free-space velocity of light to the velocity of light through a specific material.

**Optical density**  A property of a material that is a measure of the speed of light through a specific material.

**Snell's law**  The ratio of the sine of the angle of incidence to the sine of the angle of refraction is equal to the ratio of the velocity of light in the incident medium to the velocity of light in the refracted medium.

**Dispersion**  The separation of light into its component wavelengths, usually accomplished with a prism.

**Critical angle**  The limiting angle of incidence in a denser medium that results in an angle of refraction equal to 90°.

**Total internal reflection**  The process by which light incident from a denser medium is reflected back into that medium on striking the boundary between two media.

**Apparent depth**  The depth an object appears to have in water when viewed from the air above.

## CONCEPTS AND EXAMPLES

**Refraction**  Refraction is the process by which light bends as it passes from one medium to another. The amount of bending is determined by the index of refraction for a material, which is defined as the ratio of the free-space velocity $c$ of light to the velocity $v$ of light in that material:

$$n = \frac{c}{v} \qquad c = 3 \times 10^8 \text{ m/s} \qquad \textbf{index of refraction}$$

Consider the case illustrated by Fig. 35-1 where light enters from medium 1 and is refracted into medium 2. The angle of incidence is $\theta_1$, and the angle of refraction is $\theta_2$. Both angles are measured with respect to a line perpendicular to the surface at the point of incidence. Snell's law for refraction may be written in terms of the indices of refraction or in terms of the velocities in the two media:

$$n_1 \sin \theta_1 = n_2 \sin \theta_2 \qquad \frac{v_1}{v_2} = \frac{\sin \theta_1}{\sin \theta_2}$$

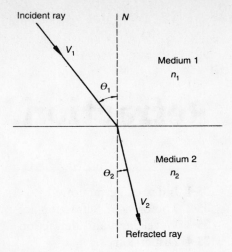

**Fig. 35-1** Snell's law of refraction.

**EXAMPLE 35-1**    A beam of light passes from water ($n_w = 1.33$) into glass so that the beam makes an angle of 30° with the glass interface. If the angle of refraction is 50°, what is the index of refraction for the glass?

*Solution*
The angle of incidence at the glass is 90° − 30° = 60°. Remember, for Snell's law, the angles are measured with respect to the normal to the surface. We write Snell's law for the two media and solve for the index of refraction $n_g$:

$$n_w \sin \theta_w = n_g \sin \theta_g \qquad n_g = \frac{n_w \sin \theta_w}{\sin \theta_g}$$

So

$$n_g = \frac{1.33 \sin 60°}{\sin 50°} = \boxed{1.50}$$

Thus, the index of refraction for glass is 1.5.

**Total Internal Reflection**    When light passes from one medium to an optically denser medium, the light may be reflected totally into the incident medium. The smallest angle of incidence for which such total internal reflection occurs is called the *critical angle* $\theta_c$, which is given by

$$\sin \theta_c = \frac{n_2}{n_1} \qquad \text{critical angle}$$

In this relation, $n_1$ is the index for the optically denser medium, or the incident medium.

**EXAMPLE 35-2**    The critical angle for a cut of diamond surrounded by water is 35°. What is the index of refraction for this diamond?

**Solution**
We first solve for $n_1$ and then substitute known quantities:

$$n_d = \frac{n_w}{\sin 35°} = \frac{1.33}{0.574} = \boxed{2.32}$$

Note that the critical angle of 35° applies when the diamond is surrounded by water. This angle would be around 25° if the surrounding medium were air.

**Refraction and Wavelengths**   We have seen that when light travels from one medium to another, the velocities are different when the indices of refraction are different. It is also true that the wavelength changes as light travels from one medium to another. Such a decrease is in the same proportion as the decrease in velocity. The following equivalent ratios are useful in solving many physical problems involving refracted light:

$$\frac{\lambda_1}{\lambda_2} = \frac{v_1}{v_2} = \frac{n_2}{n_1} = \frac{\sin \theta_1}{\sin \theta_2}$$

By setting any two of these ratios equal, you may obtain as many as six useful formulas to apply to given problems.

**EXAMPLE 35-3**    A monochromatic beam of light whose wavelength is 600 nm in air passes into another medium. If the velocity is decreased to two-thirds of its value in air, what is the wavelength in the medium?

**Solution**
We know that $\lambda_a/\lambda_m = v_a/v_m$. Thus, if the ratio of velocities is $\frac{2}{3}$, then the wavelength will be reduced by the same fraction. Hence,

$$\lambda_m = \frac{2}{3}\lambda_a = \frac{2(600 \text{ nm})}{3} = \boxed{400 \text{ nm}}$$

**Summary**

- The concepts of refraction, critical angle, dispersion, and internal reflection play important roles in the operation of many instruments.
- The index of refraction of a particular material is the ratio of the free-space velocity of light $c$ to the velocity $v$ of light through the medium.
- Snell's law states that if the index of refraction decreases, the angle will *increase*.

- When light enters medium 2 from medium 1, its wavelength is changed by the fact that the index of refraction is different.
- The critical angle $\theta_c$ is the maximum angle of incidence from one medium that will still produce refraction (at 90°) into a bordering medium.
- Refraction causes an object in one medium to be observed at a different depth when viewed from above in another medium.

| **Key Equations by Section** | **Section** | **Topic** | **Equation** | **Equation No.** |
|---|---|---|---|---|
| | 35-1 | Index of refraction $n$ given as the ratio of the speed of light in a vacuum $c$ to the speed of light in a new medium $v$ | $n = \dfrac{c}{v}$ | 35-1 |
| | 35-2 | Snell's law: The ratio of the sine of the angle of incidence to the sine of the angle of refraction equals the ratio of the speed of light in the two mediums | $\dfrac{\sin \theta_1}{\sin \theta_2} = \dfrac{v_1}{v_2}$ | 35-2 |
| | 35-2 | Snell's law using the index of refraction of each medium | $n_1 \sin \theta_1 = n_2 \sin \theta_2$ | 35-3 |
| | 35-3 | The speed of a wave equals the frequency times the wavelength | $c = f\lambda_a$ and $v_m = f\lambda_m$ | 35-4 |
| | 35-3 | The wavelength in the new medium $\lambda_m$ given by the wavelength in air $\lambda_a$ divided by the index of refraction of the medium | $\lambda_m = \dfrac{\lambda_a}{n_m}$ | 35-5 |
| | 35-3 | Relationships of incidence and refraction | $\dfrac{\sin \theta_1}{\sin \theta_2} = \dfrac{v_1}{v_2} = \dfrac{n_2}{n_1} = \dfrac{\lambda_1}{\lambda_2}$ | 35-6 |
| | 35-5 | Conditions for total internal reflection, at the critical angle $\theta_c$ | $\sin \theta_c = \dfrac{n_2}{n_1}$ | 35-7 |

| Section | Topic | Equation | Equation No. |
|---|---|---|---|
| 35-9 | Apparent depth of an underwater object viewed from air | $$\frac{\sin \theta_1}{\sin \theta_2} = \frac{n_2}{n_1} = \frac{q}{p}$$ $$\boxed{\frac{\text{Apparent depth } q}{\text{Actual depth } p} = \frac{n_2}{n_1}}$$ | 35-10 |

---

## True-False Questions

**T  F  1.** When light enters from a medium of lower refractive index into a medium of larger refractive index, the path of the light bends toward the normal.

**T  F  2.** When light crosses a boundary between medium 1 and medium 2, the ratio of $v_1$ to $v_2$ will be the same as the ratio of $n_1$ to $n_2$.

**T  F  3.** Objects of higher optical density have greater critical angles.

**T  F  4.** When white light is dispersed by a prism, the smaller-wavelength components are deviated the most.

**T  F  5.** For total internal reflection, the angle of incidence is equal to the angle of reflection.

**T  F  6.** When an object in air is viewed from a position under water, the object appears to be farther away than it actually is.

**T  F  7.** Whenever light enters a denser medium, both the velocity and frequency of the light are reduced.

**T  F  8.** Whenever light passes through any number of parallel media and finally returns to the original medium, the angle of incidence in the original medium is equal to the angle of emergence.

**T  F  9.** The critical angle for right-angle prisms should not exceed 45°.

**T  F  10.** The lateral displacement of light as it passes through a pane of glass is greater when the optical density of the glass is large.

---

## Multiple-Choice Questions

**1.** For refraction of light from medium 1 to medium 2, the ratio of $\sin \theta_1$ to $\sin \theta_2$ is equal to
   (a) $v_1/v_2$     (b) $n_1/n_2$     (c) $\lambda_2/\lambda_1$     (d) $\theta_1/\theta_2$

**2.** The index of refraction for a substance is
   (a) constant
   (b) constant for a given wavelength
   (c) variable with the speed of light
   (d) never constant

**3.** Total internal reflection occurs when the angle of incidence is
   (a) greater than the angle of refraction
   (b) equal to the critical angle
   (c) greater than the critical angle
   (d) greater than 45°

**4.** The refractive index of benzene is 1.5. The velocity of light in benzene is approximately
   (a) $1.5 \times 10^8$ m/s     (b) $1.75 \times 10^8$ m/s
   (c) $2 \times 10^8$ m/s     (d) $2.5 \times 10^8$ m/s

5. Light passes at an angle of incidence of 37° from water to air. The angle of refraction in the air is approximately
   (a) 27°   (b) 53°   (c) 45°   (d) 60°

6. Monochromatic green light has a wavelength of 520 nm. The wavelength of this light inside glass of refractive index 1.5 is approximately
   (a) 300 nm   (b) 340 nm   (c) 520 nm   (d) 780 nm

7. The critical angle for diamond ($n = 2.42$) surrounded by air is approximately
   (a) 24°   (b) 35°   (c) 45°   (d) 66°

8. A fish is located 2 m from the surface of a small pond. The apparent depth is
   (a) 1.0 m   (b) 1.5 m   (c) 2.0 m   (d) 2.66 m

9. Light is incident from water at an angle of 60° into a transparent medium. If the angle of refraction is 30°, the index for the medium is
   (a) 1.15   (b) 1.5   (c) 2.3   (d) 3.1

10. Light travels from air at an angle of incidence of 41°. What will the speed of light in the second medium be if the angle of refraction is 30°?
    (a) $1.5 \times 10^8$ m/s      (b) $2.1 \times 10^8$ m/s
    (c) $2.29 \times 10^8$ m/s      (d) $2.5 \times 10^8$ m/s

**Completion Questions**

1. The _____ of a material is the ratio of the free-space velocity of light to the velocity of light through the material.

2. For refraction, the _____, the _____, and the _____ all lie in the same plane.

3. Whenever light enters glass from air, the _____ remains unchanged, but the _____ and the _____ are reduced.

4. Whenever white light enters a prism, it separates into its component _____. This phenomenon is known as _____.

5. The _____ is the limiting angle of incidence in a denser medium that results in an angle of refraction of 90°.

6. Total internal reflection occurs when a light ray from a certain medium is incident upon another medium of _____ optical density at an angle of incidence greater than the _____.

7. According to Snell's law, the ratio of the _____ to the _____ is equal to the ratio of the _____ in the incident medium to the _____ in the refracted medium.

8. Right-angle prisms make use of the principle of _____ to deviate the path of light along right angles.

9. When objects under water are viewed from the air above the water, the apparent depth is equal to the actual depth times the ratio of _____ to the _____.

10. The minimum value of the index of refraction is _____.

# Chapter
# 36

# Lenses and Optical Instruments

**DEFINITIONS OF KEY TERMS**

**Lens**   A transparent object that alters the shape of a wavefront that passes through it.

**Converging lens**   A lens that refracts and converges parallel light to a point focus beyond the lens.

**Diverging lens**   A lens that refracts and diverges parallel light from a point located in front of the lens.

**Focal length**   The distance from the optical center of a lens to either focus.

**Lensmaker's equation**   The physical relationship among the focal length, surface curvatures, and refractive index that guides the construction of lenses.

**Magnification**   The ratio of image size to object size for images formed by lenses.

**Telescope**   An optical instrument designed to examine large, distant objects.

**Objective lens**   The outermost lens of an optical instrument that first receives light from the object being viewed.

**Eyepiece**   The lens of an optical instrument through which light leaves the instrument and enters the eye.

**Spherical aberration**   A lens defect that shows its inability to focus light of different colors (wavelengths) to the same point.

**Achromatic lens**   A combination lens that is made of convergent and divergent lens materials and removes chromatic aberration.

**Chromatic aberration**   A lens defect in which different wavelengths of light (i.e., colors) are focused at different places.

**Diaphragm**   An apparatus that allows light through a central aperture that can be adjusted in size.

**Lensmaker's equation**   A relationship between focal length, the radii of the two lens surfaces, and the index of refraction of the lens material.

**Meniscus lens**   A lens whose shape is like a meniscus or surface of a liquid in a cylinder.

**Virtual focus**   A focus that does not have actual light rays converging on it but is found by tracing the rays backward to a crossing point.

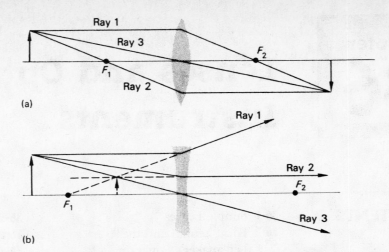

(a)

(b)

**Fig. 36-1** (a) Ray tracing for a converging lens, (b) ray tracing for a diverging lens.

**Image Construction** Ray-tracing techniques are illustrated in Fig. 36-1. Remember that the first focal point $F_1$ is the one on the same side of the lens as the incident light. The second focal point $F_2$ is located on the far side.

**Ray 1** A ray parallel to the axis passes through the second focal point $F_2$ of a converging lens or appears to come from the first focal point $F_1$ of a diverging lens.

**Ray 2** A ray that passes through $F_1$ of a converging lens or proceeds toward $F_2$ of a diverging lens is refracted parallel to the lens axis.

**Ray 3** A ray that passes through the geometrical center of a lens will not be deviated.

**Lensmaker's Equation** The equation used in the construction of lenses relates the focal length $f$ to the index of refraction $n$ and the radii of the two surfaces $R_1$ and $R_2$:

$$\frac{1}{f} = (n - 1)\left(\frac{1}{R_1} + \frac{1}{R_2}\right) \qquad \textbf{lensmaker's equation}$$

The meaning of each symbol is seen in Fig. 36-2. Remember that $R_1$ or $R_2$ is positive if the outside surface is convex or negative if the surface is concave. The focal length $f$ is positive for converging lenses and negative for diverging lenses.

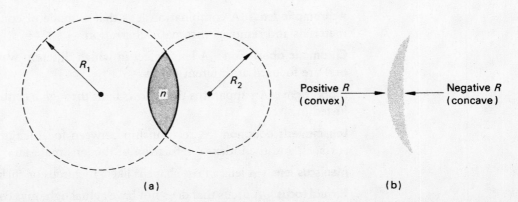

(a)

(b)

**Fig. 36-2**

**EXAMPLE 36-1**  A lens for eyeglasses has a concave surface of radius 30 cm and a convex surface of radius 50 cm. If the lens is constructed of glass whose refractive index is 1.5, what is the focal length?

*Solution*
According to sign convention, the concave surface is negative ($-30$ cm). Substitution yields

$$\frac{1}{f} = (1.5 - 1)\left(\frac{1}{50 \text{ cm}} + \frac{1}{-30 \text{ cm}}\right)$$

$$= (0.5)(0.02 - 0.03) = -0.0667 \text{ cm}^{-1}$$

Now solving for $f$ gives

$$f = -150 \text{ cm} \qquad \text{diverging meniscus lens}$$

Problems such as Example 36-1 are solved easily by using the reciprocal button on your calculator.

**Lens Equation**  The equations for object and image locations and for the magnification are the same as for the mirror equations:

$$\frac{1}{p} + \frac{1}{q} = \frac{1}{f} \qquad p = \frac{qf}{q - f} \qquad q = \frac{pf}{p - f} \qquad f = \frac{pq}{p + q}$$

$$\text{Magnification} = \frac{\text{image size}}{\text{object size}} \qquad M = \frac{y'}{y} = \frac{-q}{p}$$

$p$ or $q$ is positive for real and negative for virtual

$y$ or $y'$ is positive if erect and negative if inverted

**EXAMPLE 36-2**  What are the nature, size, and location of the image formed of a 6-cm object located 200 cm from the diverging meniscus lens of Example 36-1?

*Solution*
Remember that $f = -150$ cm and $p = 200$ cm. So

$$q = \frac{pf}{p - f} = \frac{(200 \text{ cm})(-150 \text{ cm})}{200 \text{ cm} - (-150 \text{ cm})}$$

$$= \frac{-30,000 \text{ cm}^2}{350 \text{ cm}} = -85.7 \text{ cm}$$

The image is virtual (negative) and appears to be located a distance of 85.7 cm from the lens on the near side. Now the image size is found from the magnification equation. Solving for $y'$, we obtain

$$y' = \frac{-qy}{p} = \frac{-(-85.7 \text{ cm})(6 \text{ cm})}{200 \text{ cm}}$$

$$= 2.57 \text{ cm} \qquad \text{erect (positive) and diminished}$$

## Summary

- A lens is a transparent device that converges or diverges light to or from a focal point.
- Image formation by thin lenses can be understood more easily through ray-tracing techniques for converging lenses and diverging lenses.
  - **a. Ray 1** A ray parallel to the axis that passes through the second focal point $F_2$ of a converging lens or appears to come from the first focal point $F_1$ of a diverging lens.
  - **b. Ray 2** A ray that passes through $F_1$ of a converging lens or proceeds toward $F_2$ of a diverging lens is refracted parallel to the lens axis.
  - **c. Ray 3** A ray that passes through the geometrical center of a lens and will not be deviated.
- The lensmaker's equation is a relationship between the focal length, the radii of the surfaces of the two lens, and the index of refraction of the lens material.
- The equations for object and image locations and for the magnification are the same as for the mirror equations.

## Key Equations by Section

| Section | Topic | Equation | Equation No. |
|---------|-------|----------|--------------|
| 36-2 | The lensmaker's equation | $\dfrac{1}{f} = (n - 1)\left(\dfrac{1}{R_1} + \dfrac{1}{R_2}\right)$ | 36-1 |
| 36-4 | The lens equation, which is the same as the mirror equation | $\dfrac{1}{p} + \dfrac{1}{q} = \dfrac{1}{f}$ | 36-2 |
| 36-4 | Lens equation alternative forms | $p = \dfrac{fq}{q - f} \qquad q = \dfrac{fp}{p - f}$ $f = \dfrac{qp}{p + q}$ | 36-3 |
| 36-4 | Definition of the magnification of a lens | $M = \dfrac{y'}{y} = -\dfrac{q}{p}$ | 36-4 |
| 36-5 | Overall magnification of more than one lens | $M = M_1 M_2$ | 36-5 |

## True-False Questions

**T   F   1.** A lens that is thinner in the middle than it is at the edges will be a converging lens.

**T   F   2.** A plano-concave lens has a virtual focus.

**T   F   3.** Both surfaces of a converging meniscus lens should be reckoned as positive, according to convention.

**T   F   4.** Virtual images are formed on the same side of the lens as the object.

**T   F   5.** The overall magnification of a compound optical instrument is equal to the product of the magnifications of the component lenses.

**T  F  6.** Chromatic aberration is a lens defect in which the extreme rays are brought to a focus nearer the lens than those rays entering near the optical center of the lens.

**T  F  7.** According to convention, the object distance is reckoned as negative when it is measured to a virtual object.

**T  F  8.** All images formed by diverging lenses are virtual, diminished, and erect.

**T  F  9.** In a simple microscope, the greatest magnification occurs as the object gets closer and closer to the lens surface.

**T  F  10.** Whenever the object is beyond the focal point of a converging lens, the magnification will always be negative.

## Multiple-Choice Questions

1. Images formed from real objects by diverging lenses are always
   (a) virtual     (b) enlarged   (c) inverted     (d) real

2. A diverging lens may not have
   (a) a negative focal length     (b) a positive focal length
   (c) one plane surface     (d) one convex surface

3. For a compound microscope, the image formed by the eyepiece is
   (a) real          (b) inverted   (c) erect          (d) diminished

4. A negative magnification always means that the image is
   (a) erect          (b) real          (c) virtual          (d) inverted

5. Which of the following is *not* characteristic of images formed by real objects located inside the focal point of a converging lens?
   (a) Virtual       (b) Erect         (c) Real          (d) Enlarged

6. A meniscus lens has a convex surface of curvature 20 cm and a concave surface of curvature $-30$ cm. If the lens is constructed from glass ($n = 1.5$), the focal length will be
   (a) $-4$ cm       (b) $+4$ cm       (c) $-120$ cm    (d) $+120$ cm

7. An object is located 10 in. from a thin converging lens whose focal length is 30 in. The image distance is approximately
   (a) $-7.5$ in.    (b) $+7.5$ in.    (c) 15 in.          (d) $-15$ in.

8. A diverging meniscus lens has a focal length of $-20$ cm. If the lens is held 10 cm from the object, the magnification is
   (a) $-0.667$      (b) $+0.667$      (c) $-2$          (d) $+2$

9. A plano-convex lens is ground from glass ($n = 1.5$). If the focal length is to be 20 cm, the radius of the curved surface should be
   (a) 10 cm       (b) 20 cm       (c) 30 cm         (d) 40 cm

10. A 6-ft-high image is projected on a screen located 40 ft from a converging lens. If the object size is 0.2 ft, the focal length must be
   (a) 0.736 ft     (b) 1.29 ft       (c) 1.38 ft        (d) 2.79 ft

## Completion Questions

1. Images formed by diverging lenses are always _____, _____, and _____ in size.

2. The _____ of a lens is the distance from the optical center of the lens to either focus.

3. An object at a distance beyond twice the focal length of a convex lens forms an image that is _____, _____, and _____ in size.

4. The object distance and the image distance are considered _____ for real images and objects and _____ for virtual images and objects.

5. A positive magnification means that the image is _____, and a negative magnification means the image is _____.

6. Three examples of converging lens are _____, _____, and _____.

7. A converging lens is _____ in the middle than at the edges, whereas a diverging lens is _____ in the middle.

8. A ray parallel to the axis passes through the _____ of a converging lens or appears to come from the _____ of a diverging lens.

9. A(n) _____ image is formed on the same side of the lens as the object; a(n) _____ image is formed on the opposite side.

10. A ray that passes through the _____ of a lens will not be deviated.

# Chapter 37

# Interference, Diffraction, and Polarization

**DEFINITIONS OF KEY TERMS**

**Diffraction**   The ability of waves to bend around obstacles in their path.

**Young's experiment**   A classic experiment that predicts the location of interference fringes produced by passing monochromatic light through two slits.

**Diffraction grating**   An optical device consisting of many parallel slits that produce a spectrum through the interference of light diffracted through them.

**Order**   A system of numbering interference fringes as they move away from the central fringe.

**Resolving power**   A measure of the ability of an instrument to produce well-defined, separate images.

**Polarization**   The process by which the oscillations of wave motion are confined to a definite pattern.

**Plane polarization**   Polarization in which the oscillations are restricted to a plane.

**Analyzer**   A secondary filter which is used to test light for polarization.

**Coherent light**   Light that is all the same frequency, traveling in the same direction.

**Huygens's principle**   Each point on a wavefront can be considered as a new source of secondary waves.

**Inference fringes**   A description of the pattern of light and dark bands projected on a screen through multiple slits due to constructive and obstructive interference.

**Polarizer**   A filter that allows only polarized light to pass.

**Polaroid sheets**   Flat transparent film or lenses that allow only polarized light to pass.

**Superposition principle**   It is possible to add two coherent waves together, to get a third composite wave.

**CONCEPTS AND EXAMPLES**

**Young's Experiment**   In Young's experiment, interference and diffraction produce bright and dark fringes that are located by the following equations (see Fig. 37-1):

Bright fringes:

$$\frac{yd}{x} = n\lambda \qquad n = 0, 1, 2, 3, \ldots$$

Dark fringes:

$$\frac{yd}{x} = \frac{n\lambda}{2} \qquad n = 1, 3, 5, 7, \ldots$$

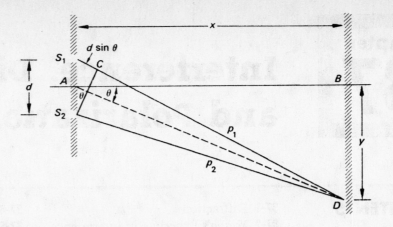

**Fig. 37-1**

In this equation $x$ is the distance from the slit to the screen and $y$ is the displacement of the $n$th fringe. The slit separation is $d$, and $\lambda$ represents the wavelength of the incident light.

---

**EXAMPLE 37-1**

In Young's experiment, monochromatic light of wavelength 560 nm passes through two slits that are 0.05 mm apart. The screen is located 2 m away from the slits. What is the displacement of the third dark fringe?

**Solution**
We solve the above equation for $y$ and substitute $n = 5$ for the dark fringe:

$$y = \frac{n\lambda x}{2d} = \frac{(5)(560 \times 10^{-9})(2 \text{ m})}{2(5 \times 10^{-5} \text{ m})}$$

$$= 0.056 \text{ m} = \boxed{56 \text{ mm}}$$

---

**Diffraction Grating**    In the diffraction grating, brighter and sharper diffraction patterns are produced by passing light through many regularly spaced, parallel slits. A central bright fringe is formed on the screen, and other bright fringes occur on either side. The first bright line formed on each side is called the first-order fringe ($n = 1$), the second bright line on either side is called the second-order fringe ($n = 2$), and so forth. The equation predicting the formation of these fringes is

$$d \sin \theta_n = n\lambda \qquad n = 1, 2, 3, \ldots$$

In this equation, $d$ is the separation of each slit (line), $\lambda$ is the wavelength of the incident light, and $\theta_n$ is the deviation angle of the $n$th bright fringe.

---

**EXAMPLE 37-2**

Light of wavelength 620 nm passes through a diffraction grating located 800 mm from a screen. The second bright fringe occurs 500 mm from the central fringe. What is the spacing of the slits on this grating?

*Solution*

We must first determine the deviation angle:

$$\tan \theta_n = \frac{500 \text{ mm}}{800 \text{ mm}} \quad \text{or} \quad \theta_n = 32.0°$$

Now, we solve the grating equation for $d$ and substitute $n = 2$ for the second-order fringe:

$$d = \frac{n\lambda}{\sin \theta_n} = \frac{(2)(620 \times 10^{-9} \text{ m})}{\sin 32.0°}$$

$$2.34 \times 10^{-6} \text{ m} = \boxed{2.34 \text{ } \mu\text{m}}$$

**Resolving Power**   The resolving power is important for optical instruments because it determines the ability to distinguish separate images. It is often expressed as the angle $\theta_0$ subtended at the opening by the objects being resolved. The minimum conditions for resolution are illustrated by Fig. 37-2. The resolving power is calculated from the wavelength $\lambda$ of the incident light and the diameter $D$ of the objective lens opening; it may be expressed also as the ratio of the object separation $s_0$ to the distance $p$ from the object to the lens:

$$\theta_0 = \frac{1.22\lambda}{D} \quad \text{or} \quad \theta_0 = \frac{s_0}{p}$$

Often the resolving power is expressed simply as $s_0$, the minimum separation of two objects that can be distinguished separately by the instrument. From the figure, the value of $s_0$ also can be determined from the focal length $f$ of the objective lens and the radius $R$ of the central maximum at the point of minimum resolution:

$$s_0 = p\theta_0 \quad \text{or} \quad s_0 = \frac{pR}{f}$$

Example 37-3 of the text shows the application of these formulas.

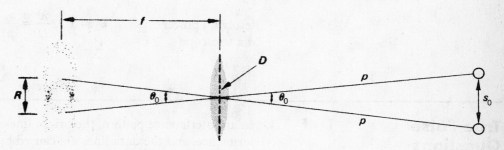

**Fig. 37-2**

---

**Summary**

- The bending of light around obstacles placed in its path is called *diffraction*.
- In Young's experiment, interference and diffraction account for the production of bright and dark fringes.
- The resolving power of an instrument is a measure of its ability to produce well-defined, separate images.

| Key Equations by Section | Section | Topic | Equation | Equation No. |
|---|---|---|---|---|
| | 37-2 | Conditions for bright fringes | $d \sin \theta = n\lambda$<br>$n = 0, 1, 2, 3, \ldots$ | 37-1 |
| | 37-2 | Conditions for dark fringes | $d \sin \theta = n\dfrac{\lambda}{2}$<br>$n = 1, 3, 5, \ldots$ | 37-2 |
| | 37-2 | Conditions for bright fringes | $\dfrac{yd}{x} = n\lambda \qquad n = 0, 1, 2, \ldots$ | 37-3 |
| | 37-2 | Conditions for dark fringes | $\dfrac{yd}{x} = n\dfrac{\lambda}{2} \qquad n = 1, 3, 5, \ldots$ | 37-4 |
| | 37-3 | Conditions for bright fringes through a diffraction grating | $d \sin \theta_n = n\lambda$<br>$n = 1, 2, 3, \ldots$ | 37-5 |
| | 37-4 | Optimum resolution given by the angle subtended by the radius of the diffraction pattern for resolution of an object through a telescope | $\theta_0 = \dfrac{1.22\lambda}{D}$ | 37-6 |
| | 37-7 | Angle, in radians, required to resolve two separate objects $s_0$ from each other and $p$ from the objective lens | $\theta_0 = \dfrac{s_0}{p}$ | 37-7 |
| | 37-7 | Resolving power of a telescope | $\theta_0 = 1.22\dfrac{\lambda}{D} = \dfrac{s_0}{p}$ | 37-8 |

## True-False Questions

**T  F  1.** In an interference pattern, the bright lines are due to constructive interference, and the dark lines are caused by destructive interference.

**T  F  2.** A diffraction grating deviates red light more than blue light.

**T  F  3.** In Young's experiment, decreasing the separation of the two slits will also decrease the separation of the interference fringes.

**T  F  4.** The resolving power of a telescope depends on the diameter of its objective lens and is not a function of the magnification.

**T  F  5.** First-order images are bright lines, and second-order images are dark lines.

T  F   **6.** The limit of resolution for two objects occurs when the central maximum of the interference pattern of one object coincides with the first dark fringe of the interference pattern from the other object.

T  F   **7.** Both transverse and longitudinal waves can be polarized by an appropriate choice of material for the polarizer.

T  F   **8.** The angular separation $\theta_0$ of two distant objects at the limit of resolution is a more accurate measure of resolving power than their linear separation $d_0$.

T  F   **9.** The greater the number of lines per inch on a diffraction grating, the greater the angle of deviation for the diffracted light.

T  F   **10.** A polarizer can be used as an analyzer in studying polarized light.

---

## Multiple-Choice Questions

**1.** The resolving power of an instrument is determined by
(a) magnification
(b) focal length of the objective lens
(c) diameter of the objective lens
(d) none of these

**2.** The main advantage of a grating over Young's apparatus is the
(a) sharpness of the bright lines
(b) absence of dark fringes
(c) absence of bright fringes
(d) greater deviation of light

**3.** Which of the following demonstrates the transverse nature of light waves?
(a) Interference          (b) Polarization
(c) Diffraction           (d) Refraction

**4.** A diffraction grating with a spacing of 15,000 lines/in. has a slit separation of
(a) $6.67 \times 10^{-6}$ in.          (b) $5.9 \times 10^{-3}$ cm
(c) $3.81 \ \mu m$                      (d) $1.69 \times 10^{-4}$ cm

**5.** In Young's experiment, the slit separation in 0.02 mm, and the screen is 1 m away. If the slit is illuminated with light of wavelength 500 nm, the second bright fringe will be displaced from the central fringe by approximately
(a) 3 cm          (b) 4 cm          (c) 5 cm          (d) 6 cm

**6.** A diffraction grating having 7000 lines/cm is illuminated by light of wavelength 589 nm. The angular separation of the second-order bright fringe is approximately
(a) 51.2°          (b) 55.5°          (c) 61.5°          (d) 65°

**7.** A parallel beam of light illuminates a diffraction grating with 15,000 lines/in. The first-order image is located 16 cm from the central image of a screen 50 cm from the grating. The wavelength of the light is approximately
(a) 515 nm          (b) 571 nm          (c) 541 nm          (d) 592 nm

**8.** A 30-in.-diameter optical telescope examines a large skylab orbiting 150 mi above the earth. The minimum separation of two points that can be resolved by the telescope, receiving light of average wavelength 500 nm, is approximately
(a) 0.56 ft          (b) 0.634 ft          (c) 0.75 ft          (d) 2.67 ft

**9.** In Question 8, the angular resolution is
(a) $8 \times 10^{-7}$ rad          (b) $7.2 \times 10^{-7}$ rad
(c) $8.1 \times 10^{-6}$ rad        (d) $7.5 \times 10^{-6}$ rad

**Completion Questions**

1. The ability of waves to bend around obstacles in their path is called _____.

2. The minimum separation of two objects that can just be distinguished as separate images by a telescope is a measure of its _____.

3. When two or more waves exist simultaneously in the same medium, the resultant _____ at any point is the sum of the _____ of the composite waves at that point.

4. Question 3 represents a statement of the _____ principle.

5. The dark lines in Young's experiment are the result of _____ interference.

6. The resolving power of an instrument for use with light of constant wavelength is determined by the _____ of the objective lens.

7. An optical device that produces a spectrum as a result of interference of light passing through thousands of parallel slits is called a(n) _____.

# Chapter 38

# Modern Physics and the Atom

**DEFINITIONS OF KEY TERMS**

**Relativistic mass**   The mass of an object that takes into account the increase in mass resulting from relative motion.

**Relativistic contraction**   The decrease in length that is due to relative motion between an observer and the object being measured.

**Time dilation**   A term that refers to the fact that a clock moving with respect to an observer ticks less rapidly, in other words, records longer time intervals, than to a person traveling with the clock.

**Einstein's first postulate**   The laws of physics are the same for all frames of reference moving with a constant velocity with respect to each other.

**Einstein's second postulate**   The free-space velocity of light is constant for all observers, independent of their state of motion.

**Work function**   The minimum energy required to eject a photoelectron from a surface.

**Line emission spectrum**   A spectrum formed by the dispersion of light emitted from an excited gas. It consists of bright lines that correspond to characteristic wavelengths.

**Bohr's first postulate**   An electron can exist only in orbits where its angular momentum is an integral multiple of $h/(2\pi)$.

**Bohr's second postulate**   If an electron changes from one stable orbit to any other, it loses energy in discrete quanta equal to the difference in energy between the initial and final states.

**Principal quantum number**   A positive integer used to describe the possible energy levels for an atomic electron.

**Energy level**   The energy state of an electron.

**Excited electron**   An electron that has a higher energy level than its normal state.

**Relativity**   According to Einstein's equations of relativity, length, mass, and time are affected by relativistic speeds. The changes become more significant as the ratio of an object's velocity $v$ to the free-space velocity of light $c$ becomes larger.

$$L = L_0\sqrt{1 - v^2/c^2} \qquad \text{relativistic contraction}$$

$$m = \frac{m_0}{\sqrt{1 - v^2/c^2}} \qquad \text{relativistic mass}$$

$$\Delta t = \frac{\Delta t_0}{\sqrt{1 - v^2/c^2}} \qquad \text{time dilation}$$

**Relativistic Energy**   The total energy of a particle of rest mass $m_0$ and speed $v$ can be written in either of the following forms:

$$E = mc^2 \qquad E = \sqrt{m_0^2 c^4 + p^2 c^2}$$

The relativistic kinetic energy is found from

$$E_K = (m - m_0)c^2 \qquad \text{relativistic kinetic energy}$$

In the above equations, $m$ is the relativistic mass, $m_0$ is the rest mass, $p$ is the momentum $mv$, and $c$ is the free-space velocity of light ($3 \times 10^8$ m/s).

---

**EXAMPLE 38-1**

The rest mass of a proton is $1.673 \times 10^{-27}$ kg. Determine the mass and the speed of protons that have a relativistic kinetic energy of 200 MeV.

*Solution*
First, by converting (1 MeV $= 1.6 \times 10^{-13}$ J), we determine the kinetic energy in joules:

$$E_K = (200 \text{ MeV})(1.6 \times 10^{-13} \text{ J/MeV}) = 3.2 \times 10^{-11} \text{ J}$$

Now, recalling the equation for relativistic kinetic energy, we may solve for $m$ in terms of $m_0$, $c$, and $E_K$:

$$E_K = (m - m_0)c^2 = mc^2 - m_0 c^2$$

From this we determine that

$$mc^2 = m_0 c^2 + E_K \qquad \text{or} \qquad m = \frac{m_0 c^2 + E_K}{c^2}$$

Thus, the relativistic mass is found by substitution:

$$m = \frac{(1.673 \times 10^{-27} \text{ kg})(3 \times 10^8 \text{ m/s})^2 + 3.2 \times 10^{-11} \text{ J}}{(3 \times 10^8 \text{ m/s})^2}$$

$$= 2.029 \times 10^{-27} \text{ kg} \qquad \text{(a 21.4 percent increase!)}$$

Now, the velocity must be found from the relativistic mass equation. If we let $\alpha = v/c$, this equation becomes

$$m = \frac{m_0}{\sqrt{1 - \alpha^2}} \quad \text{or} \quad m^2 = \frac{m_0^2}{1 - \alpha^2}$$

Solving for $\alpha$, we obtain

$$\alpha^2 = \frac{m^2 - m_0^2}{m^2} = \frac{(2.029 \times 10^{-27})^2 - (1.673 \times 10^{-27})^2}{(2.029 \times 10^{-27})^2} = 0.320$$

Now, recalling that $\alpha = v/c$ and $\alpha = \sqrt{0.320}$, we solve for $v$:

$$v = c\sqrt{0.320} = \boxed{1.70 \times 10^8 \text{ m/s}}$$

## Photoelectric Effect

According to modern quantum theory, electromagnetic waves consist of bundles of energy called quanta, each having an energy proportional to its frequency:

$$E = hf \qquad E = \frac{hc}{\lambda} \qquad h = 6.63 \times 10^{-24}\,\text{J} \cdot \text{s}$$

In the photoelectric effect, the kinetic energy of the ejected electrons is the energy of the incident radiation $hf$ less the *work function* $W$ of the surface:

$$E_K = \tfrac{1}{2}mv^2 = hf - W$$

The lowest frequency $f_0$ at which a photoelectron is ejected is the threshold frequency. It corresponds to the work-function energy $W$. The following relations apply:

$$f_0 = \frac{W}{h} \qquad W = hf_0 \qquad \text{threshold frequency}$$

**EXAMPLE 38-2**

When a beam of electromagnetic radiation strikes a metal surface, photoelectrons are ejected form the surface with a kinetic energy of $4 \times 10^{-19}$ J. If the threshold frequency for this metal is $4.3 \times 10^{14}$ J, what is the wavelength of the incident radiation?

*Solution*

First, we calculate the work function $W$ from the threshold frequency:

$$W = hf_0 = (6.63 \times 10^{-34}\,\text{J} \cdot \text{s})(4.3 \times 10^{14}\,\text{Hz}) = 2.85 \times 10^{-19}\,\text{J}$$

Now, the energy $hf$ of the incident radiation must be equal to the work function $W$ plus the kinetic energy of the ejected electrons:

$$hf = E_K + W = 4 \times 10^{-19}\,\text{J} + 2.85 \times 10^{-19}\,\text{J} = 6.85 \times 10^{-19}\,\text{J}$$

Recalling that $hf = hc/\lambda$, we may now write

$$hf = \frac{hc}{\lambda} = 6.85 \times 10^{-19}\,\text{J}$$

Solving for $\lambda$, we obtain

$$\lambda = \frac{(6.63 \times 10^{-34}\,\text{J} \cdot \text{s})(3 \times 10^8\,\text{m/s})}{6.85 \times 10^{-19}\,\text{J}} = \boxed{290 \text{ nm}}$$

## de Broglie Wavelength

By combining wave theory with particle theory, de Broglie was able to give the following equation for the wavelength of any particle whose mass $m$ and velocity $v$ are known:

$$\lambda = \frac{h}{mv} \qquad h = 6.63 \times 10^{-34} \, \text{J} \cdot \text{s}$$

## Bohr Atom and Modern Atomic Theory

The Bohr model of the atom led to two important physical postulates. Bohr's first postulate states that the angular momentum of an electron in any orbit must be a multiple of $h/2\pi$. His second postulate states that the energy absorbed or emitted by an atom is, in discrete amounts, equal to the difference in energy levels of an electron. These concepts are given as equations:

$$mvr = \frac{nh}{2\pi} \qquad hf = E_i - E_f \qquad \textbf{Bohr's postulates}$$

The subscripts $f$ and $i$ refer to final and initial energy levels.

Absorption and emission spectra for gases verify the discrete nature of radiation. The wavelength $\lambda$ or frequency $f$ that corresponds to a change in electron energy levels is given by

$$\frac{1}{\lambda} = R\left(\frac{1}{n_f^2} - \frac{1}{n_i^2}\right) \qquad f = Rc\left(\frac{1}{n_f^2} - \frac{1}{n_i^2}\right)$$

$$R = \frac{me^4}{8e_0^2 h^3 c} = 1.097 \times 10^7 \, \text{m}^{-1} \qquad \texttext{Rydberg's constant}$$

---

**EXAMPLE 38-3**

Determine the wavelength of the radiation emitted from a hydrogen atom when an electron drops from the fourth energy level to the ground state.

**Solution**
We apply Balmer's equation for $n_f = 4$ and $n_i = 1$:

$$\frac{1}{\lambda} = (1.097 \times 10^7 \, \text{m}^{-1})\left(\frac{1}{1^2} - \frac{1}{4^2}\right) = 1.028 \times 10^7 \, \text{m}^{-1}$$

from which

$$\lambda = 9.72 \times 10^{-8} \, \text{m} = \boxed{\textbf{97.2 nm}}$$

---

## Energy Levels

The total energy of a particular quantum state $n$ for the hydrogen atom is given by

$$E_n = -\frac{me^4}{8\epsilon_0^2 n^2 h^2} \qquad \text{or} \qquad E_n = -\frac{13.6 \, \text{eV}}{n^2}$$

where $\epsilon_0 = 8.85 \times 10^{-12} \, \text{C}^2/(\text{N} \cdot \text{m}^2)$

$e = 1.6 \times 10^{-19} \, \text{C}$

$m_e = 9.1 \times 10^{-31} \, \text{kg}$

$h = 6.63 \times 10^{-34} \, \text{J} \cdot \text{s}$

## Summary

- According to Einstein's equations of relativity, length, mass, and time are affected by relativistic speeds.
- The quantum theory of electromagnetic radiation relates radiation energy to its frequency $f$ or wavelength $\lambda$.
- In the photoelectric effect, the kinetic energy of the ejected electrons is the energy of the incident radiation $hf$ minus the work function of the surface $W$.
- The lowest frequency $f_0$ at which a photoelectron can be ejected is the threshold frequency.
- de Broglie combined wave theory with particle theory and gave an equation for the wavelength of any particle whose mass and velocity are known.
- Bohr's first postulate states that the angular momentum of an electron in any orbit must be a multiple of $h/2\pi$.
- Bohr's second postulate states that the energy absorbed or emitted by an atom is in discrete amounts equal to the difference in energy levels of an electron.
- Absorption and emission spectra for gases verify the discrete nature of radiation.

## Key Equations by Section

| Section | Topic | Equation | Equation No. |
|---|---|---|---|
| 38-3 | Relativistic contraction of length of objects traveling at speed $v$ close to light speed | $L = L_0 \sqrt{1 - \dfrac{v^2}{c^2}}$ | 38-1 |
| 38-3 | Time dilation, the time interval experienced by objects traveling at speed $v$ close to light speed | $\Delta t = \dfrac{\Delta t_0}{\sqrt{1 - v^2/c^2}}$ | 38-2 |
| 38-3 | Relativistic mass, the mass experienced by objects traveling at speed $v$ close to light speed | $m = \dfrac{m_0}{\sqrt{1 - v^2/c^2}}$ | 38-3 |
| 38-4 | Total energy of a particle of rest mass $m_0$ and momentum $p = mv$ | $E = \sqrt{(m_0 c^2)^2 + p^2 c^2}$ | 38-5 |
| 38-4 | Einstein's equivalence between relativistic mass and energy | $E = mc^2$ | 38-6 |
| 38-4 | Energy equation for momentum and rest energy | $E = \frac{1}{2}m_0 v^2 + m_0 c^2$ | 38-7 |

| Section | Topic | Equation | Equation No. |
|---|---|---|---|
| 38-4 | Relativistic kinetic energy of a particle | $E_K = (m - m_0)c^2$ | 38-8 |
| 38-5 | Planck's equation | $E = hf$ | 38-9 |
| 38-5 | Einstein's photoelectric equation; the kinetic energy of a photon striking a metallic surface that causes an electron to leave the surface is a maximum of the wavelength of the ejected electron ($hf$) minus the work function $W$ | $E_K = \frac{1}{2}mv^2{}_{max} = hf - W$ | 38-10 |
| 38-5 | The lowest frequency of light that can eject an electron from a metallic surface | $f_0 = \frac{W}{h}$ | 38-11 |
| 38-6 | The energy of a photon | $E = \sqrt{p^2 c^2} = pc$ | 38-12 |
| 38-6 | The wavelength of a photon | $\lambda = \frac{h}{p}$ | 38-13 |
| 38-6 | de Broglie wavelength, the wavelength of any particle | $\lambda = \frac{h}{mv}$ | 38-14 |
| 38-8 | The electrostatic force of attraction on an electron by a proton | $F_e = \frac{e^2}{4\pi\epsilon_0 r^2}$ | 38-15 |
| 38-8 | Centripetal force of an electron | $F_c = \frac{mv^2}{r}$ | 38-16 |
| 38-8 | Orbital radius $r$ of an electron as a function of its speed $v$ | $r = \frac{e^2}{4\pi\epsilon_0 mv^2}$ | 38-18 |
| 38-9 | Balmer's formula for predicting the wavelengths of the hydrogen atomic spectrum | $\frac{1}{\lambda} = R\left(\frac{1}{2^2} - \frac{1}{n^2}\right)$ | 38-19 |

| Section | Topic | Equation | Equation No. |
|---|---|---|---|
| 38-9 | General formula for predicting the wavelengths of the hydrogen atomic spectra for ultraviolet and infrared | $\dfrac{1}{\lambda} = R\left(\dfrac{1}{l^2} - \dfrac{1}{n^2}\right)$ | 38-20 |
| 38-10 | Bohr's orbital wavelengths | $n\lambda = 2\pi r \qquad n = 1, 2, 3, \ldots$ | 38-21 |
| 38-10 | Bohr's angular momentum | $mvr = \dfrac{nh}{2\pi}$ | 38-22 |
| 38-10 | Radius of an electron from the center of the nucleus | $r = n^2\dfrac{\epsilon_0 h^2}{\pi m e^2}$ | 38-23 |
| 38-10 | Velocity of an electron | $v = \dfrac{e^2}{2\epsilon_0 nh}$ | 38-24 |
| 38-10 | Kinetic energy of an electron | $E_K = \tfrac{1}{2}mv^2 = \dfrac{me^4}{8\epsilon_0^2 n^2 h^2}$ | 38-25 |
| 38-10 | Potential energy for any electron orbit | $E_p = \dfrac{1}{4\pi\epsilon_0}\dfrac{e^2}{r} = -\dfrac{me^4}{4\epsilon_0^2 n^2 h^2}$ | 38-26 |
| 38-10 | Total energy of an electron in an orbit | $E_T = -\dfrac{me^4}{8\epsilon_0^2 n^2 h^2}$ | 38-27 |
| 38-11 | The total energy of an electron at the $n$th level where $n = 1$ is the ground state | $E_n = -\dfrac{me^4}{8\epsilon_0^2 n^2 h^2}$ | 38-30 |
| 38-11 | Total energy of a ground-state electron in an inner orbit | $E_n = \dfrac{-13.6 \text{ eV}}{n^2}$ | 38-31 |

## True-False Questions

**T  F  1.** When two rocket ships $A$ and $B$ move toward each other at uniform speed, it is not possible for the astronauts on either ship to determine whether ship $A$ is moving and ship $B$ is at rest, ship $B$ is moving and ship $A$ is at rest, or both ships are moving.

**T  F  2.** Einstein's second postulate tells us that the velocity of light is always constant.

T  F  **3.** The kinetic energy of a particle traveling at relativistic speeds is equal to the difference between its total energy and its rest mass energy.

T  F  **4.** In the photoelectric effect, a surface with a large work function is likely to produce photoelectrons with higher kinetic energy.

T  F  **5.** de Broglie wavelengths can be calculated only for charged particles such as protons and electrons.

T  F  **6.** The wavelengths from an emission spectrum of an element should not be expected to coincide with the wavelengths determined from an absorption spectrum of that element.

T  F  **7.** The Balmer spectral series in an emission spectrum results from electrons in higher energy levels dropping to the ground state for the hydrogen atom.

T  F  **8.** The stable electron orbits are those that contain an integral number of de Broglie wavelengths.

T  F  **9.** The energy of an electron in the first excited states is 4 times its energy in the ground state.

T  F  **10.** Rutherford's work with the scattering of alpha particles resulted in a nuclear theory of matter.

---

## Multiple-Choice Questions

**1.** According to the modern theory of relativity, newtonian mechanics is
(a) totally incorrect
(b) approximately correct for any velocity
(c) approximately correct for speeds much less than the speed of light
(d) correct for all velocities

**2.** A clock of mass $m$ and length $L$ records a time interval $\Delta t$. If this clock moves with a velocity of $0.6c$, which of the following statements is *not* true?
(a) The mass will be larger.
(b) The length will be shorter.
(c) The time interval will be longer.
(d) The time interval will be shorter.

**3.** The rest mass of a proton is $1.673 \times 10^{-27}$ kg. The relativistic mass of a proton when its velocity is $0.6c$ is approximately
(a) $1.3 \times 10^{-27}$ kg         (b) $2.1 \times 10^{-27}$ kg
(c) $2.6 \times 10^{-27}$ kg         (d) $8.4 \times 10^{-27}$ kg

**4.** What is the relativistic mass of an electron whose kinetic energy is 1.0 MeV?
(a) $35.6 \times 10^{-31}$ kg        (b) $29.2 \times 10^{-31}$ kg
(c) $26.9 \times 10^{-31}$ kg        (d) $22.5 \times 10^{-31}$ kg

**5.** According to the Bohr theory of an atom, an electron may circle the nucleus indefinitely without radiating energy if
(a) the radius of its orbit is an integral multiple of the nuclear radius
(b) its orbital path is an integral number of de Broglie wavelengths
(c) its orbit is an integral multiple of its angular momentum
(d) the coulomb force is constant

**6.** What is the de Broglie wavelength of an electron when it is accelerated through a potential difference of 200 V?
(a) 86.8 pm      (b) 62.2 pm      (c) 23.6 pm      (d) 6.92 pm

7. For the hydrogen atom, which of the following energy-level transitions will result in the emission of a photon with the greatest frequency?
   (a) From $n = 2$ to $n = 1$?
   (b) From $n = 1$ to $n = 4$
   (c) From $n = 4$ to $n = 2$
   (d) From $n = 3$ to $n = 1$

8. The threshold frequency for a surface is known to be $5 \times 10^{14}$ Hz. What is the wavelength of light required to eject a photoelectron having a kinetic energy of 5 eV?
   (a) 120 nm
   (b) 176 nm
   (c) 211 nm
   (d) 306 nm

9. The energy of the photon emitted when an electron in the hydrogen atom drops from the $n = 3$ level to the $n = 2$ level is
   (a) 10.2 eV
   (b) 6.8 eV
   (c) 1.89 eV
   (d) 1.51 eV

---

**Completion Questions**

1. For objects traveling past an observer at relativistic speeds, the measurements of length are _____, the measurements of mass are _____, and the time intervals on the object are observed to be _____ than when at rest.

2. The _____ is constant for all observers independent of their state of motion.

3. The maximum kinetic energy of an ejected photoelectron is equal to the difference between the _____ and _____.

4. The lowest energy level in an atom is known as its _____ state.

5. Electron jumps from higher quantum levels back to the first or lowest quantum level produce the _____ series for the hydrogen atom. Jumps back to the second level produce the _____ series.

6. If light from an incandescent platinum wire passes through sodium vapor before reaching the slit of a spectroscope, the resulting spectrum lacks the wavelength's characteristic of _____. Such a spectrum is known as a(n) _____ spectrum.

7. The _____ is calculated by dividing Planck's constant by the product of the mass and velocity of a particle.

8. According to _____ first postulate, an electron may occupy only those orbits for which its angular momentum is an integral multiple of _____.

9. The minimum energy required to eject a photoelectron from a surface is known as the _____.

# Chapter 39

# Nuclear Physics and the Nucleus

**DEFINITIONS OF KEY TERMS**

**Nuclear force** The fundamental, natural force that holds the nuclear particles together in the nucleus of an atom.

**Nucleon** A general term applied to neutrons and protons, the constituent particles of an atomic nucleus.

**Atomic number** A number used to identify elements. It is equal to the number of protons in the nucleus and is represented by the symbol $Z$.

**Mass number** The total number of nucleons in an atomic nucleus. The mass number is the sum of the atomic number and the number of neutrons ($A = Z + N$).

**Atomic mass unit** A unit of mass that is equal to one-twelfth the mass of the most abundant form of the carbon atom. Its value is equal to $1.6606 \times 10^{-27}$ kg.

**Isotopes** Atoms that have the same atomic number $Z$ but different mass numbers $A$, for example, $^{12}_{6}C$ and $^{13}_{6}C$.

**Mass spectrometer** An instrument used to separate the isotopes of elements based on their differences in mass.

**Mass defect** The difference between the rest mass of a nucleus and the sum of the rest masses of its constituent nucleons.

**Binding energy** The energy required to separate a nucleus into its constituent nucleons. It is an energy equivalent to the mass defect of an atom.

**Alpha particle** The nucleus of a helium atom, which consists of two neutrons and two protons bound together by nuclear forces.

**Beta particle** A beta-minus particle is simply an electron. The beta-plus particle has a mass equivalent to the electron, but its charge is equal and opposite to that of the electron.

**Activity** The rate of disintegration of an unstable isotope. An activity of 1 Ci is equivalent to $3.7 \times 10^{10}$ disintegrations per second.

**Half-life** The time in which one-half the unstable nuclei of a radioactive sample will decay.

**Nuclear reaction**    A process by which nuclei, radiation, and/or nucleons collide to form different nuclei, radiation, and/or nucleons.

**Nuclear fission**    The process by which heavy nuclei are split into two or more nuclei of intermediate mass numbers.

**Nuclear fusion**    The process by which two lighter nuclei join to form a heavier nucleus.

**Chain reaction**    The process of one fission reaction emitting radiation that causes another fission or a fusion reaction that causes further nuclear reactions, and so on.

**Curie**    The basic unit of radiation. The activity of a radioactive material that decays at the rate of $3.7 \times 10^{10}$ disintegrations per second.

**Gamma rays**    High-energy electromagnetic waves, the most penetrating radiation emitted by radioactivity.

**Moderator**    A material that slows down high-speed neutron particles until they are slow enough to produce a nuclear reaction.

**Radioactivity**    The property of radioactive elements to emit alpha, beta, or gamma radiation as a natural atomic decay process.

**CONCEPTS AND EXAMPLES**

**Nuclear Particles**    The fundamental nuclear particles are summarized in the following table. The masses are given in atomic mass units, and the charge is in terms of the electronic charge $+e$ or $-e$, which is $1.6 \times 10^{-19}$ C. The average atomic masses of the elements are given in the text.

## Fundamental Particles

| Particle | Symbol | Mass, u | Charge |
|---|---|---|---|
| Electron/positron | $_{-1}^{0}e$, $_{-1}^{0}\beta$/$_{+1}^{0}e$, $_{+1}^{0}\beta$ | 0.00055 | $-e$/$+e$ |
| Proton | $_{1}^{1}p$, $_{1}^{1}H$ | 1.007276 | $+e$ |
| Neutron | $_{0}^{1}n$ | 1.008665 | 0 |
| Alpha particle | $_{2}^{4}\alpha$, $_{2}^{4}$He | 4.001506 | $+2e$ |

**Atomic Number and Atomic Mass**    The atomic number $Z$ of an element is the number of protons in its nucleus. The mass number $A$ is the sum of the atomic number $Z$ and the number of neutrons $N$. These numbers are used to write the nuclear symbol:

$$A = Z + N \qquad \text{symbol: } _{Z}^{A}X$$

For example, carbon-14 may be written as $_{6}^{14}$C. From the symbol it is clear that the atomic number is 6, the mass number is 14, and the number of neutrons is 8.

One *atomic mass unit* (u) is equal to one-twelfth the mass of the most abundant carbon atom, $_{6}^{12}$C. Its value in kilograms is given below. Also, since $E = mc^2$, we can write the conversion factor from mass to energy as $c^2$.

$$1 \text{ u} = 1.6606 \times 10^{-27} \text{ kg} \qquad c^2 = 931 \text{ MeV/u}$$

The following conversion factors are also useful in describing atomic energy:

$$1 \text{ MeV} = 1 \times 10^6 \text{ eV} \qquad 1 \text{ MeV} = 1.6 \times 10^{-13} \text{ J}$$

**Isotopes** It is possible for two different atoms of the same element to have nuclei containing different numbers of neutrons and, therefore, different atomic masses. Such atoms are called *isotopes*. Since the period table gives atomic masses of elements as the average of the naturally occurring isotopes, these values generally are not used in nuclear reactions involving particular isotopes. The following table provides the atomic masses of several isotopes. The value in the table is essentially the mass of a particular nucleus plus the mass of Z electrons.

| Nucleus | Atomic mass (u) | Nucleus | Atomic mass (u) |
|---|---|---|---|
| $^{1}_{1}H$ | 1.0078252 | $^{12}_{6}C$ | 12.000000 |
| $^{2}_{1}H$ | 2.014102 | $^{14}_{7}N$ | 14.003074 |
| $^{3}_{1}H$ | 3.016049 | $^{16}_{8}O$ | 15.994915 |
| $^{3}_{2}He$ | 3.016030 | $^{19}_{9}F$ | 18.998405 |
| $^{4}_{2}He$ | 4.002603 | $^{24}_{12}Mg$ | 23.985045 |
| $^{6}_{3}Li$ | 6.015126 | $^{28}_{14}Si$ | 27.976927 |
| $^{7}_{3}Li$ | 7.016930 | $^{56}_{26}Fe$ | 55.934932 |
| $^{9}_{4}Be$ | 9.012186 | $^{127}_{53}I$ | 126.90435 |
| $^{10}_{5}B$ | 10.012939 | $^{235}_{92}U$ | 235.04393 |
| $^{11}_{5}B$ | 11.009305 | $^{238}_{92}U$ | 238.050786 |

The mass spectrometer is often used to separate and determine the masses of specific isotopes of elements. The velocity $v$ and the radius $R$ of the singly ionized particles in a mass spectrometer experiment are

$$V = \frac{E}{B} \qquad R = \frac{mv}{eB} \qquad \textbf{mass spectrometer}$$

## Mass Defect and Binding Energy

The *mass defect* is the difference between the rest mass of a nucleus and the sum of the rest masses of its nucleons. The *binding energy* is obtained by multiplying the mass defect by $c^2$:

$$E_B = (Zm_H + Nm_n - M)c^2 \qquad \textbf{binding energy}$$

where $m_H = 1.0078252$ u

$m_n = 1.0086654$ u

$c^2 = 931$ MeV/u

$M =$ atomic mass (u)

$N =$ number of neutrons $(A - Z)$

$Z =$ atomic number

To understand this formula, we reason as follows. Since a hydrogen atom contains a proton and an electron and since $N$ is the number of neutrons, the total mass of the individual parts of an atom is given by $Zm_H + Nm_n$. Subtracting the atomic mass $M$ of the isotope, we obtain the mass defect that can then be converted to the binding energy. Remember that $c^2 = 931$ MeV/u.

**EXAMPLE 39-1** Determine the mass defect and binding energy of the isotope boron-11 ($^{11}_5$B).

*Solution*

We recognize that $Z = 5$, $N = 11 - 5 = 6$, and $M = 11.009305$ from the table of isotopes. (Do not take the mass $M$ from the periodic table.) The mass defect is

$$m_D = Zm_H + Nm_n - M$$

$$= 5(1.0078252) + 6(1.0086654) - 11.009305$$

$$= 11.091118 - 11.009305 = 0.0818134 \text{ u}$$

Now the binding energy is found by multiplying by $c^2$:

$$E_B = m_D c^2 = (0.0818134 \text{ u})(931 \text{ MeV/u})$$

$$= \boxed{76.17 \text{ MeV}}$$

**Radioactive Decay** Several general equations are useful for the description of radioactive decay:

$$^A_Z X \rightarrow ^{A-4}_{Z-2} Y + ^4_2 \alpha + \textbf{energy} \qquad \textbf{alpha decay}$$

$$^A_Z X \rightarrow ^A_{Z+1} Y + ^0_{-1} \beta + \textbf{energy} \qquad \textbf{beta-minus decay}$$

$$^A_Z X \rightarrow ^A_{Z-1} Y + ^0_{+1} \beta + \textbf{energy} \qquad \textbf{beta-plus decay}$$

The activity of $R$ of a sample is the rate at which the radioactive nuclei decay. It is generally expressed in curies (Ci):

$$1 \text{ Ci} = 3.7 \times 10^{10} \text{ disintegrations per second}$$

The *half-life* of a sample is the time $T$ in which one-half the unstable nuclei will decay. The number of radioactive nuclei remaining after a time $t$ depends on the number $n$ of half-lives that have passed. If $N_0$ nuclei exist at time $t = 0$, then a number $N$ exists at time $t$. We have

$$N = N_0(\tfrac{1}{2})^n \qquad \textbf{where} \qquad n = \frac{t}{T_{1/2}}$$

Similar relations exist for the activity $R$ and mass $m$ of a radioactive sample remaining after $n$ half-lives:

$$R = R_0(\tfrac{1}{2})^n \qquad m = m_0(\tfrac{1}{2})^n$$

**Summary**

- Protons and neutrons are held together in the nucleus by strong nuclear forces that are active only within the nucleus.
- For massive nuclei energy results from tearing these nuclei apart.
- The atomic number $Z$ of an element is the number of protons in its nucleus.

- One atomic mass unit (1 u) is equal to one-twelfth the mass of the most abundant carbon isotope.
- The *mass defect* is the difference between the rest mass of a nucleus and the sum of the rest masses of its nucleons.
- The activity $R$ of a sample is the rate at which the radioactive nuclei decay.
- The half-life of a sample is the time $T_{1/2}$ in which one-half the unstable nuclei will decay.

## Key Equations by Section

| Section | Topic | Equation | Equation No. |
|---|---|---|---|
| 39-2 | Calculation of mass, number from atomic number $Z$ and the number of neutrons $N$ | $A = Z + N$ | 39-1 |
| 39-3 | Definition of the atomic mass unit u | $1\ u = 1.6606 \times 10^{-27}\ kg$ | 39-2 |
| 39-3 | Conversion factor from mass units u to energy units MeV based on the speed of light squared | $c^2 = 931\ MeV/u$ | 39-3 |
| 39-4 | Velocity at which an isotope will be separated from differing isotopes based on the strength of the electric $E$ and magnetic $B$ fields | $v = \dfrac{E}{B}$ | 39-4 |
| 39-4 | The radius of a semicircular ion path used to determine the mass of the isotopes | $R = \dfrac{mv}{eB}$ | 39-5 |
| 39-5 | The masses of the hydrogen atom and the neutron | $m_H = 1.007825\ u$ <br> $m_n = 1.008665\ u$ | 39-6 |
| 39-5 | Binding energy | $E_B = [(Zm_H + Nm_n) - M]c^2$ | 39-7 |
| 39-7 | Radioactive decay of an alpha particle | ${}^{A}_{Z}X \rightarrow {}^{A-4}_{Z-2}Y + {}^{4}_{2}\alpha + energy$ | 39-8 |

| Section | Topic | Equation | Equation No. |
|---|---|---|---|
| 39-7 | Radioactive decay of a negative beta particle | $^A_ZX \rightarrow {^A_{Z+1}}Y + {^{\ 0}_{-1}}\beta + \text{energy}$ | 39-9 |
| 39-7 | Radioactive decay of a positive beta particle | $^A_ZX \rightarrow {^A_{Z-1}}Y + {^{\ 0}_{+1}}\beta + \text{energy}$ | 39-10 |
| 39-8 | The radioactive activity $R$ based on the number of undecayed nuclei $N$ and time | $R = \dfrac{-\Delta N}{\Delta t}$ | 39-11 |
| 39-8 | Definition of the Curie (Ci), the basic unit of radiation | $1 \text{ Ci} = 3.7 \times 10^{10} \text{ s}^{-1}$ | 39-12 |
| 39-8 | Number of nuclei present at any time $t$ based on the number originally present and the number $n$ of half-lives | $N = N_0(\frac{1}{2})^n$ | 39-13 |
| 39-8 | Activity of a radioactive sample at a time $t$ | $R = R_0(\frac{1}{2})^{t/T_{1/2}}$ | 39-14 |

## True-False Questions

T  F  **1.** The diameter of an atom is approximately 10,000 times the diameter of its nucleus.

T  F  **2.** The difference between the mass number of an isotope and its atomic number is equal to the number of nucleons in the nucleus.

T  F  **3.** An element may have more than one mass number, but the mass number for a stable isotope is fixed.

T  F  **4.** The radioactive half-life of a substance is one-half of the time required for all the unstable atoms in that substance to decay.

T  F  **5.** In alpha decay, the mass number of an unstable isotope is reduced by 4 and the atomic number is reduced by 2.

T  F  **6.** When $^{27}_{13}\text{Al}$ is bombarded by a neutron, the collision produces $^{27}_{12}\text{Mg}$ and a beta-plus particle.

T  F  **7.** The binding energy of an element is equivalent to the product of the mass defect and the square of the velocity of light.

T  F  **8.** In a balanced nuclear equation, the sum of the atomic numbers and the sum of the mass numbers must be the same on both sides of the nuclear equation.

**T   F   9.** In nuclear fission energy is emitted, whereas in nuclear fusion energy is absorbed.

**T   F   10.** The function of the moderator in a nuclear reactor is to slow down the nuclear fission process and thereby control the release of energy.

## Multiple-Choice Questions

**1.** Isotopes are atoms that have the same
    (a) number of neutrons        (b) atomic number
    (c) number of nucleons       (d) atomic mass

**2.** The process by which a nucleus with a large mass number splits into lighter nuclei is called
    (a) fission         (b) fusion        (c) alpha decay      (d) beta decay

**3.** In the mass spectrometer, the distance from the slit to the impact on the plate is 10 cm for isotope $A$ and 12 cm for isotope $B$. The ratio of their masses $M_A/M_B$ is approximately
    (a) 0.83         (b) 0.93         (c) 1.2          (d) 1.33

**4.** If two light nuclei are fused in a nuclear reaction, the average energy per nucleon
    (a) increases                (b) remains the same
    (c) decreases              (d) cannot be determined

**5.** A sample of radioactive material contains $N$ radioactive nuclei at a given instant. If the half-life is 20 s, how many unstable nuclei remain after 1 h?
    (a) $N/2$         (b) $N/4$         (c) $N/6$        (d) $N/8$

**6.** Which of the following nuclear reactions is possible?
    (a) $^{1}_{1}H + ^{3}_{2}He \rightarrow ^{4}_{2}He$        (b) $^{4}_{2}He + ^{27}_{13}Al \rightarrow ^{30}_{15}P + ^{1}_{0}n$
    (c) $^{2}_{1}H + ^{31}_{15}P \rightarrow ^{29}_{13}Al + ^{4}_{2}He$     (d) $^{224}_{88}Ra \rightarrow ^{219}_{86}Rn + ^{4}_{2}He$

**7.** When a beta-plus particle encounters an electron, they cancel each other and both disappear. This *annihilation* of matter produces two photons, each having an energy of approximately
    (a) 0.466 MeV     (b) 0.511 MeV     (c) 0.931 MeV     (d) 1.02 MeV

**8.** A deuteron is a particle consisting of a neutron and a proton bound together by nuclear forces. If the rest mass of a deuteron is $3.34313 \times 10^{-27}$ kg, the binding energy is approximately
    (a) 2.22 MeV     (b) 3.11 MeV     (c) 4.44 MeV     (d) 6.22 MeV

**9.** In a nuclear reactor, which of the following is used to slow down the fast neutrons released in the fission process?
    (a) Moderator             (b) Control rods
    (c) Radiation shielding     (d) Heat exchanger

**10.** The binding energy per nucleon for $^{235}_{92}U$, whose mass is 235.043925 u, is approximately
    (a) 7.11 MeV     (b) 6.40 MeV     (c) 7.59 MeV     (d) 7.92 MeV

## Completion Questions

1. In general, the _____ is defined as the energy required to break up a nucleus into its constituent protons and neutrons.

2. When a proton bombards $^{27}_{13}$Al in a nuclear reaction, the unstable nucleus _____ is formed, which results in the production of _____ and an alpha particle.

3. The half-life for alpha decay for deuterium is 10.2 s. One-fourth of the unstable atoms will remain after _____ s.

4. An activity of _____ is equal to $3.7 \times 10^{10}$ disintegrations per second.

5. The _____ is the total number of nucleons in the nucleus. It can be computed by adding the number of _____ to the number of _____.

6. Basic components of a nuclear reactor are _____, _____, _____, _____, and _____.

7. In a nuclear fission, the fission fragments have a(n) _____ mass number and hence a(n) _____ binding energy per nucleon.

8. Isotopes are atoms that have the same _____ but different _____.

9. A mass defect of 1 u is equivalent to an energy of _____.

10. Three quantities that must be conserved in a nuclear reaction are _____, _____, and _____.

# Chapter 40

# Electronics

## DEFINITIONS OF KEY TERMS

**Thermionic emission**   The emission of electrons from a solid body as a result of elevated temperature.

**Diode**   An electronic device consisting of two elements. In a vacuum tube, these elements are the plate and the filament; in the semiconductor diode, the two elements are formed by a PN junction.

**Triode**   An electronic device that consists of three elements. In a vacuum tube triode, the elements are the plate, the filament, and the grid; the semiconductor triode may be an NPN or a PNP transistor.

**Cathode-ray tube**   An evacuated tube in which an electron beam is focused to a small cross-section on a luminescent screen and can be varied in intensity and position to produce a pattern.

**X-ray tube**   A vacuum tube in which high-energy electromagnetic rays (X rays) are produced by bombarding a target with high-velocity electrons.

**Semiconductor**   A solid or liquid material with a resistivity between that of a conductor and an insulator.

**Conduction band**   A partially filled energy band in which electrons can move freely under the influence of a potential difference.

**Donor**   An impurity atom that tends to give up an electron to a material, enhancing its semiconductor properties. Donors are used in producing N-type semiconductors.

**Acceptor**   An impurity atom that is deficient in valence electrons. When it is added to a semiconductor crystal, it accepts an electron from a neighboring atom, leaving an electron hole in the lattice structure. Acceptors are used in producing P-type semiconductors.

**N-type semiconductor**   A semiconductor material consisting of donor impurities and free electrons (negative).

**P-type semiconductor**  A semiconductor material consisting of acceptor atoms and electron holes (positive).

**PN junction**  The union of a P-type crystal with an N-type crystal in such a way that current is conducted in only one direction. It thus forms a semiconductor diode.

**Rectifier**  An electronic device that has the property of converting alternating current to direct current.

**Zener diode**  A semiconductor device that exhibits a sudden rise in current if a certain reverse voltage is applied. If the diode is forward-biased, it acts as an ordinary rectifier. When reverse-biased, however, it exhibits a sharp break in its voltage-current graph.

**Transistor**  A semiconductor device consisting of three or more electrodes, usually the emitter, the collector, and the base.

**Integrated circuit**  A combination of interconnected circuit elements inseparably associated on or within an underlying material called a substrate.

**CONCEPTS AND EXAMPLES**

**General**  Most of the material in this chapter is presented for introductory and informational purposes. A careful reading and close attention to definitions are important, but the only problem-solving approaches discussed are relative to transistors and their use in simple circuits. A summary of the basic equations used in this regard is given.

**Transistor Applications**  For a common-base transistor amplifier, the emitter current is equal to the base current plus the collector current:

$$I_e = I_b + I_c \qquad \alpha = \frac{I_c}{I_e} \qquad \text{current gain}$$

The current gain for other connections is

$$\beta = \frac{\alpha}{1 - \alpha} \qquad \text{common-emitter amplifier}$$

$$A_i = \frac{1}{1 - \alpha} \qquad \text{common-collector amplifier}$$

The voltage gain $A_v$ and the power gain $G$ are given by

$$A_v = \frac{V_{\text{out}}}{V_{\text{in}}} \qquad \text{voltage gain}$$

$$G = \frac{\text{power out}}{\text{power in}} = \frac{V_{\text{out}}I_c}{V_{\text{in}}I_e}$$

$$G = \alpha \frac{V_{\text{out}}}{V_{\text{in}}} = \alpha A_v \qquad \text{power gain}$$

**EXAMPLE 40-1**    A transistor in a common-base amplifier circuit (Fig. 40-1) has an input resistance of 400 Ω and an output resistance of 320 kΩ. The current gain $\alpha$ is 0.97 when the emitter current is 2.5 mA. What are the input and output voltages? What are the voltage amplification and the power gain?

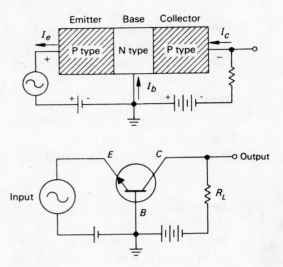

**Fig. 40-1** Common-base amplifier circuit.

**Solution**
First, we determine the collector current from the fact that $\alpha = I_c/I_e$. Solving for $I_c$, we obtain

$$I_c = \alpha I_e = 0.97(2.5 \text{ mA}) = 2.425 \text{ mA}$$

Now, the input and output voltages are found from Ohm's law:

$$V_{\text{in}} = I_e R_{\text{in}} = (2.5 \times 10^{-3} \text{ A})(400 \text{ }\Omega) = \boxed{1.0 \text{ V}}$$

$$V_{\text{out}} = I_c R_{\text{out}} = (2.425 \times 10^{-3} \text{ A})(320,000 \text{ }\Omega) = \boxed{776 \text{ V}}$$

The voltage amplification is $V_{\text{out}}/V_{\text{in}}$, or

$$A_v = \frac{776 \text{ V}}{1.0 \text{ V}} = \boxed{776}$$

Note that the current changes little ($\alpha = 0.97$), whereas the voltage gain is 776. The gain is due to the output resistance, which can be made quite large, resulting in considerable amplification.

The power gain depends on both the voltage and the current, and it is equal to the product of current gain and voltage gain ($\alpha A_v$):

$$G = \alpha A_v = 0.97(776) = \boxed{753}$$

Thus, the power gain is only slightly less than the voltage gain—again because of little difference between the input and output currents.

## Summary

- A *diode* is an electronic device that conducts electricity in only one direction.
- In a vacuum tube a diode is made of two elements, the plate and the filament.
- A semiconductor diode is formed by a PN junction.
- A *semiconductor* is a solid or liquid material with a resistivity between that of a conductor and an insulator.
- An N-type semiconductor contains *donor* impurities and free electrons.
- A P-type semiconductor consists of *acceptor* atoms and electron holes.
- A PN *junction* is the union of a P-type crystal with an N-type crystal in such a way that current is conducted in only one direction.
- A *transistor* is a semiconductor device consisting of three or more electrodes, usually the emitter, the collector, and the base.

## Key Equations by Section

| Section | Topic | Equation | Equation No. |
|---------|-------|----------|--------------|
| 40-1 | The current in a transistor | $I_e = I_b + I_c$ | 40-1 |
| 40-1 | Definition of $\alpha$, the ratio of collector current to emitter current | $\alpha = \dfrac{I_c}{I_e}$ | 40-2 |
| 40-1 | Base current of a transistor | $I_b = I_e(1 - \alpha)$ | 40-3 |
| 40-11 | Voltage amplification factor for a transistor, also called voltage gain | $A_v = \dfrac{V_{out}}{V_{in}}$ | 40-4 |
| 40-11 | Power gain of a transistor | $G = \dfrac{\text{power out}}{\text{power in}} = \dfrac{V_{out}I_c}{V_{in}I_e}$ | 40-5 |
| 40-12 | Power gain of a transistor | $G = \alpha\dfrac{V_{out}}{V_{in}} = \alpha A_v$ | 40-6 |
| 40-12 | Current gain | $A_i = \dfrac{I_{out}}{I_{in}} = \dfrac{\alpha I_e}{(1 - \alpha)I_e} = \dfrac{\alpha}{1 - \alpha}$ | 40-7 |
| 40-12 | Relation of current gain to $\alpha$ of a transistor | $\beta = \dfrac{\alpha}{1 - \alpha}$ | 40-8 |

## True-False Questions

T   F   **1.** Current flow in a vacuum-tube diode occurs only when the plate is negative with respect to the cathode.

T   F   **2.** Semiconductors are materials in which the valence and conduction bands overlap.

**T  F  3.** N-type semiconductors are produced by doping with donor atoms.

**T  F  4.** For a reverse-biased PN junction, both holes and electrons move away from the junction and there is no current flow across the junction.

**T  F  5.** PN junctions are semiconductor diodes, and simple transistors are semiconductor triodes.

**T  F  6.** The bridge rectifier uses four diodes to obtain half-wave rectification of alternating current.

**T  F  7.** The zener diode is useful as a voltage-regulating device.

**T  F  8.** In a transistor, the ratio of base current to emitter current is sometimes referred to as current gain $\alpha$.

---

## Multiple-Choice Questions

**1.** A vacuum-tube triode amplifier in a radio circuit might be replaced in a miniature circuit by
  (a) a PN junction                    (b) a zener diode
  (c) an NPN transistor                (d) a thermistor

**2.** Which of the following is *not* an application of diodes?
  (a) A filter                         (b) Bridge rectifier
  (c) Full-wave rectifier              (d) Half-wave rectifier

**3.** In an NPN transistor, the N-type part of the forward-biased section is called the
  (a) emitter        (b) collector        (c) base        (d) donor

**4.** A transistor for which $\alpha = 0.98$ has the following $\beta$ value:
  (a) 29              (b) 38              (c) 49              (d) 56

**5.** The amplifier circuit shown in Fig. 40-2 is
  (a) a common-emitter amplifier
  (b) a common-base amplifier
  (c) a common-collector amplifier
  (d) none of these

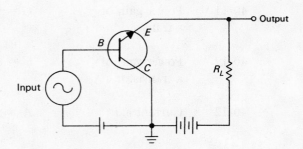

**Fig. 40-2**

**6.** If $\alpha = 0.96$ for the arrangement in Fig. 40-2, the current gain will be approximately
  (a) 0.96           (b) 1.06           (c) 24           (d) 25

**7.** Which of the following amplifier circuits act as a signal inverter?
  (a) Common emitter                   (b) Common collector
  (c) Common base                      (d) None of these

8. In a common-base amplifier circuit for which $\alpha = 0.98$ and the emitter current is 20 mA, the output resistance is 500 k$\Omega$ and the input resistance is 500 $\Omega$. The voltage gain is approximately

(a) 1000      (b) 1020      (c) 1040      (d) 1060

9. Which of the following is *not* an indication of current gain in a transistor amplifier?

(a) $\alpha$      (b) $\dfrac{\alpha}{1 - \alpha}$      (c) $\dfrac{1}{1 - \alpha}$      (d) $\dfrac{I_e}{I_c}$

---

**Completion Questions**

1. The atom impurities used in producing N-type semiconductors are called _____, whereas P-type semiconductors are produced through doping with _____.

2. A PN junction is _____ biased when the positive terminal of a battery is connected to the P side and the negative terminal to the N side.

3. An electronic device that converts alternating-current to direct-current pulses in a manner that utilizes only half of the incoming signal is called _____.

4. The highest current gain is obtained with a common-_____ amplifier, the highest power gain is obtained with a common-_____ amplifier, and the highest voltage gain is obtained with a common-_____ amplifier.

5. The process of planting impurity atoms into semiconductor crystals is known as _____.

# Answers to True-False, Multiple-Choice, and Completion Questions

## CHAPTER 1

### True-False Questions

| | | | |
|---|---|---|---|
| 1. F | 5. F | 8. F | 11. F |
| 2. T | 6. T | 9. F | 12. F |
| 3. T | 7. T | 10. F | 13. T |
| 4. F | | | |

## CHAPTER 2

### True-False Questions

| | | |
|---|---|---|
| 1. T | 5. T | 8. F |
| 2. F | 6. T | 9. F |
| 3. T | 7. F | 10. F |
| 4. T | | |

### Multiple-Choice Questions

| | | |
|---|---|---|
| 1. b | 5. a | 8. d |
| 2. c | 6. c | 9. d |
| 3. a | 7. d | 10. b |
| 4. d | | |

### Completion Questions

1. negative
2. two
3. radical
4. inverse
5. formula
6. power, exponent
7. hypotenuse
8. adjacent
9. 65°
10. equal

## CHAPTER 3

### True-False Questions

| | | |
|---|---|---|
| 1. F | 5. T | 8. T |
| 2. T | 6. F | 9. T |
| 3. T | 7. T | 10. T |
| 4. T | | |

### Multiple-Choice Questions

| | | |
|---|---|---|
| 1. b | 5. d | 8. b |
| 2. c | 6. a | 9. a |
| 3. c | 7. a | 10. c |
| 4. c | | |

### Completion Questions

1. dimensions
2. displacement, velocity, force
3. scalar
4. components
5. vector
6. polygon
7. resultant force
8. $x$ components
9. concurrent
10. negative

## CHAPTER 4

### True-False Questions

| | | |
|---|---|---|
| 1. F | 5. T | 8. F |
| 2. F | 6. T | 9. T |
| 3. F | 7. F | 10. T |
| 4. T | | |

### Multiple-Choice Questions

| | | |
|---|---|---|
| 1. c | 5. a | 8. c |
| 2. b | 6. c | 9. a |
| 3. c | 7. b | 10. b |
| 4. c | | |

### Completion Questions

1. Newton's third law
2. kinetic friction
3. equilibrium, zero
4. free-body diagram
5. along, normal to
6. the coefficient of static friction
7. acted on by a resultant force
8. 10 lb
9. equilibrant
10. angle of repose

# CHAPTER 5

## True-False Questions

| | | |
|---|---|---|
| 1. T | 5. T | 8. F |
| 2. T | 6. F | 9. T |
| 3. F | 7. T | 10. T |
| 4. F | | |

## Multiple-Choice Questions

| | | |
|---|---|---|
| 1. c | 5. b | 8. d |
| 2. c | 6. a | 9. a |
| 3. c | 7. a | 10. c |
| 4. c | | |

## Completion Questions

1. perpendicular, moment arm
2. center of gravity
3. torques, zero
4. negative, positive
5. zero
6. force, torque
7. newton-meter
8. the geometric center
9. sum of the torques due to each component
10. a wheelbarrow, wrench, and pliers

# CHAPTER 6

## True-False Questions

| | | | |
|---|---|---|---|
| 1. T | 6. F | 11. F | 16. T |
| 2. F | 7. T | 12. T | 17. F |
| 3. F | 8. T | 13. T | 18. F |
| 4. F | 9. T | 14. T | 19. T |
| 5. T | 10. F | 15. T | 20. F |

## Multiple-Choice Questions

| | | | |
|---|---|---|---|
| 1. b | 6. d | 11. b | 16. d |
| 2. b | 7. a | 12. a | 17. b |
| 3. a | 8. c | 13. c | 18. b |
| 4. c | 9. a | 14. b | 19. c |
| 5. d | 10. b | 15. a | 20. d |

## Completion Questions

1. average speed
2. uniformly accelerated motion
3. three; $v_f$, $v_0$, $a$, $t$, and $s$
4. the change in speed
5. acceleration
6. negative
7. ft/s$^2$, 9.8
8. below the point of release
9. 16, 19.6, 144
10. second, 4 ft/s
11. gravity
12. horizontal, vertical

13. $v_y = v_{0y} + \frac{1}{2}gt^2$
14. $y = v_{0y}t + \frac{1}{2}gt^2$
15. $x = v_{0x}t$
16. acceleration
17. weight
18. vertical, velocity, acceleration
19. horizontal, velocity
20. $y = \frac{1}{2}gt^2$

# CHAPTER 7

## True-False Questions

| | | |
|---|---|---|
| 1. F | 5. F | 8. T |
| 2. T | 6. T | 9. T |
| 3. T | 7. T | 10. F |
| 4. F | | |

## Multiple-Choice Questions

| | | |
|---|---|---|
| 1. b | 5. d | 8. c |
| 2. a | 6. a | 9. d |
| 3. c | 7. b | 10. a |
| 4. d | | |

## Completion Questions

1. resultant force, mass
2. its weight, acceleration of gravity
3. slug, pounds
4. kilogram, newton
5. force
6. weight, 32 ft/s$^2$
7. mass, weight
8. weight, acceleration
9. mass, decrease
10. newton, m/s$^2$

# CHAPTER 8

## True-False Questions

| | | |
|---|---|---|
| 1. F | 5. F | 8. T |
| 2. T | 6. F | 9. F |
| 3. T | 7. T | 10. F |
| 4. T | | |

## Multiple-Choice Questions

| | | |
|---|---|---|
| 1. d | 5. a | 8. c |
| 2. c | 6. b | 9. a |
| 3. a | 7. d | 10. a |
| 4. c | | |

## Completion Questions

1. a compressed spring, an elevated mass, a cocked rifle
2. kinetic energy
3. kinetic and potential energies, constant
4. work, J

**5.** power
**6.** resultant
**7.** energy
**8.** negative
**9.** applied force, displacement
**10.** power

# CHAPTER 9

*True-False Questions*

| | | |
|---|---|---|
| **1.** F | **5.** T | **8.** F |
| **2.** T | **6.** F | **9.** T |
| **3.** T | **7.** T | **10.** F |
| **4.** T | | |

*Multiple-Choice Questions*

| | | |
|---|---|---|
| **1.** d | **5.** b | **8.** a |
| **2.** b | **6.** d | **9.** c |
| **3.** d | **7.** b | **10.** b |
| **4.** b | | |

*Completion Questions*

**1.** momentum, kinetic energy
**2.** momentum, kilogram-meters per second
**3.** coefficient of restitution
**4.** impulse, change in momentum
**5.** 1, elastic, 0, inelastic
**6.** completely inelastic
**7.** pounds per second, slug-feet per second
**8.** momentum, momentum
**9.** coefficient of restitution
**10.** elasticity

# CHAPTER 10

*True-False Questions*

| | | |
|---|---|---|
| **1.** F | **5.** T | **8.** F |
| **2.** F | **6.** F | **9.** T |
| **3.** T | **7.** T | **10.** F |
| **4.** F | | |

*Multiple-Choice Questions*

| | | |
|---|---|---|
| **1.** c | **5.** a | **8.** d |
| **2.** b | **6.** b | **9.** a |
| **3.** b | **7.** b | **10.** b |
| **4.** c | | |

*Completion Questions*

**1.** speed, direction
**2.** frequency, period
**3.** 4
**4.** masses, square, distance
**5.** centripetal, $mv^2/R$
**6.** $\tan \theta = v^2/(gR)$

**7.** centrifugal
**8.** friction
**9.** critical
**10.** $\text{lb} \cdot \text{ft}^2/\text{slug}^2$

# CHAPTER 11

*True-False Questions*

| | | |
|---|---|---|
| **1.** F | **5.** T | **8.** F |
| **2.** T | **6.** F | **9.** F |
| **3.** T | **7.** T | **10.** T |
| **4.** F | | |

*Multiple-Choice Questions*

| | | |
|---|---|---|
| **1.** c | **5.** d | **8.** c |
| **2.** b | **6.** a | **9.** a |
| **3.** c | **7.** d | **10.** b |
| **4.** a | | |

*Completion Questions*

**1.** radians, radius
**2.** torques, momentum, angular momentum
**3.** torque, angular displacement
**4.** angular acceleration, torque, moment of inertia
**5.** $\text{slug} \cdot \text{ft}^2$
**6.** angular acceleration
**7.** angular velocity, acceleration, time
**8.** radius of gyration, moment of inertia
**9.** angular velocity
**10.** torque, velocity

# CHAPTER 12

*True-False Questions*

| | | |
|---|---|---|
| **1.** F | **5.** F | **8.** T |
| **2.** F | **6.** T | **9.** F |
| **3.** T | **7.** T | **10.** T |
| **4.** T | | |

*Multiple-Choice Questions*

| | | |
|---|---|---|
| **1.** b | **5.** d | **8.** b |
| **2.** c | **6.** b | **9.** c |
| **3.** c | **7.** a | **10.** d |
| **4.** a | | |

*Completion Questions*

**1.** work output, work input
**2.** actual mechanical advantage, ideal mechanical advantage, input, output
**3.** nutcracker, wheelbarrow, crowbar
**4.** greater
**5.** planetary, bevel, helical
**6.** length, thickness
**7.** work against friction, work output

8. number of teeth, number of teeth
9. pitch
10. twice, 2

# CHAPTER 13

### True-False Questions

| | | |
|---|---|---|
| 1. F | 5. T | 8. T |
| 2. F | 6. T | 9. T |
| 3. T | 7. F | 10. T |
| 4. T | | |

### Multiple-Choice Questions

| | | |
|---|---|---|
| 1. c | 5. d | 8. a |
| 2. a | 6. b | 9. c |
| 3. d | 7. b | 10. a |
| 4. b | | |

### Completion Questions

1. tensile, compressure, shearing
2. modulus of elasticity
3. strain
4. $\tan \phi$
5. compressibility, $k$
6. applied force, surface area
7. greater than
8. elastic limit, applied stress, Hooke's
9. stress
10. bulk modulus

# CHAPTER 14

### True-False Questions

| | | |
|---|---|---|
| 1. T | 5. T | 8. T |
| 2. T | 6. T | 9. T |
| 3. F | 7. T | 10. F |
| 4. F | | |

### Multiple-Choice Questions

| | | |
|---|---|---|
| 1. b | 5. b | 8. d |
| 2. a | 6. a | 9. b |
| 3. a | 7. b | 10. c |
| 4. b | | |

### Completion Questions

1. reference circle
2. velocity, acceleration
3. displacement
4. restoring force, acceleration
5. displacement, acceleration
6. moment of inertia, torsion constant
7. torsion constant
8. displacement, opposite
9. mass
10. amplitude or angle

# CHAPTER 15

### True-False Questions

| | | | |
|---|---|---|---|
| 1. F | 6. T | 11. F | 16. T |
| 2. T | 7. F | 12. T | 17. T |
| 3. F | 8. T | 13. T | 18. T |
| 4. F | 9. T | 14. F | 19. F |
| 5. T | 10. F | 15. T | 20. T |

### Multiple-Choice Questions

| | | | |
|---|---|---|---|
| 1. b | 6. b | 11. c | 16. a |
| 2. b | 7. c | 12. a | 17. b |
| 3. c | 8. b | 13. b | 18. a |
| 4. a | 9. d | 14. b | 19. d |
| 5. c | 10. b | 15. c | 20. c |

### Completion Questions

1. absolute, atmosphere
2. perpendicular
3. depth, density
4. shape, area
5. weight, volume, no
6. buyout, weight, Archimedes'
7. gauge, manometer
8. rise to the surface and float
9. hydraulic, Pascal
10. area
11. rate of flow, $ft^3/s$
12. streamline flow
13. venturi
14. pressure, velocity, density, height
15. depth, midpoint
16. depth, acceleration of gravity, Torricelli's
17. cross-sectional area, velocity
18. turbulent
19. absolute, mass
20. $P$, $pgh$, $\frac{1}{2}pv^2$

# CHAPTER 16

### True-False Questions

| | | |
|---|---|---|
| 1. F | 5. T | 8. T |
| 2. T | 6. F | 9. T |
| 3. T | 7. T | 10. T |
| 4. F | | |

### Multiple-Choice Questions

| | | |
|---|---|---|
| 1. d | 5. d | 8. c |
| 2. b | 6. c | 9. d |
| 3. b | 7. b | 10. a |
| 4. a | | |

### Completion Questions

1. absolute zero
2. average kinetic energy, thermal equilibrium

3. 0.00006
4. length, length, temperature
5. thermometer
6. Celsius, $\frac{9}{5}$
7. ice point, steam point
8. 100
9. kinetic energy, potential energy
10. density

# CHAPTER 17

## True-False Questions

| | | |
|---|---|---|
| **1.** F | **5.** T | **8.** F |
| **2.** F | **6.** F | **9.** F |
| **3.** T | **7.** T | **10.** F |
| **4.** T | | |

## Multiple-Choice Questions

| | | |
|---|---|---|
| **1.** b | **5.** a | **8.** c |
| **2.** b | **6.** b | **9.** d |
| **3.** c | **7.** a | **10.** b |
| **4.** c | | |

## Completion Questions

1. British thermal unit
2. temperature, one gram, Celsius degree
3. specific heat capacity
4. quantity of heat, a unit mass, solid, liquid, melting point
5. mass, liquid, vapor, boiling point
6. heat, combustion
7. 540, 970
8. sublimation
9. mechanical equivalent of heat
10. heat lost, heat gained

# CHAPTER 18

## True-False Questions

| | | |
|---|---|---|
| **1.** T | **5.** T | **8.** T |
| **2.** F | **6.** F | **9.** T |
| **3.** T | **7.** F | **10.** T |
| **4.** T | | |

## Multiple-Choice Questions

| | | |
|---|---|---|
| **1.** b | **5.** a | **8.** c |
| **2.** d | **6.** d | **9.** d |
| **3.** b | **7.** b | **10.** a |
| **4.** c | | |

## Completion Questions

1. natural, forced
2. convection, conduction, polished aluminum surfaces
3. radiation

4. fourth, absolute temperature
5. emissivity, 0, 1
6. geometry, floor, ceiling
7. Btu · in./(ft² · h · F°), kcal/(m · s · C°)
8. aluminum, specific heat, copper, thermal conductivity
9. convection
10. thermal conductivity

# CHAPTER 19

## True-False Questions

| | | |
|---|---|---|
| **1.** F | **5.** T | **8.** F |
| **2.** T | **6.** F | **9.** T |
| **3.** F | **7.** T | **10.** F |
| **4.** T | | |

## Multiple-Choice Questions

| | | |
|---|---|---|
| **1.** c | **5.** d | **8.** a |
| **2.** c | **6.** d | **9.** b |
| **3.** a | **7.** c | **10.** c |
| **4.** c | | |

## Completion Questions

1. mass, temperature, volume, Boyle's
2. *PV*, *nT*, universal gas
3. critical temperature
4. evaporation, boiling, sublimation
5. dew point
6. absolute, relative, actual vapor pressure, saturated vapor pressure
7. mole
8. mass, pressure, volume, temperature
9. $6.023 \times 10^{23}$, Avogadro's number
10. L · atm/(mol · K)

# CHAPTER 20

## True-False Questions

| | | |
|---|---|---|
| **1.** F | **5.** T | **8.** T |
| **2.** F | **6.** T | **9.** T |
| **3.** T | **7.** F | **10.** F |
| **4.** T | | |

## Multiple-Choice Questions

| | | |
|---|---|---|
| **1.** d | **5.** a | **8.** b |
| **2.** b | **6.** d | **9.** d |
| **3.** d | **7.** b | **10.** b |
| **4.** c | | |

## Completion Questions

1. work output, work input
2. isochoric, isothermal
3. adiabatic

4. work done on or by the system
5. pressure, volume, temperature
6. first law of thermodynamics
7. refrigerator, coefficient of performance
8. compressor, condenser, evaporator, throttling valve
9. second law of thermodynamics
10. Carnot

5. infrasonic, ultrasonic
6. odd, third
7. intensity
8. Doppler effect, higher, lower
9. constructive, destructive
10. B

# CHAPTER 21

*True-False Questions*

| | | | | | |
|---|---|---|---|---|---|
| 1. | F | 5. | F | 8. | F |
| 2. | T | 6. | F | 9. | T |
| 3. | T | 7. | T | 10. | T |
| 4. | T | | | | |

*Multiple-Choice Questions*

| | | | | | |
|---|---|---|---|---|---|
| 1. | c | 5. | b | 8. | a |
| 2. | c | 6. | d | 9. | d |
| 3. | a | 7. | c | 10. | b |
| 4. | b | | | | |

*Completion Questions*
1. transverse
2. tension, linear density
3. wavelength
4. frequency, amplitude
5. fourth
6. displacement, displacements, superposition
7. nodes, antinodes
8. frequency, wavelength
9. nodes, antinodes
10. harmonic

# CHAPTER 22

*True-False Questions*

| | | | | | |
|---|---|---|---|---|---|
| 1. | T | 5. | T | 8. | T |
| 2. | F | 6. | F | 9. | T |
| 3. | F | 7. | F | 10. | F |
| 4. | T | | | | |

*Multiple-Choice Questions*

| | | | | | |
|---|---|---|---|---|---|
| 1. | d | 5. | c | 8. | a |
| 2. | b | 6. | c | 9. | a |
| 3. | d | 7. | b | 10. | c |
| 4. | b | | | | |

*Completion Questions*
1. node, antinode
2. bulk modulus, density
3. intensity, frequency, waveform
4. hearing threshold, $10^{-16}$

# CHAPTER 23

*True-False Questions*

| | | | | | |
|---|---|---|---|---|---|
| 1. | T | 5. | T | 8. | F |
| 2. | T | 6. | F | 9. | T |
| 3. | T | 7. | F | 10. | F |
| 4. | T | | | | |

*Multiple-Choice Questions*

| | | | | | |
|---|---|---|---|---|---|
| 1. | b | 5. | c | 8. | a |
| 2. | a | 6. | c | 9. | b |
| 3. | c | 7. | c | 10. | a |
| 4. | a | | | | |

*Completion Questions*
1. negatively, negatively
2. conductor, insulator
3. $10^{-6}$
4. repel, attract
5. electrons, cloth, rod
6. induction
7. Coulomb's distance
8. $6.25 \times 10^{18}$
9. electron, $1.6 \times 10^{-19}$
10. electroscope

# CHAPTER 24

*True-False Questions*

| | | | | | |
|---|---|---|---|---|---|
| 1. | F | 5. | F | 8. | F |
| 2. | T | 6. | T | 9. | T |
| 3. | T | 7. | T | 10. | T |
| 4. | F | | | | |

*Multiple-Choice Questions*

| | | | | | |
|---|---|---|---|---|---|
| 1. | a | 5. | a | 8. | c |
| 2. | b | 6. | d | 9. | a |
| 3. | b | 7. | b | 10. | d |
| 4. | c | | | | |

*Completion Questions*
1. electric charge, electric force
2. electric field intensity, positive charge
3. magnitude of the charge, square of the distance from the charge

4. vector sum
5. far apart, close together
6. electric field lines, electric field
7. away from, toward
8. charge, Gauss's law
9. Gauss's law, surface
10. $N \cdot m^2/C^2$

# CHAPTER 25

### True-False Questions

| | | |
|---|---|---|
| 1. F | 5. T | 8. F |
| 2. F | 6. T | 9. F |
| 3. T | 7. F | 10. T |
| 4. T | | |

### Multiple-Choice Questions

| | | |
|---|---|---|
| 1. b | 5. b | 8. b |
| 2. c | 6. d | 9. a |
| 3. a | 7. c | 10. b |
| 4. d | | |

### Completion Questions

1. potential
2. positive, negative
3. algebraic sum
4. 1 C, 1 J
5. field intensity, plate separation
6. electronvolt
7. $q$, $V_A - V_B$
8. electric field intensity, newtons per coulomb
9. negative, against; or positive, by
10. increases, decreases

# CHAPTER 26

### True-False Questions

| | | |
|---|---|---|
| 1. T | 4. T | 6. T |
| 2. F | 5. F | 7. T |
| 3. T | | |

### Multiple-Choice Questions

| | | |
|---|---|---|
| 1. c | 4. c | 7. c |
| 2. d | 5. b | 8. a |
| 3. a | 6. a | 9. b |

### Completion Questions

1. area, their separation
2. dielectric strength
3. small plate separation, increased capacitance, higher breakdown voltage
4. $C/C_0$, $V_0/V$, $E_0/E$
5. permittivity

6. less than
7. dielectric constant
8. $\frac{1}{2}QV$, $\frac{1}{2}CV^2$, $Q^2/(2C)$
9. parallel

# CHAPTER 27

### True-False Questions

| | | |
|---|---|---|
| 1. F | 5. F | 8. F |
| 2. T | 6. T | 9. T |
| 3. F | 7. T | 10. F |
| 4. F | | |

### Multiple-Choice Questions

| | | |
|---|---|---|
| 1. c | 5. c | 8. d |
| 2. b | 6. b | 9. c |
| 3. d | 7. a | 10. c |
| 4. a | | |

### Completion Questions

1. volt, joule, work
2. voltage, resistance, Ohm's law
3. length, area, temperature material
4. resistance, resistance, temperature
5. resistance, current
6. positive charges
7. circular mils, diameter
8. length, area, resistivity
9. rheostats, ammeters, voltmeters
10. mechanical or chemical electrical

# CHAPTER 28

### True-False Questions

| | | |
|---|---|---|
| 1. F | 5. F | 8. T |
| 2. F | 6. T | 9. F |
| 3. T | 7. F | 10. F |
| 4. T | | |

### Multiple-Choice Questions

| | | |
|---|---|---|
| 1. b | 5. a | 8. b |
| 2. a | 6. c | 9. a |
| 3. b | 7. b | 10. d |
| 4. b | | |

### Completion Questions

1. parallel
2. current, voltage
3. terminal potential difference
4. increase in internal resistance
5. $\Sigma I_{entering} = \Sigma I_{leaving}$; $\Sigma E = \Sigma IR$
6. $E + Ir$
7. negative

8. direction
9. EMF, resistance, internal
10. Wheatstone bridge

## CHAPTER 29

*True-False Questions*

| | | |
|---|---|---|
| 1. T | 5. F | 8. T |
| 2. F | 6. T | 9. F |
| 3. F | 7. F | 10. T |
| 4. F | | |

*Multiple-Choice Questions*

| | | |
|---|---|---|
| 1. d | 5. c | 8. a |
| 2. a | 6. a | 9. b |
| 3. b | 7. d | 10. c |
| 4. a | | |

*Completion Questions*

1. magnetism, magnetizing force
2. current, the magnetic field
3. into the paper
4. permeability, vacuum
5. magnetic saturation
6. tesla, weber
7. current vector, flux-density vector, magnetic force
8. newton, coulomb, meter per second
9. diamagnetic, paramagnetic
10. retentivity

## CHAPTER 30

*True-False Questions*

| | | |
|---|---|---|
| 1. F | 5. T | 8. F |
| 2. T | 6. F | 9. T |
| 3. F | 7. T | 10. T |
| 4. T | | |

*Multiple-Choice Questions*

| | | |
|---|---|---|
| 1. c | 5. b | 8. c |
| 2. a | 6. d | 9. b |
| 3. d | 7. b | 10. a |
| 4. a | | |

*Completion Questions*

1. flux density, area, current, cosine
2. zero
3. galvanometer
4. multiplier, galvanometer
5. shunt, parallel, galvanometer
6. commutator
7. current, voltage
8. low, high
9. magnet, coil, pointer
10. parallel, series

## CHAPTER 31

*True-False Questions*

| | | |
|---|---|---|
| 1. T | 5. T | 8. T |
| 2. T | 6. F | 9. T |
| 3. F | 7. F | 10. T |
| 4. T | | |

*Multiple-Choice Questions*

| | | |
|---|---|---|
| 1. b | 4. b | 7. a |
| 2. c | 5. c | 8. a |
| 3. b | 6. c | 9. d |

*Completion Questions*

1. weber per second, volt, turn
2. applied voltage, back emf
3. field magnets, armature, slip rings, brushes
4. series-wound motor, shunt motor, compound motor
5. starting
6. power input
7. less than
8. primary coil, secondary coil, a soft-iron core
9. shunt-wound motor

## CHAPTER 32

*True-False Questions*

| | | |
|---|---|---|
| 1. F | 5. F | 8. T |
| 2. T | 6. T | 9. F |
| 3. F | 7. T | 10. T |
| 4. F | | |

*Multiple-Choice Questions*

| | | |
|---|---|---|
| 1. b | 5. c | 8. c |
| 2. b | 6. c | 9. b |
| 3. b | 7. a | 10. c |
| 4. a | | |

*Completion Questions*

1. power, 1 ampere
2. inductive, time constant
3. current, voltage, negative
4. resonant frequency
5. resistance, impedance
6. in phase
7. henry, volt, ampere per second
8. decrease
9. resistance
10. $-155$, $+155$ V

## CHAPTER 33

*True-False Questions*

| | | | |
|---|---|---|---|
| 1. F | 3. T | 5. T | 7. F |
| 2. F | 4. F | 6. T | 8. T |

## True-False Questions (continued)

| | | | |
|---|---|---|---|
| **9.** F | **12.** F | **15.** F | **18.** T |
| **10.** T | **13.** T | **16.** T | **19.** T |
| **11.** F | **14.** T | **17.** T | **20.** T |

## Multiple-Choice Questions

| | | | |
|---|---|---|---|
| **1.** b | **6.** b | **11.** b | **16.** c |
| **2.** c | **7.** c | **12.** b | **17.** d |
| **3.** a | **8.** a | **13.** d | **18.** c |
| **4.** d | **9.** d | **14.** a | **19.** d |
| **5.** b | **10.** a | **15.** c | **20.** b |

## Completion Questions

1. wavelets, Huygens's principle
2. angstrom
3. ultraviolet
4. $3 \times 10^8$, 186,000
5. Planck's constant
6. nanometer
7. photoelectric effect
8. violet, blue, green, yellow, orange, red
9. (a) long radio waves
   (b) short radio waves
   (c) infrared region
   (d) visible region
   (e) ultraviolet region
   (f) X rays
   (g) gamma rays
   (h) cosmic photons
10. electric, magnetic
11. lumen, 555 nm
12. umbra, penumbra
13. luminous intensity, candelas
14. flux, area, luminous intensity, distance
15. intensity
16. steradian, radius
17. illumination, inverse-square
18. illumination, perpendicular
19. isotropic
20. luminous flux

# CHAPTER 34

## True-False Questions

| | | |
|---|---|---|
| **1.** F | **5.** F | **8.** T |
| **2.** T | **6.** F | **9.** F |
| **3.** F | **7.** T | **10.** T |
| **4.** T | | |

## Multiple-Choice Questions

| | | |
|---|---|---|
| **1.** d | **5.** a | **8.** b |
| **2.** b | **6.** c | **9.** a |
| **3.** c | **7.** a | **10.** a |
| **4.** d | | |

## Completion Questions

1. one-half the radius of curvature
2. virtual, erect, diminished
3. positive, negative
4. wavelengths, dispersion
5. virtual, erect, enlarged
6. image size, object size, positive, negative
7. parallel
8. parallel to the mirror axis
9. virtual
10. small, spherical aberration

# CHAPTER 35

## True-False Questions

| | | |
|---|---|---|
| **1.** T | **5.** T | **8.** T |
| **2.** F | **6.** T | **9.** T |
| **3.** F | **7.** F | **10.** T |
| **4.** T | | |

## Multiple-Choice Questions

| | | |
|---|---|---|
| **1.** a | **5.** b | **8.** b |
| **2.** b | **6.** b | **9.** c |
| **3.** c | **7.** a | **10.** c |
| **4.** c | | |

## Completion Questions

1. index of refraction
2. incident ray, refracted ray, normal to the surface
3. frequency, wavelength, velocity
4. wavelengths, dispersion
5. critical angle
6. lower, critical angle
7. sine of the angle of incidence, sine of the angle of refraction, velocity of light, velocity of light
8. total internal reflection
9. index of refraction in air, index of refraction in water
10. 1

# CHAPTER 36

## True-False Questions

| | | |
|---|---|---|
| **1.** F | **5.** T | **8.** T |
| **2.** T | **6.** F | **9.** F |
| **3.** F | **7.** T | **10.** T |
| **4.** T | | |

## Multiple-Choice Questions

| | | |
|---|---|---|
| **1.** a | **5.** c | **8.** b |
| **2.** b | **6.** d | **9.** a |
| **3.** b | **7.** d | **10.** b |
| **4.** d | | |

## Completion Questions
1. virtual, erect, diminished
2. focal length
3. real, inverted, diminished
4. positive, negative
5. erect, inverted
6. double convex, plane-convex, converging meniscus
7. thicker, thinner
8. second focal point, first focal point
9. virtual, real
10. center

# CHAPTER 37

## True-False Questions
| | | |
|---|---|---|
| 1. T | 5. F | 8. T |
| 2. T | 6. T | 9. T |
| 3. F | 7. F | 10. T |
| 4. T | | |

## Multiple-Choice Questions
| | | |
|---|---|---|
| 1. c | 4. d | 7. a |
| 2. a | 5. c | 8. b |
| 3. b | 6. b | 9. a |

## Completion Questions
1. diffraction
2. resolving power
3. amplitude, amplitudes
4. superposition
5. destructive
6. diameter
7. diffraction grating

# CHAPTER 38

## True-False Questions
| | | |
|---|---|---|
| 1. T | 5. F | 8. T |
| 2. F | 6. F | 9. T |
| 3. T | 7. F | 10. T |
| 4. F | | |

## Multiple-Choice Questions
| | | |
|---|---|---|
| 1. c | 5. b | 8. b |
| 2. d | 6. a | 9. c |
| 3. b | 7. d | |
| 4. c | | |

## Completion Questions
1. shorter, larger, longer
2. free-space velocity of photon, Einstein's second postulate
3. energy of the incident light, work function of the surface

4. ground
5. Lyman, Balmer
6. sodium vapor, absorption
7. de Broglie wavelength
8. Bohr's, $h/(2\pi)$
9. work function

# CHAPTER 39

## True-False Questions
| | | |
|---|---|---|
| 1. T | 5. T | 8. T |
| 2. F | 6. F | 9. F |
| 3. T | 7. T | 10. F |
| 4. F | | |

## Multiple-Choice Questions
| | | |
|---|---|---|
| 1. b | 5. d | 8. a |
| 2. a | 6. b | 9. a |
| 3. a | 7. b | 10. c |
| 4. c | | |

## Completion Questions
1. binding energy
2. $^{28}_{14}Si$, $^{24}_{12}Mg$
3. 20.4
4. 1 curie
5. mass number, protons, neutrons
6. nuclear core, control rods, heat exchanger, moderator, radiation shielding
7. smaller, larger
8. atomic number, mass number
9. 931 MeV
10. charge, nucleons, energy

# CHAPTER 40

## True-False Questions
| | | |
|---|---|---|
| 1. F | 4. T | 7. T |
| 2. F | 5. T | 8. F |
| 3. T | 6. F | |

## Multiple-Choice Questions
| | | |
|---|---|---|
| 1. c | 5. c | 8. b |
| 2. a | 6. d | 9. c |
| 3. a | 7. a | 10. d |
| 4. c | | |

## Completion Questions
1. donors, acceptors
2. forward
3. half-wave rectifier
4. collector, emitter, base
5. doping

# Experiments for Physics

**Reading Assignments in *Physics,* Sixth Edition**

| Experiment number and title | Chapters referenced | Sections assigned |
|---|---|---|
| 1. Vernier and Micrometer Calipers | 3. Technical Measurements and Vectors | 3-5 and 3-6 |
| 2. Addition of Force Vectors | 3. Technical Measurements and Vectors | 3-7 to 3-13 |
| 3. Friction | 4. Translational Equilibrium and Friction | 4-6 |
| 4. Accelerated Motion | 6. Uniform Acceleration | 6-1 to 6-6 |
| 5. Acceleration Due to Gravity | 6. Uniform Acceleration | 6-7 |
| 6. Range of a Projectile | 6. Uniform Acceleration <br> 7. Newton's Second Law | 6-7 to 6-9 <br> 7-1 and 7-2 |
| 7. Newton's Second Law | 7. Newton's Second Law | 7-1 to 7-3 |
| 8. Conservation of Energy | 8. Work, Energy, and Power | 8-5 to 8-7 |
| 9. Conservation of Momentum | 9. Impulse and Momentum | 9-1 and 9-2 |
| 10. Kepler's Laws | 10. Uniform Circular Motion | 10-7 to 10-10 |
| 11. Pulleys | 12. Simple Machines | 12-4 and 12-5 |
| 12. Hooke's Law and Simple Harmonic Motion | 14. Simple Harmonic Motion <br> 15. Fluids | 13-1 and 13-3 <br> 14-1 and 14-4 |
| 13. Archimedes' Principle | 15. Fluids | 15-6 |
| 14. Specific Heat | 17. Quantity of Heat | 17-1 to 17-4 |
| 15. Standing Waves in a Vibrating String | 21. Mechanical Waves | 21-7 and 21-8 |
| 16. Investigating Static Electricity | 23. The Electric Force | 23-1 to 23-5 |
| 17. The Capacitor | 26. Capacitance | 26-1 to 26-4 |
| 18. Ohm's Law | 27. Current and Resistance | 27-1 and 27-4 |
| 19. Series Resistance | 28. Direct-Current Circuits | 28-1 |
| 20. Parallel Resistance | 28. Direct-Current Circuits | 28-2 |
| 21. Principles of Electromagnetism | 29. Magnetism and the Magnetic Field | 29-3 and 29-8 |
| 22. Electromagnetic Induction | 29. Magnetism and the Magnetic Field | 29-8 to 29-10 |
| 23. Planck's Constant | 33. Light and Illumination | 33-3 and 33-4 |
| 24. Reflection of Light | 34. Reflection and Mirrors | 34-1 and 34-2 |
| 25. Concave and Convex Mirrors | 34. Reflection and Mirrors | 34-3 and 34-7 |
| 26. Snell's Law | 35. Refraction | 35-1 to 35-3 |
| 27. Convex and Concave Lenses | 36. Lenses and Optical Instruments | 36-1 to 36-5 |
| 28. Double-Slit Interference | 37. Interference, Diffraction, and Polarization | 37-1 to 37-4 |
| 29. Semiconductor Properties | 40. Electronics | 40-1, 40-5 to 40-7 |

# EXPERIMENT 1 · Vernier and Micrometer Calipers

## Purpose

Use vernier and micrometer calipers to measure lengths and diameters.

## Concept and Skill Check

Often it is necessary to measure the dimensions of objects to a greater accuracy than is allowed by rulers or meter sticks. Such devices have a failing in that one must estimate the distance between two marks separated by only one millimeter. For example, if one measures the length of a pencil using a meter stick, the result may be recorded as 141.2 mm. Since the smallest scale division is one millimeter, each end of the pencil can be located with a precision of ±0.5 mm. The measured length is the difference between the two readings, so that the *uncertainty* in the length of the meter stick is ±1 mm. We might write this result as follows:

Length = 141.2 mm ± 1 mm.

The true length of the pencil is, therefore, said to be between 140.2 mm and 142.2 mm.

The uncertainty in the measurement of length can be lessened by using two scales. A fixed scale measures the length in the usual way and a moving scale permits one to estimate with greater accuracy the spaces between divisions. The vernier caliper, shown in Figure 1, has a small scale, called *a vernier*. The main or fixed scale is divided into millimeters with numbers to represent centimeters. The moving scale is marked so that 10 divisions on it equal 9 divisions on the fixed scale. In other words, each division on the vernier scale is 9/10 of a division of the fixed scale. When the jaws of the caliper are completely closed, the zero line on the vernier scale (called the index) should coincide with the zero line on the main scale (refer to Figure 2a). When the jaws are open, the vernier reading is taken by noting which linear line coincides with a main scale division line. Note that the correct readings in parts b and c of Figure 2 should be 2.44 cm and 1.07 cm, respectively. The uncertainty in these measurements is only ±0.01 cm, which is an improvement over that experienced for the meter stick. The last reading might be written as follows:

Length = 1.07 cm ± 0.01 cm.

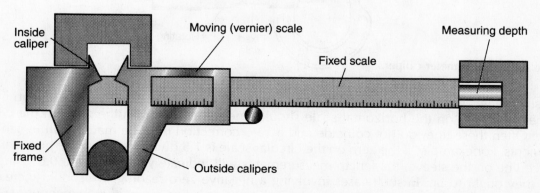

**Figure 1. The vernier caliper.**

Labels: Inside caliper · Moving (vernier) scale · Measuring depth · Fixed scale · Fixed frame · Outside calipers

# Vernier and <sub>NAME</sub> Micrometer Calipers

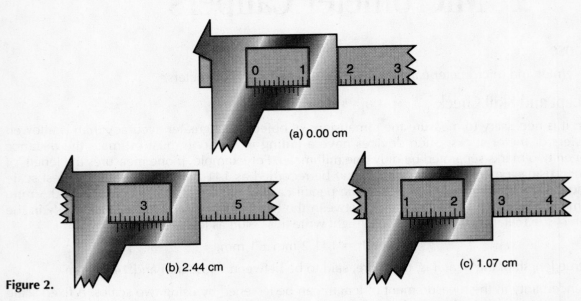

(a) 0.00 cm

(b) 2.44 cm

(c) 1.07 cm

**Figure 2.**

Even greater precision can be obtained by using a micrometer caliper which is shown in Figure 3. It can measure length to one more decimal place than the vernier caliper. Examine a micrometer, compare it with the figure, and identify each element of the instrument. The object whose length or diameter is to be measured is placed between the anvil and the spindle. The ratchet is turned until contact is made with the object to be measured. The slip gear in the ratchet allows consistent pressure to be exerted and prevents damage to the caliper. The scale on the sleeve of the micrometer is graduated in millimeters (mm) and the circular scale is divided into 50 divisions, each of which indicates 0.010 mm. One complete turn of the circular scale will advance the sleeve a distance of 0.5 mm. The uncertainty in measurements is equal to the smallest division on the circular scale or ±0.01 mm. Note that this precision is one decimal place better than possible with the vernier caliper.

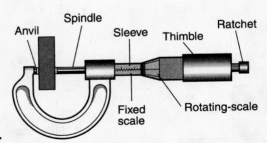

Anvil  Spindle  Sleeve  Thimble  Ratchet

Fixed scale  Rotating-scale

**Figure 3. The micrometer caliper.**

Measurements are made by closing the space between the spindle and the anvil by turning the ratchet. The zero on the horizontal scale should line up exactly with the zero on the circular scale. Often these lines do not coincide and a zero correction must be made to all length measurements. For example, if the zero on the circular scale is 1.5 divisions below the horizontal reference line on the sleeve, the length measurements will indicate values that are 0.015 mm *less* than they ought to be. In such cases involving a negative zero reading, we should increase all measurements by this amount. Conversely, if the zero reading is positive, we should subtract

# Vernier and Micrometer Calipers

from the indicated values. Remember to subtract algebraically the zero reading (change the sign and add).

The length or diameter of an object is measured by reading the horizontal scale to find the whole number of millimeters and then reading the circular scale to indicate the hundredths of a millimeter. The last digit to be estimated is the distance between divisions of the circular scale and will then be in thousandths of a millimeter. Figure 4 shows three example readings. In part (a), the edge of the circular scale has passed 17 mm, but it is not yet beyond the 17.5 mm mark. The circular scale has not completed its first revolution, and the reading is somewhere between 17.0 mm and 17.5 mm. Now, looking at the reading of 35 on the circular scale, we know that we must add 0.35 mm to the initial reading. Thus, the correct length is reported as 17.350 mm. The last "0" digit is added to indicate that we are *estimating* that the horizontal line coincides exactly with the number 35. Most of the time the horizontal line will be between two marks on the circular scale. Each reading made by the micrometer caliper is recorded in thousandths of a millimeter with an uncertainty of ±0.01 mm. In other words, we may write this measurement as follows:

$$\text{Length} = 17.350 \text{ mm} \pm 0.01 \text{ mm}.$$

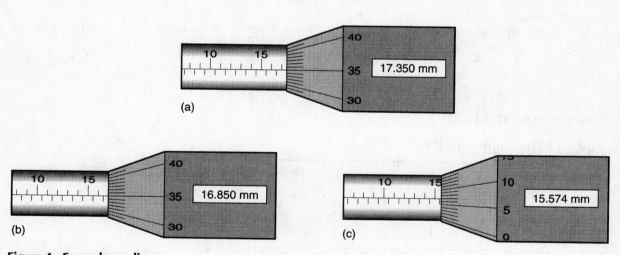

**Figure 4. Example readings.**

In Figure 4b, the horizontal scale is between 16.5 mm and 17.0 mm and the circular scale indicates the number 35. Since the edge of the circular scale has moved beyond the ½ mm mark, we must add 0.350 mm to 16.5 to obtain the correct reading: 16.850 mm. Finally, you should verify that the correct reading in Figure 4c is 15.574 mm. For the last digit, we estimate that the horizontal line on the fixed scale is positioned about ⁴⁄₁₀ of the distance between two marks on the circular scale.

## Materials

meter stick

micrometer caliper

wooden block

vernier caliper

laboratory manual

metal cylinder

# EXPERIMENT 1 Vernier and Micrometer Calipers

## Procedure A

1. Measure the length and width of your laboratory manual using the meter stick. Estimate your measurements to the nearest tenth of a millimeter and note the number of significant figures.
2. Record your measurements in Table 1 along with the number of significant digits and the uncertainty of the measurement.
3. Calculate the area by multiplying the length times the width. Be careful to include only those digits that are significant for the product of length and width.

## Procedure B

1. Measure the length, width, and height of the wooden block using the vernier caliper.
2. Record your measurements in Table 2, indicating the number of significant digits and the uncertainty for each measurement.
3. Calculate the volume of the wooden block by multiplying length times width times height. Record your answer in cubic centimeters and indicate the number of significant digits.

## Procedure C

1. Use the micrometer caliper to measure the length and diameter of the metal cylinder.
2. Record your answers in Table 3 giving the number of significant digits and the uncertainty of each measurement.
3. Recall that the volume of a right circular cylinder in terms of its diameter $D$ and its length $L$ is given by
$$V = \frac{\pi D^2 L}{4}.$$
4. Calculate the volume of the metal cylinder and record your result.

## Observations and Data

### Table 1  The meter stick

| Laboratory Manual | Measurement | Number of Significant Digits | Uncertainty (±) |
|---|---|---|---|
| Length (mm) | | | |
| Width (mm) | | | |
| Area (mm²) | | | |

### Table 2  The venier caliper

| Wooden Block | Measurement | Number of Significant Digits | Uncertainty (±) |
|---|---|---|---|
| Length (cm) | | | |
| Width (cm) | | | |
| Height (cm) | | | |
| Volume (cm³) | | | |

# 1 Vernier and Micrometer Calipers

NAME ————————————————

## Table 3  The micrometer caliper

| Metal Cylinder | Measurement | Number of Significant Digits | Uncertainty (±) |
|---|---|---|---|
| Length (mm) | | | |
| Diameter (mm) | | | |
| Volume (mm³) | | | |

## Analysis

1. The uncertainty in the measurement of length is ±0.01 cm for the vernier caliper and ±0.01 mm for the micrometer caliper. Suppose we measure the diameter of a rod first with the vernier caliper and then with the micrometer caliper and record the following measurements:

    *Vernier caliper*:   1.12 cm ± 0.01 cm        *Micrometer*:   11.223 mm ± 0.01 mm

    There is only one decimal place difference in precision. How do you explain that there are 5 significant figures for the micrometer caliper and only three for the vernier caliper?

2. Suppose the micrometer reading on the sleeve was +1.000 mm when the spindle was closed against the anchor. The zero reading correction must be made each time this micrometer is used. What would be the correct reading if a person used a flawed instrument to record 21.000 mm when (a) the zero reading is +1.000 mm and (b) the zero reading is −1.000 mm?

    (a)_____mm        (b)_____mm

3. For the three measurements described in the following table, what instrument would you choose to measure the indicated length, and what would be the uncertainty in the measurement?

| Measurement | Instrument of choice | Uncertainty of measurement |
|---|---|---|
| Depth of a small cup | | |
| Height of a table | | |
| Diameter of a wire | | |

## Application

Physics is a quantitative science which often requires precise measurements. The ability to use precision instruments such as vernier and micrometer calipers is essential for the proper design and construction of a large variety of machines and tools.

# EXPERIMENT

# 2 : Addition of Force Vectors

## Purpose

Apply the laws of vector addition to resolve forces in equilibrium.

## Concept and Skill Check

When two or more forces act at the same time on an object and their vector sum is zero, the object is in equilibrium. Each of the arrangements shown in Figure 1 illustrates three concurrent forces acting on point **P**. Since point **P** is not moving, these three forces are producing no net force on point **P**, and the diagrammed systems are in equilibrium. In this experiment, you will determine the vector sum of two of the concurrent forces, called the resultant, and investigate the relationship of the resultant to the third force.

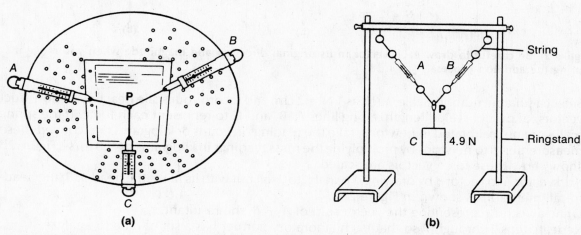

(a)

(b)

**Figure 1. The apparatus used to measure forces varies slightly. Your apparatus will probably be similar to one of these.**

## Materials

Method 1—force table and 3 spring scales
Method 2—2 spring scales, 2 ringstands,
   cross support, and 500-g mass

metric ruler          protractor
heavy string          paper
pencil

## Procedure

Your teacher will demonstrate the apparatus you will be using—either Method 1 or Method 2.
**Method 1**
1. Set up the apparatus as indicated in Figure 1a. Check each spring scale to be sure that the needle points to zero when no load is attached. Attach the spring scales to the force table so that each scale registers a force at approximately mid-range.
2. Place a piece of paper beneath the spring scale arrangement. Using a sharp pencil, mark several points along the line (the string) of action of each force.
3. Remove the paper and, using the points that you marked, construct lines **A**, **B**, and **C**, each representing the direction of force action for scales A, B, and C.
4. Record the reading of each spring scale next to its corresponding line, as shown in Figure 2a.

# 2 Addition of Force Vectors

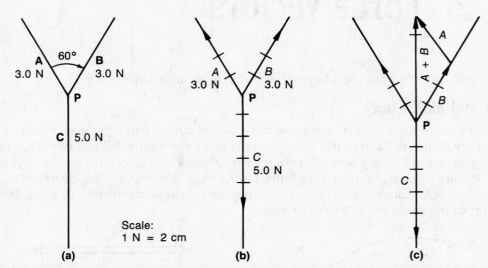

**Figure 2. Be careful to draw each vector in its original direction and magnitude when you move it during the addition process.**

5. Select a suitable number scale, such as 1 N = 2 cm, and record your scale near line **C**. Construct vectors, of proper scaled length, along lines **A**, **B**, and **C** to represent each force. If the spring scales are not calibrated in newtons, take the readings in grams or kilograms and convert these measurements to newtons by multiplying the mass readings in kilograms by 9.8 m/s². Figure 2b shows how these force vectors are scaled.

6. Add vector A to vector B by drawing A parallel to itself but with its tail at the head of B (the head-to-tail method), as shown in Figure 2c.

7. Draw a vector representing the vector sum of A + B, the resultant.

8. Repeat Steps 1 through 7 so that each laboratory partner has a set of data to analyze.

**Method 2**

1. Set up the apparatus as indicated in Figure 1b. With a protractor, measure each of the three angles at the intersection of the three strings.

2. Using these angle measurements, construct a diagram on paper of the forces acting on point **P** by drawing three lines to represent the lines of action of the three forces. Label the lines **A**, **B**, and **C**, as shown in Figure 2a.

3. Record the values of the two spring-scale readings and the weight in newtons of the 500-g mass next to lines **A**, **B**, and **C** on your paper. If the scales give readings in mass, convert the mass readings in kilograms, to weight in newtons by multiplying the mass by 9.8 m/s².

4. Select a suitable number scale and record your scale near line **C**. Using your scale, construct vectors along lines **A**, **B**, and **C** to represent the forces acting along each line of force, as shown in Figure 2b.

5. Add vector A to vector B by reproducing A parallel to itself but with its tail at the head of B (the head-to-tail method), as shown in Figure 2c.

6. Draw a vector representing the vector sum A + B, the resultant.

7. Repeat Steps 1 through 6 so that each laboratory partner has a set of data to analyze.

## Observations and Data

On a separate sheet of paper, draw the vectors obtained from your experimental procedure. Follow the form shown in Figure 2.

# 2 Addition of Force Vectors

## Analysis

1. What was the number scale you selected for your model? Compute the resultant using your number scale.

2. Compare the magnitude and direction of the computed resultant force of $A + B$ with the measured or known magnitude of force $C$. Explain your findings. Calculate the relative error in the magnitudes using force $C$ as the reference value.

3. On a separate sheet of paper, reconstruct vectors $A$, $B$, and $C$ and add them. Do this by placing a piece of paper over your first diagram and tracing the vectors. Place the tail of $B$ at the head of $A$, and then place the tail of $C$ at the head of $B$ (the head-to-tail method). Label your vectors.

4. Explain the results of the graphical addition of $A + B + C$, which you performed as requested in Question 3.

5. Suppose that you had added $B$ to $C$. What result would you expect?

6. What result would you expect if you added $C$ to $A$?

7. On the same sheet you used for Question 3, add your three vectors in the order $C + B + A$. What result do you obtain?

# 2 Addition of Force Vectors

## Application

A sky diver jumps from a plane and, after falling for 11 seconds, reaches a terminal velocity (constant speed) of 250 km/h. Changing her body configuration, she then accelerates to a different terminal velocity of 320 km/h. Finally, after releasing her parachute, she again accelerates and reaches a final terminal velocity of 15 km/h. Explain how it is possible to obtain the different terminal velocities.

## Extension

Using a different method, repeat the investigation, setting the angle between *A* and *B* at some angle other than 90°. Solve for *C* both mathematically and graphically.

EXPERIMENT

# 3 Friction

## Purpose

Investigate friction and measure the coefficients of friction.

## Concept and Skill Check

An object placed on an inclined plane may or may not slide. If the object is at rest, the force of friction is opposing the tendency of the object to slide down the plane. When the plane has been tilted at a certain angle $\theta$ with the horizontal, the object begins to slide down the inclined plane. If the object slides down the inclined plane at a constant speed, then the force of friction $F_f$ is equal to the force down the plane $F_{\parallel}$. The force down the plane is the same as the component of the object's weight parallel to the plane, as shown in the figure. The parallel component of weight is described by $F_{\parallel} = F_w \sin \theta$, where $F_w$ is the weight of the object. The perpendicular component of weight is described by $F_{\perp} = F_w \cos \theta$. When sliding just begins and the object is moving at a constant speed, the coefficient of friction $\mu$ is given by

$$\mu = \frac{F_f}{F_{\perp}} = \frac{F_{\parallel}}{F_{\perp}} = \frac{F_w \sin \theta}{F_w \cos \theta} = \tan \theta.$$

## Materials

spring scale (with capacity sufficient to
   measure the weight of the object)
object: book, chalkboard eraser,
   2-by-4 block, or similar object
flat board
string (1 m)
masking tape
protractor

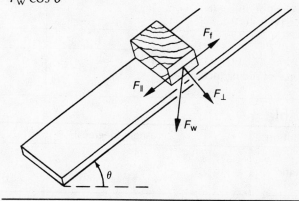

## Procedure

1. Select an object and a flat board for this experiment. Describe the object and the surface of the board in Table 1.
2. Tape the string to the end of the object. Hang the object by the string from the spring scale. Measure the weight of the object. Record this value in Table 2.
3. Place the flat board on a horizontal surface. Hold the spring scale and, with the string held parallel to the level board, pull the object along the board at a constant speed. With the spring scale, measure the amount of force required to keep the object moving at a uniform rate. Repeat this procedure several times, average your results, and record this value in Table 2 as the force of sliding friction between the surface of the board and the surface of the object.
4. Detach the string from the object. Place the object on the flat board. Slowly lift one end of the board. Continue increasing the angle of the board with the horizontal until the object starts to slide. Use the protractor to measure this angle. Record the value of this angle in Table 3 as the angle for static friction. The tangent of this angle is the coefficient of static friction.

# 3 Friction

5. Move the object to one end of the board. Again, slowly lift this end of the board while your lab partner lightly taps the object. Adjust the angle of the board until the object slides at a constant speed after it has received an initial light tap. Use the protractor to measure this angle and record it in Table 3 as the angle for sliding friction. The tangent of this angle is the coefficient of sliding friction.
6. Repeat the experiment with a different object so that each lab partner has data to analyze.

## Observations and Data

### Table 1

| | Description |
|---|---|
| Object | |
| Surface | |

### Table 2

| Weight of object (N) | | | | |
|---|---|---|---|---|
| | 1 | 2 | 3 | Average |
| Force of sliding friction (N) | | | | |

### Table 3

| Motion | Angle (degrees) | $\mu = \tan \theta$ |
|---|---|---|
| Static | | |
| Sliding | | |

## Analysis

1. From the data in Table 3, calculate the coefficients of static and sliding friction for the object used. Record these values in Table 3.

2. Explain any difference between the values for the coefficients of static and sliding friction.

3. Using the data in Table 2, calculate the coefficient of sliding friction. Show your work.

4. Are your values for $\mu_{sliding}$ from Questions 1 and 3 equal? Explain any differences.

5. A brick is positioned first with its largest surface in contact with an inclined plane. The plane is tilted at an angle to the horizontal until the brick just begins to slide, and the angle, $\theta$, of the plane with the horizontal is measured. Then the brick is turned on one of its narrow edges, the plane is tilted, and $\theta$ is again measured. Predict whether there will be a difference in these measured angles. Explain your answer in terms of the equation for the force of friction. Is the coefficient of static friction affected by the area of contact between the surfaces?

6. A brick is placed on an inclined plane, which is tilted at an angle to the horizontal until the brick just begins to slide. The angle, $\theta$, of the plane with respect to the horizontal is measured. Then the brick is wrapped in waxed paper and placed on a plane. The plane is tilted, and $\theta$ is again measured. Predict whether there will be a difference in these measured angles. Explain.

# 3 Friction

7. From your answers to the previous two questions, determine the factors that influence the force of friction.

## Application

While looking for a set of new tires for your car, you find an advertisement that offers two brands of tires, brand **X** and brand **Y**, at the same price. Brand **X** has a coefficient of friction on dry pavement of 0.90 and on wet pavement of 0.15. Brand **Y** has a coefficient of friction on dry pavement of 0.88 and on wet pavement of 0.45. If you live in an area with high levels of precipitation, which tire would give you better service? Explain.

## Extension

Using another object, measure its weight and the force necessary to pull it along a horizontal surface at a constant speed. Record the information in a data table similar to Table 2. Calculate the coefficient of sliding friction. Since $\mu = \tan\theta$, find the angle $\theta$. Estimate the angle at which sliding at a constant speed would begin. Using this as your prediction, slowly raise the flat board while lightly tapping the object. Record the angle at which the object begins sliding at a constant speed. Compare your prediction to your experimental results.

# EXPERIMENT 4 : Accelerated Motion

## Purpose

Observe and analyze the motion of a uniformly accelerated body moving in a straight line.

## Concept and Skill Check

The recording timer can be used to record the movement of a small cart as it is pulled across a table top by a falling mass. The resulting timer tape measures displacement of the moving cart per interval of time. From Experiment 7, you know that average velocity is equivalent to displacement for a given interval of time. The ratio of a change in velocity to a change in time yields acceleration $[a = (v_2 - v_1)/(t_2 - t_1)]$; this is the equation for the slope of a velocity versus time graph. With the data from this experiment, you will construct three graphs to analyze the motion of the cart: displacement versus time, velocity versus time, and acceleration versus time.

## Materials

laboratory cart
recording timer with
    necessary power supply
carbon discs

timer tape
500-g mass
masking tape
heavy string (1 m)

pulley
2 C-clamps
graph paper

## Procedure

1. Set up the apparatus as shown in Figure 1, placing the recording timer about 1 m from the edge of the table. Attach a 1-m length of timer tape to one end of the cart with a piece of masking tape. Tie a 1-m length of string to the opposite end of the cart and thread the string through the pulley. Do not attach the 500-g mass to the string at this time.

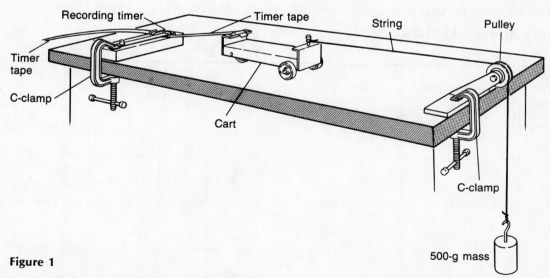

**Figure 1**

2. While one student holds the cart still, another student will attach the 500-g mass to the loose end of the string that is hanging over the table.
3. While still preventing the cart from moving, turn on the recording timer. Now release the cart, allowing the 500-g mass to pull the cart across the table.

4. Catch the cart at the edge of the table to prevent it from knocking the pulley loose or plunging to the floor.
5. Turn off the recording timer and inspect the timer tape. A dark dot should be visible at the beginning of the tape followed by a series of recognizable and distinct dots along the length of the tape.
6. Repeat Steps 1 through 5 so that each lab team member has a tape with recognizable dots to analyze.
7. Label the first distinguishable dot "0." Count off five dots (spaces) from 0 and mark this dot "1." Continue counting off five dots from each previous numbered one and marking them "2," "3," "4," etc., as shown in Figure 2. Each set of five spaces represents one time interval.

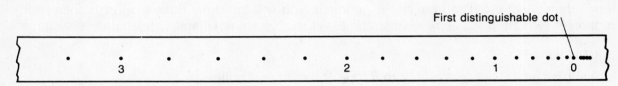

Figure 2. **Begin counting each set of five dots from the first distinguishable dot.**

8. Carefully measure the distance the cart traveled during each interval of time. Record that displacement in Table 1. Calculate the total displacement of the cart at each interval by summing the current displacement plus the displacements during the preceding intervals. Record the total displacement values in Table 1.
9. The average velocity of the laboratory cart during one interval of time has the same value as the displacement for that interval. Record the values for average velocity per interval in Table 2.

## Observations and Data

### Table 1

| Time (interval) | Displacement (cm) | Total displacement (cm) |
|---|---|---|
| 1 | | |
| 2 | | |
| 3 | | |
| 4 | | |
| 5 | | |

### Table 2

| Time (interval) | Average velocity (cm/interval) | Acceleration ($\Delta v/\Delta t$) |
|---|---|---|
| 1 | | |
| 2 | | |
| 3 | | |
| 4 | | |
| 5 | | |

# 4 Accelerated Motion

NAME _____

## Table 1 (continued)

| Time (interval) | Displacement (cm) | Total displacement (cm) |
|---|---|---|
| 6 | | |
| 7 | | |
| 8 | | |
| 9 | | |
| 10 | | |
| 11 | | |
| 12 | | |

## Table 2 (continued)

| Time (interval) | Average velocity (cm/interval) | Acceleration ($\Delta v/\Delta t$) |
|---|---|---|
| 6 | | |
| 7 | | |
| 8 | | |
| 9 | | |
| 10 | | |
| 11 | | |
| 12 | | |

## Analysis

1. On graph paper, plot the total displacement of the cart versus the time interval. Use the values from Table 1.
2. On graph paper, plot the average velocity versus the time interval. Use the values from Table 2.
3. Describe the graph of displacement versus time. What is the meaning of the graph?

4. Describe the velocity versus time graph. What is the meaning of the graph?

5. Calculate the slope of the velocity versus time graph for each time interval. Write the results of your calculations in Table 2. What is the unit for the slope?

6. On graph paper, plot acceleration versus time. Describe this graph. What does your graph show about the acceleration of the cart?

## Application

What would be the advantage of building an interplanetary spacecraft that could accelerate at 1 $g$ (9.8 m/s$^2$) for a year?

## Extension

1. At several different points along the total displacement versus time graph, draw a tangent to the curve and find the slope at each of the points. Compare the slopes to the average speeds for the various time intervals.

2. Using the actual period of your timer, convert the time unit (interval) to seconds. Compute your acceleration in m/s$^2$. How does your cart's acceleration compare to acceleration due to gravity?

# EXPERIMENT 5 : Acceleration Due to Gravity

## Purpose

Observe a falling object and determine the acceleration due to gravity.

## Concept and Skill Check

The recording timer can be used to record the displacement of a falling mass. The resulting tape is used to analyze the accelerated motion of the mass. In this experiment, you must know the period of the timer. If the period is unknown, use the procedure described in Experiment 6 to determine this value.

The average velocity during an interval of time is found by the equation

$$v = \frac{\Delta d}{\Delta t},$$

where $\Delta d$ is the distance traveled during an interval of time, $\Delta t$. A uniformly accelerated object will produce a straight (but not horizontal) line on a graph that plots velocity versus time. The slope of the velocity versus time graph is the acceleration. The slope is found from the ratio

$$\text{slope} = \frac{\text{rise}}{\text{run}} = \frac{(v_f - v_i)}{(t_f - t_i)}.$$

An object dropped from rest will travel a distance given by the following equation

$$d = \frac{1}{2} gt^2.$$

Therefore, since $d$ and $t$ have been measured, $g$ may be calculated from

$$g = \frac{2d}{t^2}.$$

The Free Fall Adapter apparatus uses the computer to measure accurately the time required for a steel marble to fall from the top of the free fall apparatus to a sensor pad. The computer software averages the times for the trials and, after the free fall distance is entered, calculates the acceleration due to gravity.

## Materials

### Part A

| | | |
|---|---|---|
| recording timer with necessary power supply | carbon discs | 1-kg mass |
| timer tape | C-clamp | meter stick |
| | masking tape | graph paper |

### Part B

| | | |
|---|---|---|
| Apple II+, IIe, or IIGS computer or IBM computer | PASCO Free Fall Adapter, #ME-9207A (Apple) or #ME-9207 (IBM) | meter stick ringstand |
| PASCO game port interface, #AI-6575 (Apple) or #CI-6588 (IBM) and IBM interface card #SE-6590 | PASCO Precision Timer III software, #SV-7401 (Apple) or #SV-7413 (IBM) | |

# 5 Acceleration Due to Gravity

## Procedure

### Part A

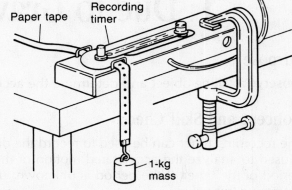

**Figure 1.**

Paper tape / Recording timer / 1-kg mass

1. Record the period of your recording timer on the line provided above Table 1.
2. Set up the apparatus as shown in Figure 1. Insert a 1.5-m strip of timer tape into the recording timer. Use the masking tape to attach the timer tape to the 1-kg mass.
3. Start the recording timer and drop the mass. Stop the recording timer when the mass hits the floor.
4. Remove the timer tape from the mass and write a zero below the first distinguishable dot. Number every subsequent dot consecutively 1, 2, 3, 4, 5, and so on. The elapsed time for the intervals can be determined by finding the product of the interval number and the period of the timer. Calculate the time for each interval and record these values in Table 1.
5. Carefully measure in meters the distance traveled during each interval of time (the space between dots). Record the displacement during each interval in Table 1.
6. The total displacement from zero to any numbered point along the timer tape is the sum of the measured distances between consecutive numbers on your tape. Record in Table 1 the total displacement of the mass during the corresponding intervals.
7. Calculate the average velocity during each interval and record these values in Table 1.

### Part B

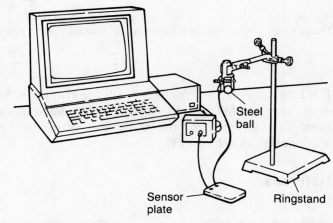

**Figure 2.**

Steel ball / Ringstand / Sensor plate

1. Set up the apparatus according to the instructions for the Free Fall Adapter and as shown in Figure 2. Allow a distance of 1 m between the sensor pad and the release mechanism.
2. Insert the ball into the Free Fall Adapter mechanism. Carefully measure in meters the distance from the bottom of the ball to the sensor pad (the distance the ball will fall). Record this distance in Table 2.
3. Release the ball. The elapsed time is automatically recorded into the computer data table. Record this time in Table 2.
4. Replace the ball in the Free Fall Adapter and drop it again. Repeat this step at least five times. Record the elapsed time for each trial in Table 2.
5. Select the Analysis section of the program. The software automatically calculates and displays on the screen the average time and standard deviation of the times. Record the average time in Table 2. Enter the free fall distance into the computer when prompted to do so. The computer calculates the value for the acceleration due to gravity. Record this value in Table 2.
6. Increase the free fall distance to about 2 m. Repeat Steps 2 through 5. Record your data in Table 3.

## Observations and Data

## Table 1

Period of recording timer _____ s

| Interval | Time (s) | Displacement (m) | Total displacement (m) | Average velocity (m/s) |
|----------|----------|------------------|------------------------|------------------------|
| 1 | | | | |
| 2 | | | | |
| 3 | | | | |
| 4 | | | | |
| 5 | | | | |
| 6 | | | | |
| 7 | | | | |
| 8 | | | | |
| 9 | | | | |
| 10 | | | | |
| 11 | | | | |
| 12 | | | | |
| 13 | | | | |
| 14 | | | | |
| 15 | | | | |

# EXPERIMENT 5 : Acceleration Due to Gravity

## Table 1 (continued)

Period of recording timer ———————————— s

| Interval | Time (s) | Displacement (m) | Total displacement (m) | Average velocity (m/s) |
|----------|----------|------------------|------------------------|------------------------|
| 16 | | | | |
| 17 | | | | |
| 18 | | | | |
| 19 | | | | |
| 20 | | | | |

## Table 2

Free fall distance ———————————— m

| Trial | Time (s) |
|-------|----------|
| 1 | |
| 2 | |
| 3 | |
| 4 | |
| 5 | |
| Average | |

$g =$ ———————————— $m/s^2$

## Table 3

Free fall distance ———————————— m

| Trial | Time (s) |
|-------|----------|
| 1 | |
| 2 | |
| 3 | |
| 4 | |
| 5 | |
| Average | |

$g =$ ———————————— $m/s^2$

## Analysis

### Part A

1. On graph paper, plot the total displacement of the mass versus time. Use the values in Table 1.
2. On graph paper, plot the average velocity versus time. Use the values in Table 1.
3. Write a brief explanation describing what the graph of total displacement versus time indicates about the motion of the falling mass.

4. Study the graph of velocity versus time. Write a brief explanation of what the graph indicates about the motion of a falling mass.

5. Calculate the slope of the velocity versus time graph. Compare your results for acceleration due to gravity to the reference value, $9.80$ m/s$^2$, by finding the relative error.

$$\text{relative error} = \frac{(\text{experimental result} - \text{reference value})}{\text{reference value}} \times 100\%$$

### Part B

1. Calculate the relative error by comparing your results for each free fall distance to $9.80$ m/s$^2$.
   a.

   b.

2. If the free fall distance is increased to about 2 m, what effect does it have on the results? Explain.

3. Which variable remained constant in each set of trials? Which variable responded in this experiment?

## Application

Would there be any potential benefit to athletes when the Olympic games are held in a country having a relatively high elevation?

EXPERIMENT
# 6 Range of a Projectile

## Purpose

Predict where a horizontally projected object will land.

## Concept and Skill Check

If an object moves in only one dimension, it is easy to describe the location of that object and to predict its final position when it comes to rest. On a calm day, an apple falls vertically from a tree limb to the ground below. However, a golf ball hit from the tee, a football thrown over the defensive line to a downfield receiver, and bullets fired from a gun move in two dimensions. Such objects, called projectiles, have horizontal and vertical components of motion that are independent of each other. The independence of vertical and horizontal motion is used to determine the location and the time of fall of projected objects. All projectiles described above had an initial vertical velocity at an angle above the horizontal. To simplify analysis of the motion of the projectile in this experiment, the steel ball will be launched horizontally; therefore, the initial vertical velocity, $v_y$, will be zero.

If you know the velocity, $v_x$, of a steel ball launched horizontally from a table and the ball's initial height, $y$, above the floor, the equations of projectile motion can be used to predict where the steel ball will land. Recall that the horizontal displacement or range, $x$, of an object with horizontal velocity, $v_x$, at time, $t$, is

$$x = v_x t.$$

The equation describing vertical displacement is that of a body falling with constant acceleration:

$$y = v_y t + \frac{1}{2} g t^2.$$

But if $v_y = 0$, then $y = \frac{1}{2} g t^2$.

The steel ball's fall time can be computed from the equation for vertical motion by solving for $t$:

$$t = \sqrt{\frac{2y}{g}},$$

where $g$ is the acceleration due to gravity and $y$ is the vertical height the projectile falls. The range of the projectile, $x$, can now be computed. In this experiment, you will try to predict the range of a projectile launched with a certain horizontal velocity.

## Materials

| | | |
|---|---|---|
| 2 50-cm pieces of U-shaped channel (track) | ringstand-support clamp | stopwatch or electronic timing apparatus with 2 |
| steel ball | meter stick | photogates and lights |
| masking tape | string (1 m) | paper cup |
| ringstand | washer | |

## Procedure

1. Set up the apparatus as shown in the figure on the next page. Construct horizontal and inclined ramps on the table top, using the U-channel, ringstand, support clamps, and masking tape. At the end of the track, use the string to suspend a washer from the edge of the table to the floor.

# EXPERIMENT 6
# Range of a Projectile

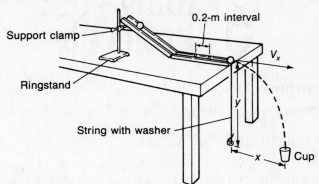

**Figure 1.**

Align the end of the horizontal ramp with the edge of the table. Make a few test rolls, but catch the ball as it rolls off the table. The steel ball should stay in the U-shaped channels when you roll it down the inclined ramp. If the steel ball jumps out of the channel at the junction of the two channel pieces, then lower the support clamp and try again.

2. Set up the electronic timing apparatus with photogates along a section of the horizontal channel. Place the photogates 0.20 m apart. Record the distance as $d$ in Table 1. If you are using a stopwatch, measure a 0.20-m interval along the horizontal channel. Use two small pieces of tape to mark the beginning and the end of the interval. Record the interval distance as $d$ in Table 1.

3. Select a point on the incline where you will release the ball. Mark this location with a small piece of tape.

4. In order to determine the ball's horizontal velocity, you must measure the time it takes for the ball to travel the known distance, $d$, along the horizontal channel. Release the ball from the tape mark on the ramp. Using either the stopwatch or the electronic timer, measure the time it takes for the ball to roll past the tape marks or photogates. Catch the ball as it rolls off the table. Record the time, $t$, in Table 1. Roll the ball twice more, releasing it for each trial from the same location on the ramp. Record these times in Table 1.

5. Place the steel ball on the channel at the edge of the table. Measure the vertical distance, in meters, from the bottom of the ball to the floor. Record this distance, $y$, on the line provided under Table 1.

## Observations and Data

### Table 1

| Trial | Distance (m) | Time (s) |
|---|---|---|
| 1 | | |
| 2 | | |
| 3 | | |
| Average | | |

Vertical distance, $y$, _____ m

## Analysis

1. Calculate the average time for the three trials. Write this value in Table 1. Compute the horizontal velocity, $v_x$, using $v_x = \Delta d / \Delta t$.

2. Calculate the time, $t$, that the ball will be falling from the table, using your value for $y$.

3. Calculate the horizontal distance, $x$, that the ball should travel, using your values for $v_x$ and $t$ from Questions 1 and 2.

4. Measure the distance, $x$, in a straight line from the horizontal channel, starting where the washer touches the floor. Place a small piece of tape at this location. Place the paper cup with the front edge of the cup over the tape and the back side of the cup toward the table. Roll the steel ball from the same height on the inclined ramp from which it was released earlier. This time, let the ball roll off the table. Where does it land?

5. Does this experiment support the premise that the horizontal and vertical components of motion are not affected by each other? Explain.

6. If a very light, but fairly large, sponge ball, or a ping-pong ball, were rolled down the ramp would you expect the same result? Explain.

## Application

A diver achieves a horizontal velocity of 3.75 m/s from a diving platform located 6.0 m above the water. How far from the edge of the platform will the diver be when she hits the water?

## Extension

1. The horizontal range, $x$, of a projectile launched at an angle, $\theta_i$, with the horizontal at an initial velocity, $v_i$, can be determined by use of the following:

$$x = \frac{v_i^2}{g} \sin 2\,\theta_i.$$

Prove that $x$ has a maximum value when $\theta_i = 45°$. Give some examples. Recall that $\sin 2\theta = 2 \sin \theta \cos \theta$.

2. Find the horizontal range of a baseball that leaves the bat at an angle of 63° with the horizontal with an initial velocity of 160 km/h. Disregard air resistance. Show all your calculations.

EXPERIMENT

# 7 : Newton's Second Law

## Purpose

Investigate Newton's second law of motion.

## Concept and Skill Check

Newton's second law of motion states that the acceleration of a body is directly proportional to the net force on it and inversely proportional to its mass. In mathematical form, this law is expressed as $a = F/m$. An object will accelerate if it has a net force acting on it, and the acceleration will be in the direction of the force. In this experiment, a laboratory cart will be accelerated by a known force, and its acceleration will be measured using a recording timer. The product of the total mass accelerated and its acceleration equals the force causing the acceleration.

In order to calculate the net force acting on the laboratory cart, frictional forces that oppose the motion must be offset. If the cart moves at a constant velocity, then the net force acting on the cart is zero because the acceleration is zero. The frictional force can be neutralized by providing enough small masses at the end of the string to make the cart move forward at a constant velocity.

If you assume that the laboratory cart experiences a uniform acceleration due to the falling mass, then the relationship

$$d = v_i t + \frac{1}{2} at^2$$

applies. If the cart has an initial velocity of zero, the equation becomes

$$d = \frac{1}{2} at^2.$$

If the displacement in a given time is known, you can solve for acceleration with

$$a = 2\frac{d}{t^2}.$$

## Materials

| | | |
|---|---|---|
| recording timer with necessary power supply | 1000-g mass | heavy string (1.5 m) |
| timer tape | set of small masses | masking tape |
| carbon-paper discs | pulley | balance |
| laboratory cart | 2 C-clamps | meter stick |

## Procedure

1. Record the period of the timer in Table 1. If you do not know the period, use the procedure described in Experiment 6 to determine the period.
2. Determine the combined mass of your laboratory cart and attached timer tape and string. Record the mass in Table 1. Record the weight, $F = mg$, of the 1000-g mass in Table 2 as the accelerating force.
3. Assemble the cart, pulley, recording timer, string, and timer tape as shown in Figure 1 on the next page. Tie a small loop in the end of the string hanging from the pulley.
4. Give the cart a small push. It should roll to a stop in a few centimeters. Attach a 10-g mass with a piece of masking tape to the end of the string that is hanging from the pulley. Give the cart a

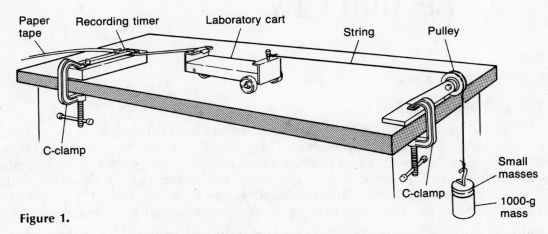

**Paper tape** · **Recording timer** · **Laboratory cart** · **String** · **Pulley** · **C-clamp** · **Small masses** · **1000-g mass** · **C-clamp**

**Figure 1.**

small push toward the pulley and observe its motion. If the cart moves at a constant velocity, then the weight of the 10-g mass is equal to the force of friction in the cart's wheels and from the timer tape sliding through the recording timer. If the cart rolled to a stop, then the total mass on the string must be increased. If the cart's velocity increased after it was released, then the total mass on the string must be decreased. Adjust the total mass until the cart moves at a constant velocity after you have given it a small push. Record in Table 1 the mass needed to equalize the friction of the cart. Leave the small masses attached to the string.

5. Move the cart next to the recording timer. Attach the timer tape to the cart. Insert the timer tape into the recording timer. Carefully hang the 1000-g mass from the loop in the string. Adjust the string length so that the mass is hanging just under the pulley. Hold the cart to prevent it from moving. Turn on the recording timer. Release the cart, allowing it to accelerate across the table top. Catch the cart before it collides with the pulley or plunges to the floor. Turn off the recording timer. Remove the timer tape and inspect it. A dark area should occur at the beginning of the tape where the timer made numerous dots close together before the cart was released. Note a series of dots, each an increasing distance from the previous one, as shown in Figure 2. It is in this interval on the tape that the cart was accelerating. Write a zero by the first distinguishable dot. Continue counting and marking dots 1, 2, 3, 4, 5, and so on. After the mass hits the floor, the dots will clump together, indicating the end of the acceleration phase. Stop counting at a dot located about 10 cm before the clump of dots.

First distinguishable dot

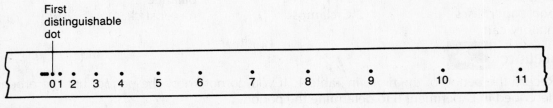

0 1 2 3 4 5 6 7 8 9 10 11

Beginning of timer tape

**Figure 2.**

6. Measure in meters the distance from the zero dot to the chosen end dot. Record this distance in Table 2. Record the number of dots to this end point in Table 2.

7. Repeat Steps 5 and 6 twice more for a total of three trials.

# 7 Newton's Second Law

## Observations and Data

### Table 1

|  | Value |
|---|---|
| Period of timer (s) |  |
| Mass of laboratory cart (kg) |  |
| Mass needed to equalize friction of cart (kg) |  |

### Table 2

| Trial | Accelerating force (N) | Distance (m) | Number of dots | Time (s) |
|---|---|---|---|---|
| 1 |  |  |  |  |
| 2 |  |  |  |  |
| 3 |  |  |  |  |

### Table 3

| Trial | Acceleration (m/s²) | Total mass (kg) | (m)(a) (N) |
|---|---|---|---|
| 1 |  |  |  |
| 2 |  |  |  |
| 3 |  |  |  |

## Analysis

1. Determine the time by multiplying the number of dots by the timer period. For example, 30 dots and a timer period of 1/60 s (30 dots × 1/60 s/dot) will yield the time value of 0.5 s. Record the times for the three trials in Table 2.

2. Calculate the acceleration of the entire system using the values for distance and time from Table 2. Show your calculations for all trials. Record the acceleration value for each trial in Table 3.

3. Are your acceleration values less than, equal to, or greater than *g*? Are these the values you would have predicted? Explain.

4. The total mass that was accelerated is equal to the mass of the cart, tape, mass and string, the small masses needed to equalize friction, and the 1000-g mass. Calculate the total mass of your system and record this value in Table 3.

5. Calculate the product of the total mass and the acceleration for each trial. Record these values in Table 3.

6. Compare the results for the product $(m)(a)$ to the accelerating force. Find the relative error using the accelerating force as the reference value.

7. Using your data, calculate the frictional force.

8. Why was it important to neutralize the effect of frictional force acting on the system?

## Application

At a specific engine rpm, an automobile engine provides a constant force that is applied to the automobile. If the car is traveling on a horizontal surface, does the car accelerate when this force is applied? Explain.

EXPERIMENT

# 8 Conservation of Energy

## Purpose

Verify the law of conservation of energy with an inclined plane.

## Concept and Skill Check

In the absence of friction, the work done to pull an object up an inclined plane is equal to the work done to lift it straight up to the same height. The change in gravitational potential energy of an object caused by raising it to a given height is independent of the path through which the object is moved to reach that height.

When an object is lifted straight up, work is done only against gravity. The work done is equivalent to the increase in the gravitational potential energy of the object, *mgh*. In order to move the object up a plane, work must be done against gravity and friction. Therefore, to compare the work required to raise the object to the same height but over different paths, you must calculate the work done pulling the object up the plane in such a way as to remove from the comparison the work done against friction.

## Materials

inclined plane
smooth wood block with hook in one end

spring scale (calibrated in newtons)
meter stick

string (1 m)
silicone spray

## Procedure

1. Weigh the block of wood and enter this value in newtons on the line provided above Table 1. If your scale reads in grams, convert the reading to kilograms and multiply by 9.80 m/s² to find the weight in newtons.
2. Clean and spray with silicone the surfaces of the plane and block to make them as smooth as possible.
3. Set the plane at an angle, $\theta$, so that the block will just slide down without being pushed. Measure the length, *d*, and the height, *h*, of the high end of the plane, as shown in Figure 1, and record these values in Table 1.

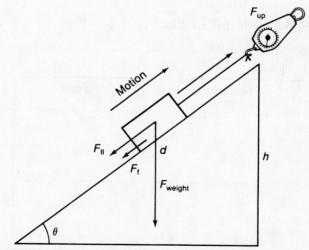

**Figure 1. Be careful to pull the block in a direction parallel to the plane.**

4. Hook the spring scale to the block and pull the block up the plane at constant velocity. While the block is moving at constant velocity, have your lab partner read the applied force on the scale. Record the value in Table 1 as $F_{up}$. The force you applied, $F_{up}$, is the equilibrant of the component of the block's weight parallel to the plane, $F_{\parallel}$, plus the force of friction, $F_f$, described by

$$F_{up} = (F_{\parallel} + F_f).$$

5. Place the block at the top of the incline with the spring scale attached to the hook by a piece of string. Let the block slide down the plane at constant velocity and have your lab partner read and record in Table 1 the force, $F_{down}$. The frictional force has the same magnitude as before, but is exerted in the opposite direction. Thus, the force you exert is the equilibrant of the parallel component of the weight minus the frictional force described by

$$F_{down} = (F_{\parallel} - F_f).$$

6. Repeat Steps 3 through 5 twice more using different angles $\theta$. Keep the height constant in all the trials. Each time you change the angle, $\theta$, measure the length of the plane, $d$, to the point at which the surface of the plane is at height, $h$, from the table, as shown in Figure 2. Note that as the angle becomes larger, the length of the plane becomes shorter.

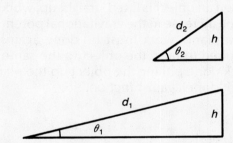

Figure 2. Vary the length, $d$, but retain the same height, $h$, of the plane for Step 6.

## Observations and Data

Weight of the block _____ N

## Table 1

| Trial | Length, $d$ (m) | Height, $h$ (m) | $F_{up}$ (N) | $F_{down}$ (N) |
|-------|-----------------|-----------------|--------------|----------------|
| 1     |                 |                 |              |                |
| 2     |                 |                 |              |                |
| 3     |                 |                 |              |                |

## Table 2

| Trial | $F_\parallel$ (N) | Work done without friction, $F_\parallel d$ (J) | Potential energy $mgh$ (J) | Work input $F_{up}d$ (J) |
|-------|------------------|------------------------------------------------|----------------------------|--------------------------|
| 1     |                  |                                                |                            |                          |
| 2     |                  |                                                |                            |                          |
| 3     |                  |                                                |                            |                          |

## Analysis

1. Add the two equations for $F_{up}$ and $F_{down}$ from Steps 4 and 5 and solve for $F_\parallel$.

2. Calculate the value of $F_\parallel$ for each trial and enter these values in Table 2.

3. For each trial, calculate and record in Table 2 the work done to pull the block up a frictionless plane. Use the calculated $F_\parallel$ and the length of the plane to solve $W = F_\parallel d$.

4. Calculate and record the potential energy, $mgh$, of the block at height $h$.

5. Compare the work required to pull the block up the plane set at various angles with the work required to lift the block straight up. Is energy conserved?

## Application

A 60.0-kg student skis down an icy, frictionless hill with a vertical drop of 10.0 m. Explain how the law of conservation of energy applies to this situation and calculate how fast she will be going when she reaches the bottom of the hill.

## Extension

Calculate the angle, $\theta$, for each trial. Using your calculated angle, calculate the theoretical values for $F_{\parallel}$. Explain any differences between these theoretical values and those you obtained in Table 2.

# EXPERIMENT

# 9 Conservation of Momentum

## Purpose

Investigate the conservation of momentum for a collision in two dimensions.

## Concept and Skill Check

Recall that the law of conservation of momentum states: The momentum of any closed, isolated system does not change. This law is true regardless of the number of objects or the directions of the objects before and after they collide. When there is a collision in two dimensions, the sum of the momentum components in the horizontal direction before the collision equals the sum of the momentum components in the horizontal direction after the collision, and, likewise, the sum of the momentum components in the vertical direction before the collision equals the sum of the momentum components in the vertical direction after the collision.

In this investigation, a steel ball rolls down an incline and collides with another steel ball of equal mass that is at rest. The force of gravity accelerates each steel ball vertically at an equal rate, and they strike the floor at the same time. The horizontal distance that each ball travels during the time of fall is the distance from a point on the floor just below the initial position of the target ball to the point where it lands. These horizontal distances can be measured directly along the floor.

When steel balls of equal mass are used for the incident ball ($m$) and the target ball ($m'$), the mass of each can be designated one mass unit ($m = m' = 1$). Since the time required for each ball to reach the floor is the same, this interval can be designated one time unit ($t = t' = 1$). Recall that $v = \Delta d/\Delta t$; but if $\Delta t = 1$, then $v = \Delta d$. Thus, the horizontal velocity of a steel ball during this time interval is equal to the horizontal distance it travels. This same horizontal distance can also represent the momentum of each ball because $p = mv$, but since $m = 1$, $p = v$. Therefore, the momentum vector for each ball can be represented by its horizontal distance and direction as measured on the floor.

## Materials

2 steel balls of equal mass
C-clamp
apparatus for a collision in two dimensions
meter stick
4 sheets carbon paper
4 sheets tracing paper
masking tape
string (1.5 m)
washer

## Procedure

1. Assemble the apparatus as shown in Figure 1. If your apparatus does not have an attached plumb line, a washer on a string suspended from the apparatus can serve as a plumb line. The set screw at the bottom of the inclined ramp has a small depression in

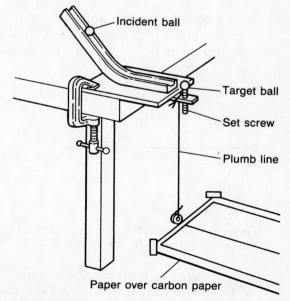

**Figure 1.**

Labels in figure: Incident ball; Target ball; Set screw; Plumb line; Paper over carbon paper

its top to hold the target ball. Adjust the depression in the set screw so that it is positioned about one radius of the steel ball away from the bottom of the inclined ramp. Roll a steel ball down the inclined ramp and observe the ball as it travels over the set screw. Adjust the set screw so that the steel ball is just able to pass over it.

2. Tape four pieces of carbon paper together to form one large sheet of carbon paper. Tape four pieces of tracing paper together to form one large sheet of tracing paper. Set the carbon paper on the floor, with the carbon side up, and place the tracing paper over it. Arrange the center of one edge of the paper so that it is located just below the plumb line. Tape the paper in place on the floor. The spot just below the plumb line should be marked 0.

3. If your inclined ramp has an adjustable slider, move it about two-thirds of the distance up the incline and secure it. Otherwise, use a piece of tape to mark a position at this same point on the incline. Without placing a target ball on the set screw, roll a steel ball from the marked position on the incline and let it fall onto the paper. Roll the incident ball four more times from the same location on the incline; circle and label the cluster of points on the paper where the initial incident ball lands.

4. Adjust the position of the set screw so that the incident ball will collide with the target ball at an angle of approximately 45°, as shown in Figure 2. Make sure that the two balls are at the same height above the floor at the time of collision.

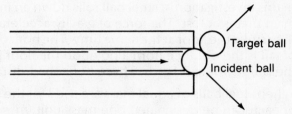

**Figure 2. Adjust the position of the target ball so that the two steel balls are deflected at an angle.**

5. Place the target ball on the set screw and release the incident ball from the marked position on the incline. Try five collisions between the incident and target balls, making sure that you release the incident ball each time from the same location on the incline. Circle and label the clusters of points where the incident and the target balls hit the paper.

6. Draw a vector from the point 0, under the plumb line, to a spot in the center of the initial incident cluster of points. This vector represents the initial momentum of the incident ball. Label this vector $p_{initial}$. Measure the magnitude of the initial momentum vector and record this value in Table 1.

(a)                                    (b)

**Figure 3. (a) Move the vectors in a parallel manner until they begin at a point 0. (b) Then add the momentum of the incident ball to the momentum of the target ball.**

7. Draw a vector from the point 0 to a spot in the center of the cluster of points where the target ball landed, as shown in Figure 3. Label this vector $p_{target}$. Measure its magnitude and record this value in Table 1.

# Conservation of Momentum

8. Draw a third vector from the point 0 to a spot in the center of the cluster of points where the incident ball landed. Label this vector $p_{incident}$. Measure its magnitude and record this value in Table 1.

## Observations and Data

### Table 1

| Magnitude of Vectors | | |
|---|---|---|
| $p_{initial}$ (cm) | $p_{target}$ (cm) | $p_{incident}$ (cm) |
| | | |

## Analysis

1. Add the magnitudes of $p_{target}$ and $p_{incident}$. Do they equal $p_{initial}$? Explain.

2. Add the vector representing the momentum of the target ball to the vector representing that of the incident ball to determine the total momentum after the collision. Label this vector $p_{final}$.

3. Measure the magnitude of $p_{final}$. Compare the initial momentum to the final momentum. Explain your results.

4. Make a generalization about the direction of $p_{final}$ and $p_{initial}$.

# Conservation of Momentum

NAME ———————————————————

## Application

While playing baseball with your friends, your hands begin to sting after you catch several fast balls. What method of catching the ball might prevent this stinging sensation?

## Extension

1. If the incident ball, a, had hit two target balls, b and c, at an angle, predict what would have happened to the total momentum of the system after the collision. Explain your answer and write an equation that proves your answer.

2. In this experiment, you found that the velocity vectors formed a closed triangle (ideally). When an incident ball, a, collides at an angle with a target ball, b, of equal mass that is initially at rest, the two balls always move off at right angles to each other after the collision. Use a familiar equation for a right triangle to show that this statement is true. *Hint:* Since the collision is elastic, kinetic energy is conserved and

$$\frac{1}{2} m v_a^2 = \frac{1}{2} m v'_a{}^2 + \frac{1}{2} m v_b^2.$$

## EXPERIMENT
# 10 : Kepler's Laws

## Purpose

Plot a planetary orbit and apply Kepler's Laws.

## Concept and Skill Check

The motion of the planets has intrigued astronomers since they first gazed at the stars, moon, and planets filling the evening sky. But the old ideas of eccentrics and equants (combinations of circular motions) did not provide an accurate accounting of planetary movements. Johannes Kepler adopted the Copernican theory that Earth revolves around the sun (heliocentric, or sun-centered, view) and closely examined Tycho Brahe's meticulously recorded observations on Mars' orbit. With these data, he concluded that Mars' orbit was not circular and that there was no point around which the motion was uniform. When elliptical orbits were accepted, all the discrepancies found in the old theories of planetary motion were eliminated. From his studies, Kepler derived three laws that apply to the behavior of every satellite or planet orbiting another massive body.

1. The paths of the planets are ellipses, with the center of the sun at one focus.
2. An imaginary line from the sun to a planet sweeps out equal areas in equal time intervals, as shown in Figure 1.
3. The ratio of the squares of the periods of any two planets revolving about the sun is equal to the ratio of the cubes of their respective average distances from the sun. Mathematically, this relationship can be expressed as

$$\frac{T_a^2}{T_b^2} = \frac{r_a^3}{r_b^3}.$$

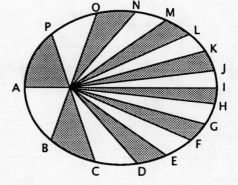

**Figure 1. Kepler's law of areas.**

In this experiment, you will use heliocentric data tables to plot the positions of Mercury on polar graph paper. Then you will draw Mercury's orbit. The distance from the sun, the radius vector, is compared to Earth's average distance from the sun, which is defined as 1 astronomical unit or 1 AU. The angle, or longitude, between the planet and a reference point in space is measured from the zero degree point, or vernal equinox.

## Materials

polar graph paper        sharp pencil        metric ruler

## Procedure

1. Orient your polar graph paper so that the zero degree point is on your right as you view the graph paper. The sun is located at the center of the paper. Label the sun without covering the center mark. Move about the center in a counter-clockwise direction as you measure and mark the longitude.
2. Select an appropriate scale to represent the values for the radius vectors of Mercury's positions. Since Mercury is closer to the sun than is Earth, the value of the radius vector will always be less than 1 AU. In this step, then, each concentric circle could represent one-tenth of an AU.

EXPERIMENT

# 10 Kepler's Laws

3. Table 1 provides the heliocentric positions of Mercury over a period of several months. Select the set of data for October 1 and locate the given longitude on the polar graph paper. Measure out along the longitude line an appropriate distance, in your scale, for the radius vector for this date. Make a small dot at this point to represent Mercury's distance from the sun. Write the date next to this point.
4. Repeat the procedure, plotting all given longitudes and associated radius vectors.
5. After plotting all the data, carefully connect the points of Mercury's positions and sketch the orbit of Mercury.

## Observations and Data

## Table 1

### Some Heliocentric Positions for Mercury for 0ʰ Dynamical Time*

| Date | Radius vector (AU) | Longitude (degrees) | Date | Radius vector (AU) | Longitude (degrees) |
|---|---|---|---|---|---|
| Oct. 1, 1990 | 0.319 | 114 | Nov. 16 | 0.458 | 280 |
| 3 | 0.327 | 126 | 18 | 0.452 | 285 |
| 5 | 0.336 | 137 | 20 | 0.447 | 291 |
| 7 | 0.347 | 147 | 22 | 0.440 | 297 |
| 9 | 0.358 | 157 | 24 | 0.432 | 304 |
| 11 | 0.369 | 166 | 26 | 0.423 | 310 |
| 13 | 0.381 | 175 | 28 | 0.413 | 317 |
| 15 | 0.392 | 183 | 30 | 0.403 | 325 |
| 17 | 0.403 | 191 | Dec. 2 | 0.392 | 332 |
| 19 | 0.413 | 198 | 4 | 0.380 | 340 |
| 21 | 0.423 | 205 | 6 | 0.369 | 349 |
| 23 | 0.432 | 211 | 8 | 0.357 | 358 |
| 25 | 0.440 | 217 | 10 | 0.346 | 8 |
| 27 | 0.447 | 223 | 12 | 0.335 | 18 |
| 29 | 0.453 | 229 | 14 | 0.326 | 29 |
| 31 | 0.458 | 235 | 16 | 0.318 | 41 |
| Nov. 2 | 0.462 | 241 | 18 | 0.312 | 53 |
| 4 | 0.465 | 246 | 20 | 0.309 | 65 |
| 6 | 0.466 | 251 | 22 | 0.307 | 78 |
| 8 | 0.467 | 257 | 24 | 0.309 | 90 |
| 10 | 0.466 | 262 | 26 | 0.312 | 102 |
| 12 | 0.464 | 268 | 28 | 0.319 | 114 |
| 14 | 0.462 | 273 | 30 | 0.327 | 126 |

*Adapted from *The Astronomical Almanac for the Year 1990*, U.S. Government Printing Office, Washington, D.C., 20402, p. E9.

EXPERIMENT

# 10 Kepler's Laws

## Analysis

1. Does your graph of Mercury's orbit support Kepler's law of orbits?

2. Draw a line from the sun to Mercury's position on December 20. Draw a second line from the sun to Mercury's position on December 30. The two lines and Mercury's orbit describe an area swept by an imaginary line between Mercury and the sun during the ten-day interval of time. Lightly shade this area. Over a small portion of an ellipse, the area can be approximated by assuming the ellipse is similar to a circle. The equation that describes this value is

$$\text{area} = (\theta/360°)\pi r^2,$$

where $r$ is the average radius for the orbit.

Determine $\theta$ by finding the difference in degrees between December 20 and December 30. Measure the radius at a point midway in the orbit between the two dates. Calculate the area in AUs for this ten-day period of time.

3. Select two additional ten-day periods of time at points distant from the interval in Question 2 and shade these areas. Calculate the area in AUs for each of these ten-day periods.

4. Find the average area for the three periods of time from Questions 2 and 3. Calculate the relative error between each area and the average. Does Kepler's law of areas apply to your graph?

5. Calculate the average radius for Mercury's orbit. This can be done by averaging all the radius vectors or, more simply, by averaging the longest and shortest radii that occur along the major axis. The major axis is shown in Figure 2. Recall that the sun is at one focus; the other focus is located at a point that is the same distance from the center of the ellipse as the sun, but in the opposite direction.

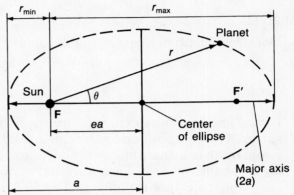

**Figure 2. The major axis passes through the two foci (F and F′) and the center of the ellipse. The value *ea* determines the location of the foci; *e* is the eccentricity of the orbit. If *e* = 0, the orbit is a circle, and the foci merge at one, central point.**

From Table 1, find the longest radius vector. Then, align a metric ruler so that it describes a straight line passing through the point on the orbit that represents the longest radius vector and through the center of the sun to a point opposite on the orbit. Find the shortest radius vector by reading the longitude at this opposite point and consulting Table 1 for the corresponding radius vector. Average these two radius vector values. Using the values for Earth's average radius (1.0 AU), Earth's period (365.25 days), and your calculated average radius of Mercury's orbit, apply Kepler's third law to find the period of Mercury. Show your calculations.

6. Refer again to the graph of Mercury's orbit that you plotted. Count the number of days required for Mercury to complete one orbit of the sun; recall that this orbital time is the period of Mercury. Is there a difference in the two values (from Questions 5 and 6) for the period of Mercury? Calculate the relative difference in these two values. Are the results from your graph consistent with Kepler's law of periods?

# 10 Kepler's Laws

## Application

There has been some discussion about a hypothetical planet **X** that is on the opposite side of the sun from Earth and that has an average radius of 1.0 AU. If this planet exists, what is its period? Show your calculations.

## Extension

Using the data in Table 2, plot the radius vectors and corresponding longitudes for Mars. Does the orbit you drew support Kepler's law of ellipses? Select three different areas and find the area per day for each of these. Does Kepler's law of areas apply to your model of Mars?

## Table 2

| | Some Heliocentric Positions for Mars for $0^h$ Dynamical Time* | | | | | |
|---|---|---|---|---|---|---|
| Date | Radius vector (AU) | Longitude (degrees) | | Date | Radius vector (AU) | Longitude (degrees) |
| Jan.  1, 1990 | 1.548 | 231 | | July  12 | 1.382 | 343 |
| 17 | 1.527 | 239 | | 28 | 1.387 | 353 |
| Feb.  2 | 1.507 | 247 | | Aug.  13 | 1.395 | 3 |
| 18 | 1.486 | 256 | | 29 | 1.406 | 13 |
| Mar.  6 | 1.466 | 265 | | Sept. 14 | 1.420 | 23 |
| 22 | 1.446 | 274 | | 30 | 1.436 | 32 |
| Apr.  7 | 1.429 | 283 | | Oct.  16 | 1.455 | 42 |
| 23 | 1.413 | 293 | | Nov.  1 | 1.474 | 51 |
| May  9 | 1.401 | 303 | | 17 | 1.495 | 60 |
| 25 | 1.391 | 313 | | Dec.  3 | 1.516 | 68 |
| June 10 | 1.384 | 323 | | 19 | 1.537 | 76 |
| 26 | 1.381 | 333 | | Jan.  4, 1991 | 1.557 | 84 |

*Adapted from *The Astronomical Almanac for the Year 1990*, U.S. Government Printing Office, Washington, D.C., 20402, p. E12.

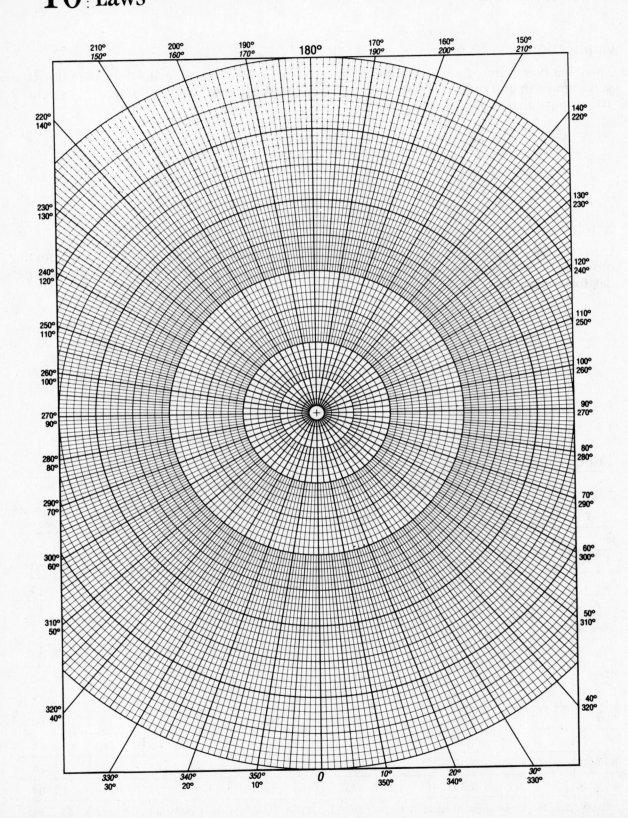

# EXPERIMENT
# 11 Pulleys

## Purpose

Find the mechanical advantage and the efficiency of several different pulley systems.

## Concept and Skill Check

Pulleys are simple machines that can be used to change the direction of a force, to reduce the force needed to move a load through a distance, or to increase the speed at which the load is moving, but that do not change the amount of work done. However, if the required effort force is reduced, the distance the load moves is decreased in proportion to the distance the force moves. Pulley systems may contain a single pulley or a combination of fixed and movable pulleys.

In an ideal machine, one lacking friction, all the energy is transferred, and the work input of the system equals the work output. The work input equals the force times the distance that the force moves, $F_e d_e$. The work output equals the output force (load) times the distance it is moved, $F_r d_r$. The ideal mechanical advantage, $IMA$, of the pulley system can be found by dividing the distance the force moves by the distance the load moves. Thus $IMA = d_e/d_r$. The ideal machine has a 100% efficiency. In the real world, however, the measured efficiencies are less than 100%. Efficiency is found by the following:

$$\text{Efficiency} = \frac{\text{Work output}}{\text{Work input}} \times 100\%.$$

## Materials

2 single pulleys            spring scale            string (2 m)
2 double pulleys            pulley support          meter stick
set of hooked metric masses

## Procedure

1. Set up the single fixed pulley system, as shown in Figure 1a.

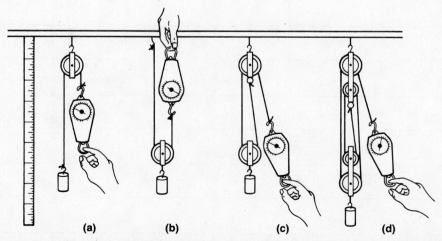

**Figure 1. In (a), (c), and (d), the spring scale's weight acts as part of the force raising the load. Therefore, the scale must be used upside down so that it will register its own weight.**

# 11 Pulleys

2. Select a mass that can be measured on your spring scale. Record the value of its mass in Table 1. Determine the weight, in newtons, of the mass to be raised by multiplying its mass in kilograms by the acceleration due to gravity. Recall that $W = mg$.

3. Carefully raise the mass by pulling on the spring scale. Measure the height, in meters, that the mass is lifted. Record this value in Table 1. Calculate the work output of the mass by multiplying its weight by the height it was raised. Record this value in Table 2.

4. Using the spring scale, raise the mass to the same height it was raised in Step 3. Ask your lab partner to read, directly from the spring scale, the force, in newtons, required to lift the mass. (If your spring scale is calibrated in grams, rather than newtons, calculate the force by multiplying the reading expressed in kilograms by the acceleration due to gravity.) Record this value in Table 1 as the force of spring scale. As you are lifting the load with the spring scale, pull upward at a slow, steady rate, using the minimum amount of force necessary to move the load. Any excess force will accelerate the mass and cause an error in your calculations.

5. Measure the distance, in meters, through which the force acted to lift the load to the height it was raised. Record this value in Table 1 as the distance, $d$, through which the force acts. Determine the work input in raising the mass by multiplying the force reading from the spring scale by the distance through which the force acted. Record the value for the work input in Table 2.

6. Repeat Steps 2 through 5 for a different load.

7. Repeat Steps 2 through 6 for each of the different pulley arrangements in Figures 1b, 1c, and 1d. Be sure to include the mass of the lower pulley(s) as part of the mass raised.

8. Count the number of lifting strands of string used to support the weight or load for each arrangement, (a) through (d). Record these values in Table 2.

## Observations and Data

### Table 1

| Pulley arrange-ment | Mass raised (kg) | Weight (W) of mass (N) | Height (h) mass is raised (m) | Force (F) of spring scale (N) | Distance (d) through which force acts (m) |
|---|---|---|---|---|---|
| (a) | | | | | |
| | | | | | |
| (b) | | | | | |
| | | | | | |
| (c) | | | | | |
| | | | | | |
| (d) | | | | | |
| | | | | | |

# 11 Pulleys

## Table 2

| Pulley arrange-ment | Work output (Wh) (J) | Work input (Fd) (J) | IMA $(d_e/d_r)$ | MA Number of lifting strands | Efficiency % |
|---|---|---|---|---|---|
| (a) | | | | | |
| | | | | | |
| (b) | | | | | |
| | | | | | |
| (c) | | | | | |
| | | | | | |
| (d) | | | | | |
| | | | | | |

## Analysis

1. Find the efficiency of each system. Enter the results in Table 2. What are some possible reasons that the efficiency is never 100%.

2. Calculate the ideal mechanical advantage, IMA, for each arrangement by dividing $d_e$ by $d_r$. Enter the results in Table 2. What happens to the force, F, as the mechanical advantage gets larger?

3. How does increasing the load affect the ideal mechanical advantage and efficiency of a pulley system?

4. How does increasing the number of pulleys affect the ideal mechanical advantage and efficiency of a pulley system?

5. The mechanical advantage may also be determined from the number of strands of string supporting the weight or load. Compare the calculated *IMA* from Question 2 with the number of strands of string you counted. Do the two results agree?

6. Explain why the following statement is false: A machine reduces the amount of work you have to do. Tell what a machine actually does.

## Application

In the space provided below, sketch a pulley system that can be used to lift a boat from its trailer to the rafters of a garage, such that the effort force would move a distance of 60 m while the load will move 10 m.

EXPERIMENT

# 12 Hooke's Law and Simple Harmonic Motion

## Purpose

To study the vibratory motion of a coiled spring oscillating with simple harmonic motion.

## Concept and Skill Check

Periodic motion occurs when a body moves back and forth over a fixed path, returning to each position and velocity after a definite interval of time. Several examples are the balance wheel in a clock, swinging pendulums, tuning forks, and the end of a vibrating diving board. In the absence of friction, periodic motion may persist indefinitely under the influence of a restoring force that is opposite to the displacement. We refer to this regular motion as *simple harmonic motion* (SHM). The time $T$ for one complete oscillation in SHM is called the *period*. For example, suppose a mass $m$ is attached to one end of a spring and the other end is attached to the ceiling as in Figure 1. If the mass is pulled downward a distance $x$ and then released, it will oscillate up and down approximating simple harmonic motion. Suppose it periodically returns to its release position every 0.5 s. The period of the motion is said to be 0.5 s/vib. The reciprocal of the period is the number of vibrations per second which is called the *frequency f* of oscillation. In this example, the frequency would be 1/0.5 or 2.0 vib/s.

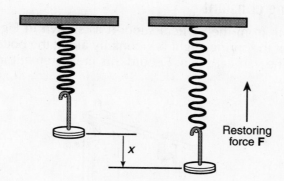

**Figure 1.**

Restoring force **F**

The acceleration $a$ in simple harmonic motion is directly proportional to the displacement $x$, but it is in the opposite direction. In other words, when the mass is displaced below its equilibrium point (negative $x$) the acceleration is *upward* due to an *upward* restoring force **F**. Similarly, when the mass is above its equilibrium point, the compressed spring produces a *downward* restoring force causing a *downward* acceleration. The magnitude of the force and the acceleration increase with increasing displacements of the spring. These observations are summarized in Hooke's law which can be expressed by the following equation:

$$F = -kx. \tag{21-1}$$

In this equation, $k$ is the spring constant which is a measure of the stiffness of the spring. Its unit is force per unit displacement such as newtons per meter (N/m). In our example of the spring stretched by a suspended mass, the spring constant can be determined by recording the change in displacement due to the addition of several new masses. A graph of force versus displacement would produce a straight line whose slope is an indication of the spring constant.

# EXPERIMENT 12

# Hooke's Law and Simple Harmonic Motion

It can be shown that the period $T$ of oscillation for a mass attached to the end of a spring is given by the following relationship:

$$T = 2\pi \sqrt{\frac{m'}{k}} \qquad (21\text{--}2)$$

where $k$ is the spring constant and $m'$ is the effective mass. Since the spring itself is part of the vibrating system, part of the spring's mass must be added to the suspended mass if accurate results are to be obtained. The effective mass $m'$ is equal to the suspended mass $m$ plus 1/3 of the spring's mass $m_s$.

$$m' = m + \frac{m_s}{3} \qquad \text{Effective mass } m'$$

## Materials

long coiled spring

meter stick

stop clock

scale and support

weight hanger and slotted masses

electronic balance

## Procedure

### A. Determine the spring constant

1. Hang the coiled spring from the vertical support as shown in Figure 2 and attach a weight hanger from the end of the spring. Sight horizontally along the bottom edge of the hanger and note the position on the vertical scale. Record this measurement in Table 1 as the zero reading position.

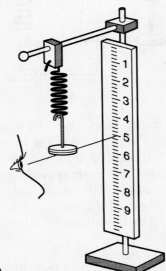

**Figure 2.**

2. Add a 100-g slotted mass to the hanger and record the new position of the bottom of the weight hanger.

# Hooke's Law and Simple Harmonic Motion

3. Place additional mass on the weight hanger and again record mass added and the new position of the weight hanger.

4. Repeat Step 3 three times until you have a total of about 500 g on the weight hanger excluding the mass of the hanger itself. You will now have five pairs of measurements of added mass and the corresponding elongations of the spring.

5. Subtract the zero reading (with just the weight hanger) from each scale reading to determine the displacement due to each increase in applied force. The force in newtons (N) is equal to the weight of the added masses, so you must multiply the mass in kilograms times gravity (9.8 m/s²).

$$W = mg \qquad g = 9.80 \text{ m/s}^2$$

6. Plot a graph of displacement versus force with force as the y-axis and displacement as the x axis. Use fine mesh graph paper for better results. Determine the slope of the graph which will be the ratio of the change in force to the change in displacement. This ratio is the spring constant.

## B. Determine the period in SHM

1. Remove the coiled spring from its support and measure its mass using the analytical balance. Record the mass in grams and then convert to kilograms. Divide this mass by 3 and record as the correction factor. Now reattach the spring to the vertical support.

2. Attach the weight hanger to the spring and add slotted masses until the total mass including the hanger is around 200 g. Convert this mass to kilograms and record it in Table 2 as the suspended mass.

3. Pull the hanger with attached masses downward about 5 cm and release it so that it vibrates with SHM. With a clock or a stop watch measure the time required for the mass to make 50 complete oscillations. Record this value and divide by 50 to obtain the experimental period.

4. Now change the mass of the vibrating system by adding additional slotted masses to the hanger. Repeat Step 3, and once again calculate period T.

5. Add the correction factor from Step 2 to the mass of the hanger and to added masses to obtain the effective mass m'. Do this calculation for each trial . Use Equation (21–2) to calculate the theoretical period T for each trial, and record these values in Table 2.

6. Compare your experimental values obtained from Steps 3 and 4 with the theoretical values determined from Step 5 by calculating the percent error. Use the theoretical value as the standard.

# 12 Hooke's Law and Simple Harmonic Motion

NAME ——————————————————————

## Observations and Data

### Table 1

| Mass added to the weight hanger (kg) | Applied force due to added mass (N) | Scale reading (cm) | Displacement $x$ of mass (m) |
|---|---|---|---|
| 0 | 0 | zero reading | 0 |
| | | | |
| | | | |
| | | | |
| | | | |
| | | | |

The slope from your graph $\dfrac{\Delta F}{\Delta x} =$ _____ N/m.    The spring constant $k =$ _____ N/m.

### Table 2

| Total mass $m$ suspended from the spring (kg) | Effective mass $m' = m + m_s/3$ (kg) | Time for 50 vibrations (s) | Period $T$ (s) | | Percent error |
|---|---|---|---|---|---|
| | | | Experimental | Calculated | |
| | | | | | |
| | | | | | |

## Analysis

1. In Procedure A you determined the spring constant by adding successive masses to the hanger and recording the stretch of the spring. The zero point for added weight was taken while the hanger was suspended. Why was it not necessary to know the weight of the hanger when you determined the spring constant graphically?

2. How would your results in Trial 1 for Procedure B change if you had ignored the correction factor for the mass of the spring? Calculate the theoretical period $T$ by ignoring the mass of the spring, and compare this value with your experimental observations.

## Application

Vibratory motion that approximates SMH is a factor in many engineering problems from simple vibrating springs to an understanding of electromagnetic theory. Some bridges have collapsed due to sympathetic vibrations due to wind. In such cases, the period of vibration is a significant consideration.

# EXPERIMENT 13 : Archimedes' Principle

## Purpose

Investigate Archimedes' principle and measure the buoyant force of a fluid.

## Concept and Skill Check

Archimedes' principle states that an object wholly or partially immersed in a fluid is buoyed up by a net force equal to the weight of the fluid that it displaces, $F_{buoyant} = F_{weight\ of\ fluid\ displaced}$. Recall that $F_{weight\ of\ fluid\ displaced} = \rho_{fluid} V_{fluid\ displaced}\ g$, where $\rho$ = density. When an object with a density less than that of the fluid is submerged, it will sink only until it displaces a volume of fluid with a weight equal to the weight of the object. At this time, the object is floating underwater, equilibrium exists, and $\rho_{fluid} V_{fluid\ displaced} = \rho_{object} V_{object}$.

If the density of an object is greater than that of the fluid, an upward buoyant force from the pressure of the fluid will act on the object, but the magnitude of the buoyant force will be too small to balance the downward weight force of the denser material. While the object will sink, its apparent weight decreases by an amount equivalent to the buoyant force.

In this experiment, you will investigate the buoyant force of water acting on an object. Recall that 1 mL of water has a mass of 1 g and a weight of 0.01 N. The buoyant force acting on the object is determined by finding the difference between the weight of the object in air and the weight of the object when it is immersed in water and is given by the following equation:

$$F_{buoyant} = F_{weight\ of\ mass\ in\ air} - F_{weight\ of\ mass\ in\ water}.$$

## Materials

| | | |
|---|---|---|
| 500-mL beaker | 500-g hooked mass | polystyrene cup |
| spring scale, 5-N capacity or greater | 100-g hooked mass | paper towel |

## Procedure

1. Pour cool tap water into the 500-mL beaker to the 300-mL mark. Carefully read the volume from the gradations on the beaker and record this value in Table 1.
2. Hang the 500-g mass from the spring scale. Measure the weight of the mass in air and record this value in Table 1.
3. Immerse the 500-g mass, suspended from the spring scale, in the water, as shown in the figure. Do not let the mass rest on the bottom of the beaker or touch the sides of the beaker and keep it suspended from the spring scale. Measure the weight of the immersed mass and record this value in Table 1.
4. Measure the volume of the water with the mass immersed. Record the new volume reading in Table 1. Remove the 500-g mass from the beaker and set it aside.
5. Measure and record in Table 2 the volume of water in the beaker. Place the 100-g mass in the beaker of water. Measure and record in Table 2 the volume of the water in the beaker with the mass immersed.
6. The polystyrene cup will serve as a "boat." Remove the mass from the water, dry it with a paper towel, and place it in the polystyrene cup. Float the cup in the beaker of water. Measure and record in Table 2 the new volume of water.

**Measure the weight of a mass submerged in water.**

## Observations and Data
### Table 1

| Weight of 500-g mass in air | |
| --- | --- |
| Weight of 500-g mass immersed in water | |
| Volume of water in beaker | |
| Volume of water in beaker with 500-g mass immersed | |

### Table 2

| Volume of water in beaker | |
| --- | --- |
| Volume of water with 100-g mass immersed | |
| Volume of water with 100-g mass in polystyrene cup | |

## Analysis

1. Calculate the buoyant force of water acting on the 500-g mass.

2. Using the values from Table 1, calculate the volume of water displaced by the 500-g mass. Calculate the weight of the water displaced. Compare the weight of the volume of water displaced with the buoyant force acting on the immersed object that you calculated in Question 1. If the values are different, describe sources of error to account for this difference.

3. What happened to the water level in the beaker when the 100-g mass was placed in the polystyrene cup (boat)? Propose an explanation, which includes density, for any difference in volume you found in Steps 5 and 6.

## Application

Tim and Sally are floating on an inflatable raft in a swimming pool. What happens to the water level in the pool if both fall off the raft and into the water?

EXPERIMENT

# 14

# Specific Heat

## Purpose

Use the law of conservation of energy to calculate the specific heat of a metal.

## Concept and Skill Check

One of several physical properties of a substance is the amount of energy that it will absorb per unit mass. This property is called specific heat, $C_s$. The specific heat of a material is the amount of energy, measured in joules, needed to raise the temperature of one kilogram of the material one Celsius degree (Kelvin).

A calorimeter is a device that can be used in the laboratory to measure the specific heat of a substance. The polystyrene cup, used as a calorimeter, insulates the water-metal system from the environment, while absorbing a negligible amount of heat. Since energy always flows from a hotter object to a cooler one and the total energy of a closed, isolated system always remains constant, the heat energy, $Q$, lost by one part of the system is gained by the other:

$$Q_{\text{lost by the metal}} = Q_{\text{gained by the water}}.$$

In this experiment, you will determine the specific heat of two different metals. The metal is heated to a known temperature and placed in the calorimeter containing a known mass of water at a measured temperature. The final temperature of the water and material in the calorimeter is then measured. Given the specific heat of water (4180 J/kg · K) and the temperature change of the water, you can calculate the heat gained by the water (heat lost by the metal) as follows:

$$Q_{\text{gained by the water}} = (m_{\text{water}})(\Delta T_{\text{water}})(4180 \text{ J/kg} \cdot \text{K}).$$

Since the heat lost by the metal is found by

$$Q_{\text{lost by the metal}} = (m_{\text{metal}})(\Delta T_{\text{metal}})(C_{\text{metal}}),$$

the specific heat of the metal can be calculated as follows:

$$C_{\text{metal}} = \frac{Q_{\text{gained by the water}}}{(m_{\text{metal}})(\Delta T_{\text{metal}})}.$$

## Materials

| | | |
|---|---|---|
| string (60 cm) | hot plate (or burner with ringstand, | balance |
| safety goggles | ring, and wire screen) | specific heat set (brass, |
| 250-mL beaker | tap water | aluminum, iron, lead, |
| polystyrene cup | thermometer | copper, etc.) |

## Procedure

1. Safety goggles must be worn for this laboratory activity. CAUTION: *Be careful when handling hot glassware, metals, or hot water.* Fill a 250-mL beaker about half full of water. Place the beaker of water on a hot plate (or a ringstand with a wire screen) and begin heating it.
2. While waiting for the water to boil, measure and record in Table 1 the mass of the metals you are using and the mass of the polystyrene cup.
3. Attach a 30-cm piece of string to each metal sample. Lower one of the metal samples, by the string, into the boiling water, as shown in the figure on the next page. Leave the metal in the boiling water for at least five minutes.

4. Fill the polystyrene cup half full of room temperature water. Measure and record in Table 1 the total mass of the water and the cup.

5. Measure and record in Table 1 the temperature of the room temperature water in the polystyrene cup and the boiling water in the beaker. The temperature of the boiling water is also the temperature of the hot metal.

6. Carefully remove the metal from the boiling water and quickly lower it into the room temperature water in the polystyrene cup.

7. Gently stir the water in the polystyrene cup for several minutes with the thermometer. CAUTION: *Thermometers are easily broken. If you are using a mercury thermometer and it breaks, notify your instructor immediately. Mercury is a poisonous liquid and vapor.* When the water reaches a constant temperature, record this value in Table 1 as the final temperature of the system.

8. Remove the metal sample and repeat Steps 3 through 7 with another metal sample.

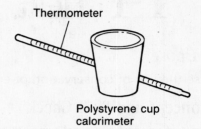

Thermometer

Polystyrene cup
calorimeter

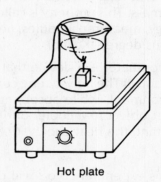

Hot plate

**The calorimeter is used to measure heat exchange by means of temperature changes.**

## Observations and Data
### Table 1

|  | Trial 1 | Trial 2 |
|---|---|---|
| Type of metal |  |  |
| Mass of calorimeter cup (kg) |  |  |
| Mass of calorimeter cup and water (kg) |  |  |
| Mass of metal (kg) |  |  |
| Initial temperature of room temperature water (°C) |  |  |
| Temperature of hot metal (°C) |  |  |
| Final temperature of metal and water (°C) |  |  |

# 14 Specific Heat

## Table 2

| | Trial 1 | Trial 2 |
|---|---|---|
| Mass of room temperature water (kg) | | |
| $\Delta T$ metal (°C) | | |
| $\Delta T$ room temperature water (°C) | | |

## Analysis

1. For each trial, calculate the mass of the room temperature water, the change in temperature of the metal, and the change in temperature of the water in the polystyrene cup. Record these values in Table 2.

2. For each trial, calculate the heat gained by the water (heat lost by the metal).

3. For each trial, calculate the specific heat of the metal. For each metal sample, use the value for heat gained by the water that you calculated in Question 2.

4. For each trial, use the values for specific heat substances found in Table B:1 of Appendix B to calculate the relative error between your value for specific heat and the accepted value for the metal sample.

5. If you had some discrepancies in your values for specific heat of the metal samples, suggest possible sources of uncertainty in your measurements that may have contributed to the difference.

## Application

The specific heat of a material can be used to identify it. For example, a 100.0-g sample of a substance is heated to 100.0°C and placed into a calorimeter cup (having a negligible amount of heat absorption) containing 150.0 g of water at 25°C. The sample raises the temperature of the water to 32.1°C. Use the values in Table B:1 in the Appendix, to identify the substance.

## Extension

Obtain a sample of an unknown metal from your teacher. Use the procedure described in this laboratory activity to identify it by the value of its specific heat.

EXPERIMENT
# 15

# Standing Waves in a Vibrating String

## Purpose

To study wave motion and its interference patterns which set up standing waves in a vibrating string.

## Concept and Skill Check

Energy propagation by means of a disturbance in a medium instead of by the motion of the medium itself is called *wave motion*. For example, if one end of a string is attached to the wall and the other end is given a sharp upward motion, a pulse is sent down the string. Thus, energy is delivered to the wall even though the individual string particles move up and down in place. Such a disturbance in which the motion of the medium particles is perpendicular to the direction of wave propagation is called a *transverse wave*. Notice in Figure 1 that a pulse moving with velocity *v* strikes the wall, is inverted, and then continues back along the string. In this manner, it is possible for returning wave pulses to coincide with arriving wave pulses in such a way to produce constructive and destructive interference. In *constructive interference,* the wave amplitude is greater than the composite waves; in *destructive interference,* the wave amplitude is less than the composite wave.

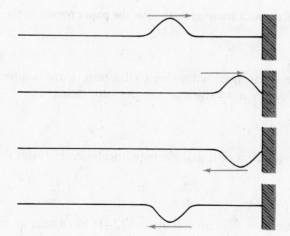

**Figure 1.**

If conditions are favorable, standing waves will occur such as those shown in Figure 2. A crest of one wave meets a trough of another wave causing regions of maximum amplitude called *antinodes A* and regions of minimum amplitude called *nodes N*. In our example, the string is vibrating in four loops. The distance from one node to another node is equal to one half of a wavelength λ/2. Thus, the four loops cover a distance of two wavelengths.

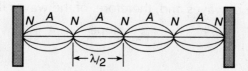

**Figure 2. Standing waves.**

# Standing Waves in a Vibrating String

It is important to understand the terms *wavelength, wave velocity,* and *wave frequency.* Consider Figure 3 in which a pen is attached to a vibrating mass. The paper moves from right to left tracing a transverse wave. Note that a wave crest occurs each time the pen reaches its highest point (one complete vibration of the mass). The distance between such crests, which is the distance traveled by the wave during one complete oscillation, is called the *wavelength* λ. The number of complete oscillations (or waves) per second is called the *frequency f* of the wave. The units of frequency are vibrations per second (1/s) which is renamed as hertz (Hz). If the wavelength is in meters and the frequency is in hertz, the product of wavelength and frequency will give units of velocity (m/s). This relationship is one of the fundamental equations satisfied by all waves.

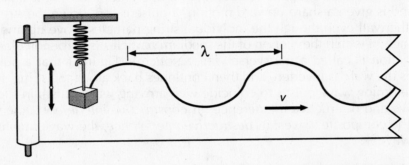

**Figure 3. A vibrating pencil draws a transverse wave as the paper moves to the right with a speed *v*. Note the wavelength λ as labeled.**

$$v = f\lambda \tag{22–1}$$

It can be shown that the velocity of waves in a vibrating string is affected by the tension $T$ in the string and by its linear density $\mu$, or the mass $m$ per unit length $L$.

$$\mu = \frac{m}{L} \tag{22–2}$$

For a transverse wave in a string of linear density and tension $T$, the wave speed is given by

$$v = \sqrt{\frac{T}{\mu}} \quad \text{or} \quad v = \sqrt{\frac{TL}{m}}. \tag{22–3}$$

Finally, we can substitute $v = f\lambda$ from Equation (22–1) to obtain an equation for calculating the frequency of vibration in terms of the tension in the string and the linear density.

$$f\lambda = \sqrt{\frac{T}{\mu}} \quad \text{or} \quad f = \sqrt{\frac{T}{\mu\lambda^2}} \tag{22–4}$$

In this experiment, you will determine the frequency of the waves along a vibrating string, which is equal to the frequency of an electromagnetic vibrator. The technique will be to produce standing waves of one, two, three, four, and five loops by varying the tension in the string. You will plot a graph of tension $T$ versus the square of the wavelength $\lambda^2$. The slope of this graph will give an experimental value for the ratio $T/\lambda^2$. By comparison with Equation (22–4) you can see that the frequency of the source of waves and, therefore, of the waves themselves can be found from

$$f = \frac{\text{slope}}{\mu}. \tag{22–5}$$

Note that the frequency of the generated waves is the same regardless of the number of loops. Thus, the slope, which is the ratio of $T/\lambda^2$, is constant for each measurement.

## Materials

| | |
|---|---|
| electromagnetic vibrator | clamp and support |
| electronic balance | meter stick |
| pulley and clamp set | weight hanger |
| slotted masses | string |

## Procedure

1. Obtain a length of string approximately 1.2 m long. Measure its mass using the electronic balance.
2. Calculate the linear density $\mu$ of the entire string using Equation (22–2). Record on the data sheet.
3. Mount the pulley to the edge of the table using the provided clamp. Attach the weight hanger to one end of the string, and tie the other end to the hole in the end of the electromagnetic vibrator. Mount the electromagnetic vibrator to its support a distance of approximately 1 m from the pulley, and pass the string over the pulley as shown in Figure 4. Make sure that the string is parallel with the table top and that the weight hanger is hanging freely.

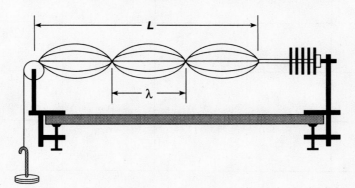

**Figure 4.**

4. After verifying that your setup is similar to that in the figure, connect the line cord from the vibrator to an ac outlet and turn on the switch.
5. Add slotted masses to the weight hanger until the string vibrates in five (5) loops. The length of each loop is equal to half of a wavelength ($\lambda/2$). Adjust the total mass by adding or removing small masses until the loops are of maximum amplitude and appear almost stationary. Assume that a node forms at the center of the pulley and at the vibrator hole. Determine the average length of one loop by dividing the length of the vibrating string by the number of loops. The length of the string is taken from the center of the pulley to the hole in the vibrator.
6. Find the tension in the string by multiplying the suspended mass (in kg) by gravity (9.80 m/s²). $W = mg$. Don't forget to include the mass of the hanger.
7. Increase the tension in the string by adding more mass to the hanger until the string vibrates in 4, 3, 2, and 1 loops, respectively. Record the length of the string, the average length of one loop, and the suspended mass for each of the five trials. Enter your data on the data sheet.

# Standing Waves in a Vibrating String

8. Using fine-mesh graph paper, plot the data for the five observations with the tension $T$ as the $y$-axis with $\lambda^2$ as the $x$-axis. With a ruler, draw a straight line which best represents the plotted data, and determine the slope of this line, including appropriate units. The slope represents an experimental determination of the ratio $T/\lambda^2$. Calculate the frequency of the wave motion from Equation (22–4). Record your answers on the data sheet.

## Observations and Data

Mass of string                                 $m =$ _____   $g =$ _____ kg

Length of string                            $L =$ _____   cm = _____ m

Linear density of string              _____   $\mu =$ _____ kg/m

| Number of loops | 5 | 4 | 3 | 2 | 1 |
|---|---|---|---|---|---|
| Length of vibrating string in meters (m) | | | | | |
| Average length of one loop (m) | | | | | |
| Wavelength $\lambda$ (m) | | | | | |
| Wavelength squared $\lambda^2$ (m$^2$) | | | | | |
| Mass suspended from string (kg) | | | | | |
| Tension in string $T$ (N) | | | | | |

Slope of your graph $T/\lambda^2$ (include units)      _____

Calculated frequency of the vibrator      _____

## Application

Understanding all forms of wave motion from simple guitar strings to mechanical waves that result in destruction of property begin with a study of wave speed, wave length, frequency, resonance, and standing waves. This experiment develops the fundamental understanding necessary for more complex applications.

EXPERIMENT

# 16 Investigating Static Electricity

## Purpose

Investigate the behavior of static electricity.

## Concept and Skill Check

Clothes removed from a clothes dryer usually cling to each other and spark or crackle with static electricity when they are separated. A fabric softener is usually added to prevent this static buildup. When two dissimilar materials are rubbed together, they can become charged. Objects can acquire static electric charges by either gaining or losing electrons. An object that gains electrons has a net negative charge and is said to be negatively charged. An object that loses electrons has a net positive charge and is said to be positively charged. Only those objects separated from a ground, or Earth, by an insulator will retain their charge for any length of time. Objects that are attached to a ground through a conductor will remain uncharged, since the charge travels into the ground and is quickly dissipated. Recall that a hard rubber rod rubbed with wool or fur will become negatively charged, while a glass rod rubbed with silk will become positively charged. In this experiment, you will charge objects and observe their interactions with other charged objects.

## Materials

hard rubber rod or vinyl strip
glass rod
plastic wrap or acrylic rod
silk pad

wool pad or cat's fur
pith ball suspended from
  holder by silk thread

leaf or vane electroscope
Leyden jar
wire with alligator clip

## Procedure

It is best to perform these activities on a cool, dry day. If it is a humid day, charged objects tend to lose their charge fairly rapidly, making it difficult to observe some phenomena. Perform each part of the experiment several times to be sure you have proper observations of the phenomena. You may have to rub the rods vigorously to charge them, particularly when you use the glass rods.

### A. Negatively Charging a Pith Ball

1. Rub the hard rubber rod with the fur or wool pad to charge it. Bring the rod close to, but not touching, a suspended pith ball, as shown in Figure 1. Observe the behavior of the pith ball and record your observations in Table 1. Touch the pith ball with your finger to remove any charge it may have.
2. Charge the rubber rod again. Bring it close to the pith ball and allow it to touch the ball. Then bring the charged rod near the charged ball and observe the behavior of the ball. Record your observations in Table 1.

### B. Positively Charging a Pith Ball

1. Rub the glass rod with a piece of silk to charge it. Bring the rod close to, but not touching, a sus-

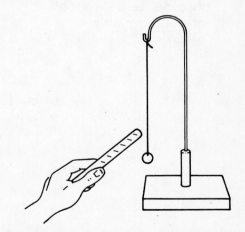

**Figure 1. Bring the charged rod near but not touching the uncharged pith ball.**

pended pith ball. Observe the behavior of the pith ball and record your observations in Table 2. Touch the pith ball with your finger to remove any charge it may have.

2. Charge the glass rod again. Bring it close to the pith ball and allow it to touch the ball. Then bring the charged rod near the charged ball and observe the behavior of the pith ball. Record your observations in Table 2.

## C. Charging an Electroscope by Conduction

1. Look at the vane electroscope or the leaf electroscope that you will be using and familiarize yourself with its response to static charges. A vane electroscope is shown in Figure 2. Charge the rubber rod with the fur or wool. Gently bring it in contact with the metal top of the vane or leaf electroscope. The electroscope is now negatively charged. Observe what changes take place in the electroscope. Record your observations in Table 3.

2. Recharge the rubber rod. Bring it near the charged electroscope top. Observe what happens to the leaves or vane. Record your observations in Table 3.

3. Recharge the electroscope with a negative charge from the rubber rod. Charge the glass rod and bring it near the top of the negatively-charged electroscope. Observe the electroscope deflection as the charged glass rod is moved close to and away from the electroscope. Record your observations in Table 3. Discharge the electroscope by momentarily touching the top of it with your finger.

4. Charge the plastic wrap or acrylic rod with a piece of silk. Touch the plastic material to the top of the electroscope to charge it. Set the plastic material aside. Charge the rubber rod and bring it near the electroscope while observing the deflection of the leaves or vane. Record your observations in Table 3. What type of charge is on the electroscope?

## D. Charging an Electroscope by Induction

1. Select one of the rods and charge it. Bring it near, within 1–2 cm, but not touching the electroscope. Observe the leaves or vane. Record your observations in Table 4.

2. With the charged rod near the electroscope, momentarily touch the top of the electroscope with your finger. Remove your finger from the top of the electroscope, and then remove the charged rod. Observe the electroscope leaves or vane. Record your observations in Table 4.

3. Following the procedure from Part C, determine the type of charge on the electroscope. Record your observations in Table 4.

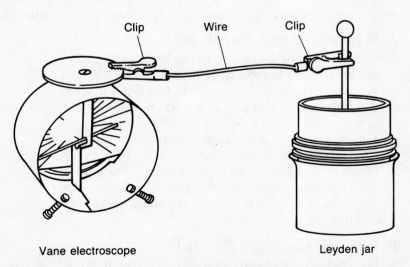

Clip        Wire        Clip

Vane electroscope                    Leyden jar

**Figure 2. The clip wire attaches the electroscope to the Leyden jar**

### E. Charging a Leyden Jar
1. Place a Leyden jar beside the electroscope.
2. Charge one of the rods and gently touch the electroscope to charge it. Repeat the procedure to place a fairly large charge on the electroscope and to cause a large deflection.
3. Discharge the electroscope by momentarily touching the top with your finger. Use the clip wire to attach the top of the electroscope to the top of the Leyden jar, as shown in Figure 2 on the previous page. Note that the Leyden jar is not grounded. Repeat Step 2 to charge the combination. Is there any difference in the ability of the electroscope to become charged? Record your observations in Table 5.

## Observations and Data
### Table 1
A. Negatively Charging a Pith Ball

Observations of pith ball with nearby charged rod:

Observations of pith ball after it has been touched with a charged rod:

### Table 2
B. Positively Charging a Pith Ball

Observations of pith ball with nearby charged rod:

Observations of pith ball after it has been touched with a charged rod:

EXPERIMENT
# 16 Investigating Static Electricity

## Table 3
### C. Charging an Electroscope by Conduction

Observations of uncharged electroscope when negatively-charged rod touches it:

_____

_____

Observations of negatively-charged electroscope when negatively-charged rod is brought near:

_____

_____

_____

Observations of negatively-charged electroscope when positively-charged rod is brought near:

_____

_____

_____

Observations of deflection of leaves or vane when plastic material charges the electroscope and negatively-charged rod is brought near:

_____

_____

_____

_____

## Table 4
### D. Charging an Electroscope by Induction

Observations of electroscope when charged rod is brought near:

_____

_____

_____

## Table 4 (continued)

Observations of electroscope after touching with finger:

_____

_____

Observations to determine the type of charge:

_____

_____

## Table 5

E. Charging a Leyden Jar

Observations of charging a Leyden jar-electroscope combination:

_____

_____

## Analysis

1. Summarize your observations of negatively charging a pith ball (Part A).

2. Summarize your observations of positively charging a pith ball (Part B).

3. Compare the negative and positive charging of a pith ball.

4. Summarize your observations of charging the electroscope by conduction (Part C).

5. Using your data from Part C, explain why the leaves of the electroscope diverged or the vane deflected.

6. Using your data from Part C, explain why the leaves (vane) remained apart (deflected) when the charged rod was removed.

7. What type of charge did the plastic wrap/plastic rod produce? How can you prove this?

8. Why did the electroscope vane or leaves move in Part D?

9. Compared to that of the charged rod, what type of charge did the electroscope receive by induction?

10. What purpose did your finger serve when it touched the electroscope?

11. Compared to that of the charging body, what is the type of charge on an electroscope when it is charged by conduction? Induction?

12. In Part E, what effect did the Leyden jar have on the ability to charge the electroscope? What is your proof of this effect?

## Application
Some long-playing records are packaged in a plastic dust cover. As the record is slid out of its cover, the record and cover rub together. What is the likely result of this action? What function do record cleaning systems perform?

EXPERIMENT

# 17 The Capacitor

## Purpose

Investigate the relationship of the flow of charge to time in a charging capacitor.

## Concept and Skill Check

A capacitor is a device that stores electrical charge. An early form of a capacitor is the Leyden jar, which you worked with in Experiment 29. Capacitors are made of two conducting plates separated by air or another insulating material, often referred to as a dielectric. The capacitance, or capacity, of a capacitor is dependent upon the nature of the dielectric substance, the area of the plates, and the distance between the plates.

The figure below is a circuit diagram of a capacitor, a battery, a switch, a resistor, a voltmeter, and an ammeter to measure the flow of charge, connected in series. The resistor is a simple device that resists the flow of charge. The flow of electrical charge over a period of time is measured in units called amperes or amps; 1 coulomb/second = 1 ampere. When the switch is open, as shown in the figure, no charge flows from the battery. However, when the switch is closed, the battery supplies electrical energy to move positive charges to one plate of the capacitor and negative charges to the other. Charge accumulates on each plate of the capacitor, but no current flows through it since the center of the capacitor is an insulator. As charge accumulates in the capacitor, the potential difference increases between the two plates until it reaches the same potential difference as that of the battery. At this point, the system is in equilibrium, and no more charge flows to the capacitor. Capacitance is measured by placing a specific amount of charge on a capacitor and then measuring the resulting potential difference. The capacitance, $C$, is found by the following relationship

$$C = q/V,$$

where $C$ is the capacitance in farads, $q$ is the charge in coulombs, and $V$ is the potential difference in volts.

In this experiment, you will charge a capacitor and measure the amount of current that flows to it over a period of time. Then you will determine the capacitance of the capacitor.

## Materials

| | |
|---|---|
| 1000-µF capacitor | DC ammeter, 0–1.0 mA |
| 10-kΩ resistor | knife switch |
| 27-kΩ resistor | connecting wire |
| voltmeter | stopwatch |
| 15-VDC source | (or watch with |
| (battery or | second hand) |
| power supply) | |

## Procedure

1. Set up the circuit as shown in the figure. The ammeter, capacitor, and battery must be wired in the proper order. Look for the + and − markings on the circuit components. The positive plate of the capacitor must be wired to the positive terminal of the battery. If the connections

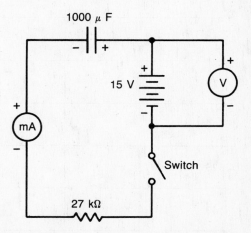

1000 µF

15 V

27 kΩ

Switch

**Carefully follow the circuit diagram as you wire the + and − connections.**

# EXPERIMENT 17

# The Capacitor

are reversed, the capacitor may be destroyed. Resistors have no + or − end. Record in Table 1 the battery voltage and the value of the capacitor.

2. With one lab partner timing and the other reading and recording the current values, close the knife switch and begin taking readings. At the instant the switch is closed, a large current will flow. Take a current reading every five seconds until the current is too small to measure. Estimate your ammeter readings as closely as possible. Record the readings in Table 2.

3. Open the knife switch. Using a spare piece of wire, connect both ends of the capacitor to discharge it.

4. Replace the 27-kΩ resistor with the 10-kΩ resistor.

5. Repeat Steps 1 through 3 with the 10-kΩ resistor. Record the readings in Table 2.

6. After all readings have been taken, dismantle the circuit. Be sure to disconnect all wires from the battery (or power supply).

## Observations and Data

### Table 1

| | |
|---|---|
| Battery voltage | |
| Capacitance | |

### Table 2

| | 27 kΩ | 10 kΩ |
|---|---|---|
| Time (s) | Current (mA) | Current (mA) |
| 0 | | |
| 5 | | |
| 10 | | |
| 15 | | |
| 20 | | |
| 25 | | |
| 30 | | |
| 35 | | |

### Table 2 (continued)

| | 27 kΩ | 10 kΩ |
|---|---|---|
| Time (s) | Current (mA) | Current (mA) |
| 40 | | |
| 45 | | |
| 50 | | |
| 55 | | |
| 60 | | |
| 65 | | |
| 70 | | |
| 75 | | |
| 80 | | |
| 85 | | |
| 90 | | |

# The Capacitor

NAME ————————————————————————————

## Analysis

1. Why did the current start at a maximum value and drop toward zero while the capacitor was charging?

2. Look at the data for the two different resistors. Explain the purpose of the resistor in the circuit.

3. Using the data in Table 2, plot two graphs for current as a function of time. Sketch in a smooth curve.

4. The area between the curve and the time axis represents the charge stored in the capacitor. Estimate the area under the curve by sketching in one or more triangles to approximate the area. Note that the current is in mA, so this should be converted to amps by using 1 mA = $1 \times 10^{-3}$ A. What is the estimated charge for the capacitor with the 27-k$\Omega$ resistor and with the 10-k$\Omega$ resistor?

5. Calculate the capacitance of the capacitor, $C = q/V$, using the value for charge from Question 4 and the measured potential difference of the power source.

6. Compare the value determined in Question 5 with the manufacturer's stated value you recorded in Table 1. Electrolytic capacitors have large tolerances, frequently on the order of 50%, so there may be a sizeable difference. Find the relative error between the two values.

7. Predict what would happen if the circuit were set up without the resistor.

8. Which variables were manipulated and which remained constant during the experiment? Describe the current versus time curve.

## Application

Describe how an *RC* circuit (a circuit that includes a resistor and a capacitor), capable of charging and discharging at a very specific and constant rate, could be applied for use in the home, an automobile, or an entertainment center.

## Extension

Repeat the experiment with another capacitor obtained from your teacher or connect several capacitors in parallel. Capacitors connected in parallel have an effective value equal to the sum of the individual capacitors. The quantity *RC* is called the time constant for the circuit and has units of seconds, as long as *R* is measured in ohms and *C* is measured in farads. The time constant is the time required for the current to drop to 37% of its original value. Before you begin, check to see that the value of *RC* can be easily measured. Follow the same procedure as before and compare the results of the graph with your earlier results.

EXPERIMENT

# 18 : Ohm's Law

## Purpose

Use Ohm's law to determine the values of resistors.

## Concept and Skill Check

Ohm's law states that as long as the temperature of a resistor, $R$, does not change, the electric current, $I$, flowing in a circuit is directly proportional to the applied voltage, $V$, and inversely proportional to the resistance. The current, $I$, is given by

$$I = \frac{V}{R} \text{ ; thus } R = \frac{V}{I} .$$

The resistance is measured in ohms ($\Omega$), where $1\,\Omega = 1\,V/1\,A$. In this experiment, you will measure the applied voltage and the current passing through a resistor and use these data to determine the value of the resistance.

## Materials

voltmeter, 0–5 VDC
power supply, 0–6 VDC
milliammeter, 0–50 mA

knife switch
connecting wires

resistors: 100 $\Omega$, 150 $\Omega$, and 220 $\Omega$, with
0.5-W or 1.0-W rating

## Procedure

1. Review Appendix C, "Rules for Using Meters." Study the scales on the meter faces until you can read them properly. Some meters may have more than one positive or negative terminal. Select the set of terminals corresponding to the voltage range you will be using, such that the meter will give readings in the middle of the scale. CAUTION: *Resistors can become hot if too high a voltage is applied to them. Do not touch resistors when current is supplied to the circuit.*

2. Study Figure 1 on the next page. Notice in Figure 1(a) the arrangement of the meters, resistor, and power supply. Compare the arrangement in Figure 1(a) with the schematic diagram of the same circuit in Figure 1(b). Select a 100-$\Omega$ resistor. Refer to the resistor color code in Appendix D to determine the corresponding values for the various color bands. Determine the tolerance range for the resistor and record this value in Table 1. Connect the circuit, knife switch open, as shown in Figure 1; use the 100-$\Omega$ resistor. Leave the knife switch open so that no current flows in the circuit until your teacher has checked your circuit and has given you permission to proceed to the next step.

3. To prevent the resistors from overheating, the knife switch should be closed only long enough to obtain readings of current and voltage. Close the switch and then adjust the voltage to about 1.5 V. If you are using a battery, there will be no adjustment of the potential difference. Quickly read the meters. Open the knife switch as soon as you complete the readings. Record these readings in Table 1.

4. Close the switch and adjust the voltage to a higher reading, such as 3.0 V. If you are using a battery, repeat Step 3, so that you have two sets of data for each resistor. Read the milliammeter current and the potential difference. Open the knife switch. Record these readings in Table 1. Convert the current from mA (milliamperes) to A (amperes) by dividing the current in mA by 1000, since 1 mA = 0.001 A. Record this converted value for current in A.

# 18 Ohm's Law

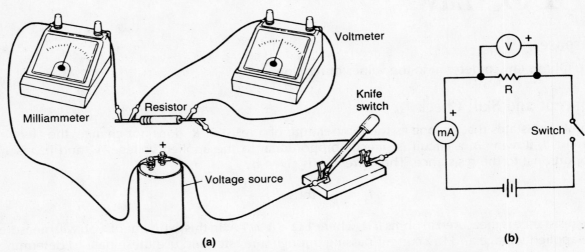

Figure 1. Be sure to connect the voltmeter in parallel with the resistor and the ammeter in series with the resistor. Note the + and − terminals of both meters in relation to the + and − terminals of the power supply. The circuit diagram for the apparatus is shown in (b).

5. Remove the 100-Ω resistor and replace it with the 150-Ω resistor. Starting with the lower voltage, repeat the procedure to obtain two sets of data for current and voltage.
6. Remove the 150-Ω resistor and replace it with the 220-Ω resistor. Starting with the lower voltage, repeat the procedure to obtain two sets of data for current and voltage.

## Observations and Data
## Table 1

| Resistor | Printed value of resistor (Ω) | Tolerance range (+/− %) | Voltage (V) | Current (mA) | Current (A) | Resistance (Ω) |
|---|---|---|---|---|---|---|
| $R_1$ | | | | | | |
| | | | | | | |
| $R_2$ | | | | | | |
| | | | | | | |
| $R_3$ | | | | | | |
| | | | | | | |

## Analysis

1. Use the data in Table 1 to calculate the resistance for each set of data, applying $R = \dfrac{V}{I}$, where $V$ is in units of volts and $I$ is in units of amperes. Record the resistances in Table 1.

2. Calculate the average of your two values for each resistor. Compare the printed values of the resistors you used with your averages of the calculated values. Determine the relative error for each resistor.

3. If your values were not within the tolerance range of the printed values, suggest some reasons for the discrepancy.

4. Describe the proper placement of an ammeter in a circuit.

5. Describe the proper placement of a voltmeter in a circuit.

6. State the relationship between the current flowing through a circuit and the voltage and resistance of the circuit.

## Application

A light bulb, after it has reached its operating temperature, has a potential difference of 120 V applied across it while 0.5 A of current passes through it. What is the resistance of the light bulb?

## Extension

From your teacher obtain a resistor of unknown value. Connect the circuit components in proper order and take several voltage and current readings through this resistor. Plot a graph of voltage (y-axis) versus current (x-axis). Sketch in a smooth line that best fits all the data. Determine the resistance and calculate the slope of the line, using measurements of potential difference in volts and of current in amperes.

EXPERIMENT

# 19 : Series Resistance

## Purpose

Investigate resistance and determine the equivalent resistance of resistances in series.

## Concept and Skill Check

A series circuit consists of resistors connected with the voltage source in such a way that all current travels through each resistor in turn. The equivalent resistance, $R$, of a series circuit is the sum of the resistances of the individual resistors in the circuit.

$$R = R_1 + R_2 + R_3$$

The potential difference across the voltage source is equal to the sum of the voltage drops across each resistor.

$$V = V_1 + V_2 + V_3$$

The total current flowing in the circuit is most easily found by calculating the equivalent resistance and then using the equation

$$I = \frac{V}{R},$$

where $I$ is the circuit current in amperes, $R$ is the equivalent resistance of the circuit in ohms, and $V$ is the voltage source in volts.

In this experiment, you will measure voltage drops across circuit resistances and apply Ohm's law to determine the equivalent circuit resistance.

## Materials

| | | |
|---|---|---|
| DC power supply or dry cells | voltmeter, 0–5 VDC knife switch | 68-$\Omega$, 82-$\Omega$, and 100-$\Omega$ resistors, 0.5 W or greater |
| 8 connecting wires | | milliammeter, 0–50 mA or 0–100 mA |

## Procedure

### A. One Resistor

1. Refer to the resistor color code in Appendix D to determine the value of your resistors. Arrange the 100-$\Omega$ resistor in series with the the ammeter, open switch, and voltage source, as shown in Figure 1(a) on the next page. Arrange the voltmeter in parallel with the resistor, as shown in the figure. Record the printed value for $R_1$ and its tolerance in Table 1. Have your teacher check the circuit before you proceed.
2. Close the switch and set the voltage source at 5 V. Read the ammeter. Open the switch as soon as your readings are complete. Do not leave the switch closed for more than a few seconds at a time. Record the meter readings in Table 1.

### B. Two Resistors

1. Arrange a second resistor in series with the first, as shown in Figure 1(b). Record the printed values for the resistors and their tolerances in Table 2.
2. Close the switch and adjust the voltage as necessary to maintain 5 V. Read the ammeter. Open the switch. Record the meter readings in Table 2.

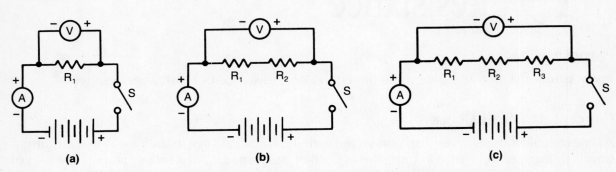

**Figure 1. Circuit diagram for measuring the current and voltage through (a) one resistor, (b) two resistors in series, and (c) three resistors in series.**

## C. Three Resistors

1. Arrange a third resistor in series with the first two, as shown in Figure 1(c). A diagram of the equipment is shown in Figure 2. Record the printed values for the resistors and their tolerances in Table 2.

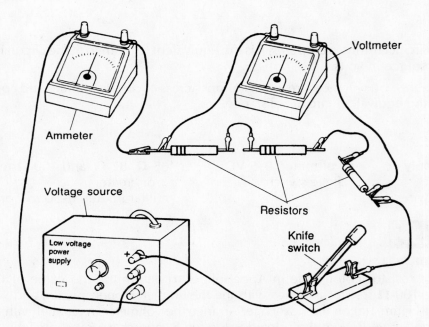

**Figure 2. The apparatus is connected for three resistors in series.**

2. Close the switch and adjust the voltage as necessary to maintain 5 V. Read the ammeter. Open the switch. The voltmeter reading is $V_{1,2,3}$ across the three resistors. Record the meter readings in Table 3.

3. With the three resistors still in series, use the voltmeter to find the voltage drop across each resistor. Touch the lead wires of the voltmeter to each end of the resistor $R_1$. Close the switch and read the voltmeter. Record the meter reading for $V_1$ in Table 3.

4. Repeat Step 3 for $R_2$ and $R_3$ and record the voltage readings of $V_2$ and $V_3$ in Table 3.

## Observations and Data
### Table 1

| | $R_1$ ($\Omega$) | Ammeter reading (mA) | Voltmeter reading (V) |
|---|---|---|---|
| | | | |
| Tolerance (%) | | | |

### Table 2

| | $R_1$ ($\Omega$) | $R_2$ ($\Omega$) | Ammeter reading (mA) | Voltmeter reading (V) |
|---|---|---|---|---|
| | | | | |
| Tolerance (%) | | | | |

### Table 3

| | $R_1$ ($\Omega$) | $R_2$ ($\Omega$) | $R_3$ ($\Omega$) | Ammeter reading (mA) | Voltmeter reading (V) | | | |
|---|---|---|---|---|---|---|---|---|
| | | | | | $V_{1,2,3}$ | $V_1$ | $V_2$ | $V_3$ |
| | | | | | | | | |
| Tolerance (%) | | | | | | | | |

## Analysis

1. Use the current and voltage data in Table 1 to calculate the measured value of resistance. Be sure to convert milliampere readings to amperes before you perform the calculations. Taking into account the tolerance, compare the printed value of the resistor with your measured value of $R_1$.

2. Use the current and voltage data in Table 2 to calculate the measured value of the equivalent resistance. The equivalent tolerance could be as large as the sum of the individual tolerances. Calculate the equivalent resistance by adding the printed values of $R_1 + R_2$. Taking into account the total tolerance, compare the calculated equivalent resistance with your measured equivalent resistance of $R_1 + R_2$.

3. Use the current and voltage ($V_{1,2,3}$) data in Table 3 to calculate the measured equivalent resistance. Calculate the equivalent resistance using the printed values for the resistors. Taking into account the total tolerance, compare the calculated equivalent resistance with your measured equivalent resistance of $R_1 + R_2 + R_3$.

4. When a circuit consists of resistors in series, how is the equivalent resistance determined?

5. Use your data in Table 3 to describe how the voltage drops across individual resistances are related to the total voltage drop in a series circuit.

6. What determines the total current in a series circuit?

## Application

A set of miniature, decorative tree lights contains 50 individual light bulbs of equal resistance, wired in series, and is designed for 120-V operation. If the set uses 1.0 A of current, what is the resistance of an individual light bulb and what is the voltage drop across each light bulb?

EXPERIMENT

# 20 Parallel Resistance

## Purpose

Measure current and voltage to determine the equivalent resistance of resistances in parallel.

## Concept and Skill Check

When resistors are connected in parallel, each resistor provides a path for current to follow and, therefore, reduces the equivalent resistance to the current. In parallel circuits, each circuit element has the same applied potential difference. In Figure 1(c), for example, three resistors are connected in parallel across the voltage source. There are three paths by which the current can pass from junction *a* to junction *b*. More current will flow between these junctions with three resistors attached than would be the case if only one or two resistors connected them. The total current, *I*, is given by

$$I = I_1 + I_2 + I_3.$$

Each time an additional resistance is connected in parallel with other resistors, the equivalent resistance decreases. The equivalent resistance of resistors in parallel can be determined by

$$\frac{1}{R} = \frac{1}{R_1} + \frac{1}{R_2} + \frac{1}{R_3} + ....$$

In this experiment, you will take numerous current and voltage readings with resistors in parallel and apply Ohm's law to verify your results. It will be necessary for you to follow very closely the circuit diagrams of Figure 1. Although the diagrams show several meters in use at one time, you probably have been provided with only one ammeter and one voltmeter. In that case, you must move the meters from position to position until all readings are obtained. You can, for example, take the total current, *I*, and total voltage *V*, readings, move the meters to positions $A_1$ and $V_1$, and so on, until you have all the required readings. Before applying Ohm's law, be sure to convert milliampere readings to amperes, where 1 mA = 0.001 A.

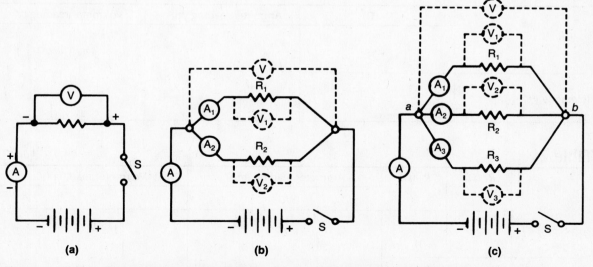

(a)  (b)  (c)

**Figure 1. Circuit diagram for resistors connected in parallel. Be sure to note the positive and negative terminals of the voltmeter and ammeter in relation to the positive and negative terminals of the voltage source.**

# EXPERIMENT 20  Parallel Resistance

## Materials

DC power supply or dry cells
3 resistors, 150–330 Ω range, 0.5 W,
  such as 180 Ω, 220 Ω, and 330 Ω

connecting wires
knife switch
voltmeter, 0–5 V

milliammeter, 0–50 mA
  or 0–100 mA

## Procedure

### A. One Resistor

1. Set up the circuit as shown in Figure 1(a), using one of the resistors. Close the switch. Adjust the power supply to a set voltage on the voltmeter, such as 3.0 V. Read the current value on the ammeter. Open the switch. Record your readings in Table 1.

### B. Two Resistors

1. Set up the circuit as indicated in Figure 1(b) by adding a second resistor. Close the switch and adjust the power supply as necessary to maintain the same voltage reading as in Part A. Read the current value on the ammeter. Open the switch. Record your readings in Table 2.
2. Move the meters as necessary to obtain the required readings of current and voltage. Record the readings in Table 2.

### C. Three Resistors

1. Set up the circuit as shown in Figure 1(c) by adding the third resistor. Close the switch, adjust the power supply, and read the meters. Open the switch. Record the readings in Table 3.
2. Move the meters as necessary to obtain all required readings. Record the readings in Table 3.

## Observations and Data

### Table 1

| | $R_1$ (Ω) | Ammeter reading (mA) | Voltmeter reading (V) |
|---|---|---|---|
| | | | |
| Tolerance (%) | | | |

### Table 2

| | $R_1$ (Ω) | $R_2$ (Ω) | Ammeter reading (mA) | | | Voltmeter reading (V) | | |
|---|---|---|---|---|---|---|---|---|
| | | | $I$ | $I_1$ | $I_2$ | $V$ | $V_1$ | $V_2$ |
| | | | | | | | | |
| Tolerance (%) | | | | | | | | |

# EXPERIMENT 20 Parallel Resistance

## Table 3

| | $R_1$ ($\Omega$) | $R_2$ ($\Omega$) | $R_3$ ($\Omega$) | Ammeter reading (mA) | | | | Voltmeter reading (V) | | | |
|---|---|---|---|---|---|---|---|---|---|---|---|
| | | | | $I$ | $I_1$ | $I_2$ | $I_3$ | $V$ | $V_1$ | $V_2$ | $V_3$ |
| | | | | | | | | | | | |
| Tolerance (%) | | | | | | | | | | | |

## Analysis

1. Use the readings from Table 1 to calculate the measured value for $R_1$, where $R_1 = \dfrac{V}{I}$. Is this within the tolerance expected for the printed value for $R_1$?

2. Use the readings from Table 2 to calculate the following values:

   a. the measured value for the equivalent resistance, $R$, where $R = \dfrac{V}{I}$.

   b. the measured current, $I_1 + I_2$.

   c. the measured resistance of $R_1$, where $R_1 = \dfrac{V_1}{I_1}$.

   d. the measured resistance of $R_2$, where $R_2 = \dfrac{V_2}{I_2}$.

   e. the calculated equivalent resistance, $R$, where $\dfrac{1}{R} = \dfrac{1}{R_1} + \dfrac{1}{R_2}$.

3.  a.  Compare the measured current sum, $I_1 + I_2$, to the measured current, $I$.



    b.  Compare the calculated equivalent resistance to the measured equivalent resistance. Was
        the measured equivalent resistance within the tolerance range for the resistors?



4.  Use the readings from Table 3 to calculate the following:

    a.  the measured equivalent resistance, $R$, where $R = \dfrac{V}{I}$.



    b.  the measured current, $I = I_1 + I_2 + I_3$.



    c.  the measured resistance of $R_1$, where $R_1 = \dfrac{V_1}{I_1}$.



    d.  the measured resistance of $R_2$, where $R_2 = \dfrac{V_2}{I_2}$.



    e.  the measured resistance of $R_3$, where $R_3 = \dfrac{V_3}{I_3}$.



    f.  the calculated equivalent resistance, $R$, where $\dfrac{1}{R} = \dfrac{1}{R_1} + \dfrac{1}{R_2} + \dfrac{1}{R_3}$.

5. a. Compare the value of $I$ with the measured current sum, $I_1 + I_2 + I_3$.

b. Compare the calculated equivalent resistance to the measured equivalent resistance. Was the measured equivalent resistance within the tolerance range for the resistors?

6. How does the current in the branches of a parallel circuit relate to the total current in the circuit?

7. How does the voltage drop across each branch of a parallel circuit relate to the voltage drop across the entire circuit?

8. As more resistors are added in parallel to an existing circuit, what happens to the total circuit current?

## Application

Tom has a sensitive ammeter that requires only 1.000 mA of current to provide a full-scale reading. The resistance of the ammeter coil is 500.0 $\Omega$. He wants to use the meter for a physics experiment that requires an ammeter capable of reading to 1.000 A. He has calculated that an equivalent resistance of 0.5000 $\Omega$ will produce the necessary voltage drop of 0.5000 V ($V = IR = 1.000 \times 10^{-3}$ A $\times$ 500.0 $\Omega$), so that only 1.000 mA of current passes through the meter. What value of shunt resistor, a resistor placed in parallel with the meter, is needed?

# Parallel Resistance

## Extension

A unique cube of 100-Ω resistors, as shown in Figure 2, is attached to a 1.0-V battery. The circuit can be reduced to combinations of parallel and series resistances. Predict the equivalent resistance. What is the total current flowing from the battery? Obtain twelve 100-Ω resistors, construct the circuit, and test your prediction.

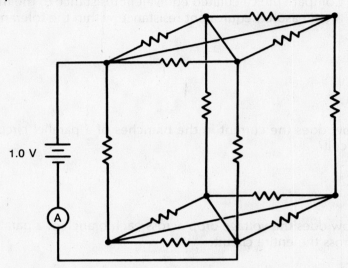

**Figure 2. All resistors are 100 Ω. What is the equivalent resistance and the current moving through the circuit?**

EXPERIMENT
# 21 Principles of Electromagnetism

## Purpose

Investigate the basic principles of electromagnetism.

## Concept and Skill Check

While experimenting with electric currents in wires, Hans Christian Oersted discovered that a nearby compass needle is deflected when current passes through the wires. This deflection indicates the presence of a magnetic field around the wire. You have learned that a compass shows the direction of the magnetic field lines. Likewise, the direction of the magnetic field can be determined with the first right-hand rule: when the right thumb points in the direction of the conventional current, the fingers point in the direction of the magnetic field.

When an electric current flows through a loop of wire, a magnetic field appears around the loop. An electromagnet can be made by winding a current-carrying wire around a soft iron core. A wire looped several times forms a coil. The coil has a field like that of a permanent magnet. A coil of wire wrapped around a core is called a solenoid. The magnetic field lines around the wire windings are collected by the iron core. The result is a strong magnet.

In this experiment, you will investigate some of the basic principles of electromagnetism while working with current-carrying wires, wire loops, and an electromagnet.

## Materials

compass
DC power supply, 0–5 A (or 3 dry cells)
ammeter, 0–5 A
14–18 gauge wire, 1.5 m
large iron nail or other piece of iron for
  use as a magnet core

knife switch
ringstand
ringstand clamp
iron filings
cardboard sheet,
  8½ × 11 inches

paper
meter stick
masking tape
box of steel paper clips

## Procedure

### A. The Field Around a Long, Straight Wire

1. Place the cardboard at the edge of the lab table. Arrange the wire so that it passes vertically through a hole in the center of the cardboard, as shown in Figure 1. Position the ringstand and clamp so that the wire can continue vertically up from the hole to the clamp. Bring the wire down the clamp and ringstand to the ammeter, and then to the positive terminal of the power supply.

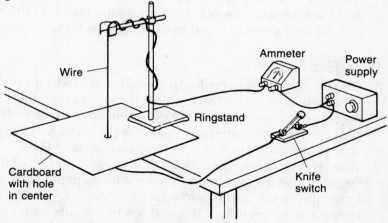

**Figure 1. Arrangement of equipment for Part A.**

# Principles of Electromagnetism

The wire passing below the cardboard should pass vertically at least 10 cm below the cardboard before it is routed along the table to the knife switch and then to the negative terminal of the power supply. Observe the proper polarity of the power supply and the ammeter as you connect the wires.

2. Close the switch and set the current to 2–3 A. Open the knife switch. Place the compass next to the wire. CAUTION: *The wire may become hot if current is permitted to flow for very long. Close the switch only long enough to make your observations.* Move the compass about the wire to map the field. In Part A of Observations and Data, record your observations with a sketch of the field around the wire.

3. Reverse the connections at the power supply and ammeter so that the current will flow in the opposite direction. Close the switch and again map the field around the wire, using the compass to determine the direction of the magnetic field. Record your observations with a sketch of the field around the wire.

## B. Strength of the Field

1. Place a piece of paper, containing a slit and a hole, on the cardboard with the wire at the center. Sprinkle some iron filings on the paper around the wire.

2. Close the switch and set the current to about 4 A. Gently tap the cardboard several times with your finger. Turn off the current. In Part B of Observations and Data, record your observations.

3. Tap the cardboard to disarrange the filings. Turn on the current and reduce it to 2.5 A. Gently tap the cardboard several times with your finger. Record your observations.

4. Tap the cardboard to disarrange the filings. Turn on the current and reduce it to 0.5 A. Gently tap the cardboard several times with your finger. Record your observations. Return the iron filings to the container.

## C. The Field Around a Coil

1. Remove the long, straight wire from the cardboard. In the center of the wire, make three loops of wire around your hand so that a coil of wire with a diameter of approximately 10 cm is formed. Place several pieces of tape on the wire to keep the loops together.

2. Connect the coil to the power supply through the ammeter and knife switch. Close the knife switch and adjust the current to 2.5 A. Hold the wire coil in a vertical plane and bring the compass near the coil, moving it through and around the coil of wire. In Part C of Observations and Data, record your observations with a sketch of the magnetic field direction around the coil. Show the positive and negative connections of your coil.

## D. An Electromagnet

1. Uncoil your large air-core loops. Wind loops of wire around the iron core until about half of the core is covered with wire. Connect the coil to the power supply, knife switch, and ammeter. Adjust the current so that 1.0 A of current flows in the coil. Bring the core near the box of paper clips and see how many paper clips the electromagnet can pick up. Open the switch and record your observations in Part D of Observations and Data.

2. Wind several more turns of wire on the iron core so that the number of turns is double that of Step 1. Close the switch and see how many paper clips the electromagnet can now pick up. Open the switch and record your observations.

3. Increase the current to 2.0 A and see how many paper clips the electromagnet can pick up. Record your observations.

4. Turn the current back on and, using the compass, determine the polarity of the electromagnet.

# 21 Principles of Electromagnetism

NAME ————————————————————

## Observations and Data

### A. The Field Around a Long, Straight Wire
Observations of direction of north pole:

### B. Strength of the Field
Observations of magnetic field with different currents:

### C. The Field Around a Coil
Observations of magnetic field around a coil of current-carrying wire:

### D. An Electromagnet
1. Several loops of wire on iron core:

2. Double the number of loops:

3. Double the current:

# 21 Principles of Electromagnetism

## Analysis

1. How does the right-hand rule apply to the current in a long, straight wire?

2. What is the effect of increasing the current strength in a wire?

3. Draw the magnetic field direction and the polarities for the current-carrying coil shown in Figure 2.

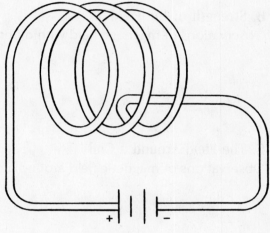

**Figure 2.**

4. What three factors determine the strength of an electromagnet?

5. Explain the difference between a bar magnet and an electromagnet.

## Application

List several solenoid applications that operate with either continuously applied or intermittently applied currents.

# EXPERIMENT 22 : Electromagnetic Induction

## Purpose

Observe the generation of an electric current when a wire cuts through a magnetic field.

## Concept and Skill Check

In 1831, Michael Faraday discovered that when a conductor moves in a magnetic field in any direction other than parallel to the field, an electric current is induced in the conductor. The strongest current is generated when the conductor moves perpendicular to the magnetic field. This process of generating an electric current is called electromagnetic induction, and the current produced is called induced current. Current is produced only when there is relative motion between the conductor and the magnetic field; it does not matter which moves. In this experiment, you will use a moving magnetic field to induce a current in a stationary conductor.

## Materials

galvanometer with zero in center of scale
1-turn coil of wire

25-turn coil of wire
100-turn coil of wire

2 bar magnets
connecting wires

## Procedure

1. Attach the single loop of wire to the galvanometer, as shown in Figure 1. Thrust one of the bar magnets through the coil. Record your observations in Item 1 of Observations and Data.

2. Move the galvanometer connections to the 25-turn coil of wire. Thrust the magnet into the coil. Record your observations in Item 2 of Observations and Data.

3. Move the galvanometer connections to the 100-turn coil of wire. Thrust the magnet into the coil. Record your observations in Item 3 of Observations and Data.

4. With the galvanometer connected to the 100-turn coil of wire, thrust the north pole of the magnet into the coil. Observe the direction of movement of the galvanometer

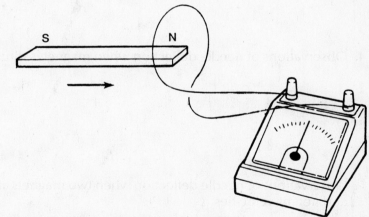

**Figure 1. Thrust the bar magnet through a single loop of wire.**

needle. Pull the magnet out of the coil. Observe the direction of movement of the galvanometer needle. Thrust the south pole of the magnet into the coil. Pull it back out. Note the movement of the galvanometer needle. Record your observations in Item 4 of Observations and Data.

5. Place two bar magnets together to make a more powerful magnet. Thrust them into the 100-turn coil of wire. Observe the deflection of the galvanometer needle. Slowly pull the two magnets out of the coil. Try different rates of speed for pushing the magnets into and out of the coil and compare the deflections of the galvanometer needle. Record your observations in Item 5 of Observations and Data.

6. Place the magnets in the 100-turn coil. While the magnets are stationary, does the galvanometer needle deflect? Move the coil back and forth while the magnets are stationary. Observe the motion of the galvanometer needle. Record your observations in Item 6 of Observations and Data.

## Observations and Data

1. Observations of needle deflection when magnet is thrust into single loop of wire:

2. Observations of needle deflection when magnet is thrust into 25-turn coil of wire:

3. Observations of needle deflection when magnet is thrust into 100-turn coil of wire:

4. Observations of needle deflection when magnet is thrust into and then pulled out of the coil:

5. Observations of needle deflection when two magnets are thrust into 100-turn coil and moved at different velocities:

6. Observations of needle deflection when magnets are stationary in 100-turn coil and when coil is moving:

# 22 Electromagnetic Induction

## Analysis

1. In Step 4, why did the galvanometer needle deflect in one direction when the magnet went into the coil and in the opposite direction when the magnet was pulled back out?

2. Summarize the factors that affect the amount of current and *EMF* induced by a magnetic field.

3. In your textbook, the equation given for the electromotive force induced in a wire by a magnetic field is *EMF* = *Blv*, where *B* is magnetic induction, *l* is the length of the wire in the magnetic field, and *v* is the velocity of the wire with respect to the field. Explain how the results of this experiment substantiate this equation.

4. What happens when a conducting wire is held stationary in or is moved parallel to a magnetic field? Explain.

5. Compare the induced electric field generated in this experiment to the electric field caused by static charges.

# Electromagnetic Induction

NAME _____

## Application

Figure 2 shows a diagram of a transformer. This device changes the AC voltage of the primary coil by inducing an increased or decreased *EMF* in the secondary coil. The value of the secondary voltage depends on the ratio of the number of turns of wire in the two coils. How is it possible for a magnetic field to move across the secondary coil and induce the *EMF*? Why does this device operate only on alternating, not direct, current?

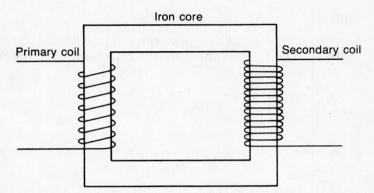

**Figure 2. A transformer contains two coils of insulated wire wound around an iron core.**

## Extension

1. The equation *EMF* = *Blv* applies only when the wire moves perpendicular to the magnetic field lines. A more general form of the equation would have to take into account movement at an angle to the magnetic field lines. Find the induced *EMF* of a wire 0.40 m long that moves through a magnetic field of magnetic strength $0.75 \times 10^{-2}$ T at a speed of 5.0 m/s. The angle $\theta$ between *v* and *B* is 45°. Show the equation you will use to solve for *EMF* and show all your calculations.

2. Write a statement that describes the effect on *EMF* of the angle $\theta$ between *v* and *B*.

3. As a conducting loop rotates in an external magnetic field, the induced current alternately increases and decreases. Describe the output produced and predict the shape of a graph that plots current against time.

# EXPERIMENT 23 Planck's Constant

## Purpose

Determine the value of Planck's constant.

## Concept and Skill Check

While studying radiation emitted from glowing, hot material, Max Planck assumed that the energy of vibration, $E$, of the atoms in a solid, could have only specific frequencies, $f$. He proposed that vibrating atoms emit radiation only when their vibrational energy is changed and that energy is quantized, changing only in multiples of $hf$. This relationship is given by $E = nhf$, where $n$ is an integer and $h$ is a constant. A light-emitting diode, or LED, is a modern application of this phenomenon.

An LED is a device made of a semiconductor wafer, which has been "doped" with two different types of impurities. A semiconductor doped with impurities that have loosely bound electrons donates those electrons and is designated an $n$-type material, while a semiconductor doped with acceptor impurities has "holes" that collect electrons and is designated a $p$-type material. The combination of $n$-type and $p$-type impurities in a semiconductor forms a $pn$ junction that acts like a vacuum tube diode, permitting current to flow in only one direction. If a DC is applied in the circuit when the $pn$ junction is reverse biased, the diode has a very high resistance and does not conduct current. When the $pn$ junction is forward biased, there is a very low resistance, and a large current can flow through the diode. An LED is a specially doped diode that emits light when current moves across a forward-biased $pn$ junction. LEDs can produce light in a wide variety of wavelengths, from the far-infrared to the near-ultraviolet region. The different wavelengths of visible light in LEDs are produced by varying the type and amount of impurity added to the semiconductor crystal structure.

When current moves across a forward-biased $pn$ junction, free electrons from the $n$-type material are injected into the $p$-type material, as shown in Figure 1. When these carriers recombine, energy is released. The energy released can be in the form of light or of heat from vibrations in the crystal lattice. The proportions of heat and light produced are determined by the recombination process taking place. As Planck proposed, the energy produced is given by $E = hf = hc/\lambda$, where $E$ is the energy in joules, $h$ is Planck's constant, $c$ is the speed of light ($c = 3.0 \times 10^8$ m/s), $f$ is the frequency of the emitted light, and $\lambda$ is the wavelength of the emitted light. In an LED, energy is supplied from a battery or DC power supply. Electrical energy supplied to the charge is given by $E = qV$, where $E$ is the energy in joules, $q$ is an elementary charge ($q = 1.6 \times 10^{-19}$ C), and $V$ is the energy per charge in volts. Equating these two relationships for $E$ yields

$$qV = \frac{hc}{\lambda}.$$

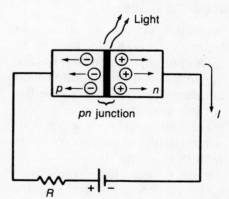

**Figure 1. A forward-biased, light-emitting diode $pn$ junction.**

# EXPERIMENT 23 Planck's Constant

Rearranging terms and solving for *h* gives

$$h = \frac{qV\lambda}{c}.$$

A typical curve for a graph of current versus voltage in a forward-biased diode is shown in Figure 2. The point at which recombination begins producing a significant amount of light, compared to that of heat, is at the knee. At the knee, the resistance drops sharply and the current increases rapidly within the diode. In this experiment, you will measure the current and various voltages across a forward-biased LED to find the voltage, *V*, where recombination begins to cause a large amount of emitted light. With this information, the value for Planck's constant can be determined.

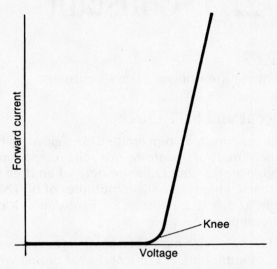

**Figure 2. A typical curve of a forward-biased LED.**

## Materials

green, red, and yellow light-emitting diodes (LEDs)
2 1.5-V batteries or 3-V power supply

battery holder
22-Ω resistor
1000-Ω potentiometer
connecting wires

ammeter, 0–50 mA DC
voltmeter, 0–5 VDC
knife switch

## Procedure

1. Wire the circuit, as shown in Figure 3. CAUTION: *Handle the LEDs carefully; their wire leads are fragile and will not tolerate much bending.* Be sure to observe the correct polarity for the meters. Have your teacher inspect your wired circuit before you continue with this activity.

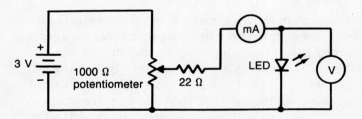

**Figure 3. Schematic diagram of a circuit to measure voltage and forward current across a light-emitting diode.**

2. Record in Table 1 the colors and the wavelengths of the LEDs supplied by your teacher.
3. CAUTION: *At no time during the experiment should the current to the LED exceed 25 mA.* Rotate the potentiometer control to its center position. Close the switch and observe the voltage on the voltmeter. Slowly adjust the voltage to approximately 2.0 V. If the LED is not glowing, open the switch and reverse the LED leads. The rotatable potentiometer forms a voltage divider network across the power supply to supply voltages from 0–3 V. Beginning at 1.50 V, and increasing in increments of 0.05–0.10 V until the current is less than or equal to 25 mA, collect current readings for the various voltage settings. Record your data in Table 2. When you have taken the last reading near, but less than, or equal to 25 mA, turn off the current to the circuit.
4. Replace the LED with one of another color and repeat Step 3. Collect data for all of the LEDs.

## Observations and Data

### Table 1

| LED | LED color | Wavelength (nm) |
|-----|-----------|-----------------|
| 1   |           |                 |
| 2   |           |                 |
| 3   |           |                 |

### Table 2

| LED 1 | | LED 2 | | LED 3 | |
|-------|-------|-------|-------|-------|-------|
| Voltage (V) | Current (mA) | Voltage (V) | Current (mA) | Voltage (V) | Current (mA) |
|  |  |  |  |  |  |
|  |  |  |  |  |  |
|  |  |  |  |  |  |
|  |  |  |  |  |  |
|  |  |  |  |  |  |
|  |  |  |  |  |  |
|  |  |  |  |  |  |
|  |  |  |  |  |  |
|  |  |  |  |  |  |
|  |  |  |  |  |  |
|  |  |  |  |  |  |
|  |  |  |  |  |  |
|  |  |  |  |  |  |
|  |  |  |  |  |  |
|  |  |  |  |  |  |

**Planck's
Constant**

## Analysis

1. Make a single graph that plots voltage on the horizontal axis and current on the vertical axis. Plot each set of LED data separately and label each curve. As the curve moves from zero current through the knee to a linear relationship, determine the point where the graph becomes linear. Hint: The current will be approximately 5–10 mA. For each LED curve, find the corresponding voltage for this point. This is the voltage at which recombination is producing a significant amount of light. List below the voltages for each LED.
   a. red voltage:

   b. yellow voltage:

   c. green voltage:

2. Calculate the value of Planck's constant for each LED. Show your work.

3. Compute the relative error for Planck's constant for each trial, using $h = 6.626 \times 10^{-34}$ J·s as the accepted value.

4. What approximate value for $V$ might be expected for an LED that produces blue light and for one that produces infrared light?

## Application

What advantages are there to using light-emitting diodes rather than a regular, incandescent light bulb?

# EXPERIMENT 24 : Reflection of Light

## Purpose

Use the law of reflection to locate the image formed by a plane mirror.

## Concept and Skill Check

When a light ray strikes a reflecting surface, the angle of reflection is equal to the angle of incidence. Both angles are measured from the normal, an imaginary line perpendicular to the surface at the point where the ray is reflected. In this laboratory activity, you will investigate and measure light rays reflected from the smooth, flat surface of a plane mirror in order to determine the apparent location of an image. This type of reflection is called regular reflection. The image of an object viewed in a plane mirror is a virtual image. By tracing the direction of the incident rays of light, you will be able to construct a ray diagram that locates the image formed by a plane mirror.

## Materials

| | | |
|---|---|---|
| thin plane mirror | 2 straight pins or | 4 thumbtacks or pieces of |
| small wooden block |    dissecting pins |    masking tape |
| metric ruler | rubber band | He-Ne Laser |
| protractor | cork board (or cardboard) | glass plate (optional) |
| 2 sheets blank paper | thick plane mirror | small blocks of wood (optional) |

## Procedure

### A. The Law of Reflection

1. Attach a sheet of paper to the cork board with the tacks or tape. Draw a line, **ML**, across the width of the paper. Attach the small block of wood to the back of the mirror with a rubber band, so that the mirror is perpendicular to the surface of the paper. Center the silvered surface (normally the back side) of the mirror along the line, **ML**, as shown in Figure 1.

2. About 4 cm in front of the mirror, make a dot on the paper with a pencil and label it point **P**. Place a pin, representing the object, upright at point **P**.

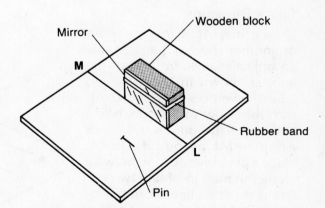

**Figure 1. Arrangement of materials for Part A.**

3. Place your ruler on the paper about 5 cm to the left of the pin. Sight along the edge of the ruler at the image of the pin in the mirror. When the edge of the ruler is lined up on the image, draw a line along it toward, but not touching, the mirror. Label this line **A**.

4. Move the ruler another 3 or 4 cm to the left and sight along it at the image of the pin in the mirror. Draw a line along the edge of the ruler toward, but not touching, the mirror. Label it line **B**.

5. Remove the pin and the mirror from the paper. Extend lines **A** and **B** to line **ML**. Using dotted lines, extend each of these lines beyond line **ML** until they intersect. Label the point of intersection point **I**. This is the position of the image. Measure the object distance from point **P** to line **ML** and the image distance from point **I** to line **ML**. Record these distances in Table 1.

6. Draw a line from point **P** to point **X**, where line **A** meets line **ML**, as shown in Figure 2. Using your protractor, construct a normal at this point and measure the angle of incidence, $i_1$, and the angle of reflection, $r_1$. Record the values of these angles in Table 1. In a similar manner, draw line **PY** from point **P** to point **Y** where line **B** meets line **ML**. Construct the normal at point **Y**. Measure the angle of incidence, $i_2$, and the angle of reflection, $r_2$. Record the values of these angles in Table 1. A line such as **PXA** or **PYB** is the path followed by a ray of light as it is transmitted from the pin (object) to the mirror, and reflected from the mirror to your eye. Drawing many such rays enables you to locate and recognize images in a mirror.

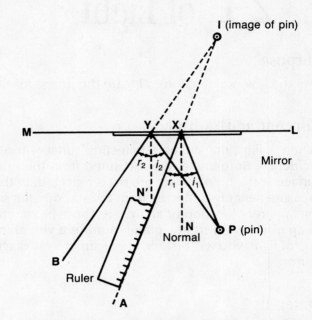

**Figure 2. Sight along the ruler at the pin's image.**

## B. Image Formation

1. Draw a line, **ML**, across the width of another sheet of paper. Draw an object triangle in front of the line, as shown in Figure 3, and label the vertices **A**, **B**, and **C**. Attach the paper to the cork board and place the mirror and block along line **ML**, as you did in Part A.

2. Place a pin in vertex **A**. Sight twice along the ruler to obtain two different sketched lines from the location of the pin at vertex **A** to line **ML**, as you did in Part A. Label these lines **A₁** and **A₂**.

3. Remove the pin from vertex **A** and place it at vertex **B**. Again, sight along the ruler to obtain two lines from the location of the pin at **B** to **ML** and label these lines **B₁** and **B₂**.

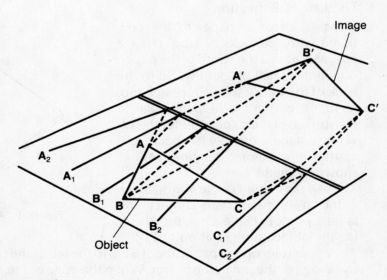

**Figure 3. Describe the front-back and right-left orientation of the image with respect to the object.**

4. Remove the pin from vertex **B** and place it at vertex **C**. Repeat the procedure and draw lines **C₁** and **C₂**.

5. Remove the mirror and the pin. Extend the pairs of lines, **A₁** and **A₂**, **B₁** and **B₂**, and **C₁** and **C₂**, beyond the mirror and locate points **A'**, **B'**, and **C'**, as shown in Figure 3. Construct the image of the triangle. Measure and record in Table 2, the distances from points **A**, **B**, **C**, **A'**, **B'**, and **C'** perpendicular to the mirror line **ML**.

## C. Laser Light Reflection

CAUTION: *Do not look directly at the laser beam or its reflection.*

1. Arrange your thin plane mirror so that the laser beam strikes it at an angle of incidence of about 60° from the normal. Reflect the laser beam onto the ceiling. In Table 3, draw your observations of the reflected image. Repeat the process using a clean, thick plane mirror (rear surfaced). In Table 3, record your observations of the reflected image.

2. With the room lights off, spray a fine mist of water in the air (or clap two chalk-filled erasers) over the reflected beam. In Table 3 record your observations.

## Observations and Data

### Table 1

| | |
|---|---|
| Object distance | |
| Image distance | |
| Angle of incidence, $i_1$, **PXN** | |
| Angle of reflection, $r_1$, **AXN** | |
| Angle of incidence, $i_2$, **PYN'** | |
| Angle of reflection, $r_2$, **BYN'** | |

### Table 2

| Point | Distance to mirror line (cm) |
|---|---|
| A | |
| B | |
| C | |
| A' | |
| B' | |
| C' | |

EXPERIMENT
## 24 | Reflection of Light

## Table 3

| | |
|---|---|
| Drawing of reflected laser light from thin plane mirror | |
| Drawing of reflected laser light from thick plane mirror | |
| Observation of reflected laser beams | |

## Analysis

1. Using your observations from Table 1, what can you conclude about the angle of incidence and the angle of reflection?

2. How far behind a plane mirror is the image of an object that is located in front of the mirror?

3. Using your observations from Table 2, compare the size and orientation of your constructed image with those of the triangle object.

4. From your observations in this experiment, summarize the general characteristics of images formed by plane mirrors.

5. Why do you think the image formed by a plane mirror is called virtual rather than real?

6. When you used the thick plane mirror, which reflected spot of the laser beam was brightest? Suggest a possible explanation for the bright beam.

7. Provide an explanation to account for the difference in patterns of the reflected laser beam from the thin and thick plane mirrors. Draw a diagram that demonstrates the pattern of laser light reflected from a thick plane mirror and your explanation for this result.

NAME _____

## Application

In many department stores, large plane mirrors have been placed high on walls or on projections from ceilings. These may be one-way mirrors that are designed to allow one-way surveillance of the store. From one side, this surface looks like a mirror, but from the other side, the activities of the shoppers can be observed. Suggest how this type of mirror works.

## Extension

Use small blocks of wood or a stand to set a piece of plate glass vertically on a line, **ML**, drawn on a sheet of paper that is attached to a cork board. The glass must be perpendicular to the paper, and you must be able to sight through the middle portion of the glass. Place one pin upright in front of the plate glass. Place a second pin upright in back of the plate glass at a point where the reflected image from the plate glass seems to be, as you sight through the glass. View the image at different angles and adjust the second pin slightly until this pin behind the plate glass is always at the position of the image, regardless of the viewing angle. Measure the distance from the object pin to the mirror line, **ML**, and compare it to the distance from the second pin to **ML**.

EXPERIMENT

# 25 : Concave and Convex Mirrors

## Purpose

Investigate positions and characteristics of images produced by curved mirrors.

## Concept and Skill Check

Spherical mirrors are portions of spheres one side of which is silvered and serves as a reflecting surface. If the inner side is the reflecting surface, the mirror is a concave mirror. If the outer side is the reflecting surface, the mirror is a convex mirror.

The center of the sphere of which the mirror is a portion is called the center of curvature (**C**) of the mirror. An imaginary line perpendicular to the center or vertex of the mirror (**A**) and that also passes through the center of curvature is called the principal axis of the mirror. The point halfway between **C** and **A** is called the focal point (**F**) of the mirror. The distance from the focal point **F** to the center of the mirror **A** is called the focal length ($f$) of the mirror. The distance of the object from the mirror, $d_o$, and the distance of the image from the mirror, $d_i$, are related to the focal length by the mirror equation

$$\frac{1}{f} = \frac{1}{d_i} + \frac{1}{d_o}.$$

Figure 1 illustrates these relationships.

Concave mirrors produce real images, virtual images, or no images. Light rays actually pass through real images, which can be projected on a screen. Convex mirrors produce only virtual images. Virtual images appear to originate behind the mirror and cannot be projected onto a screen.

Light rays that approach a concave spherical mirror from a position close to the mirror and parallel with the principal axis will, upon reflection, converge at the focal point. If all the rays approaching a concave mirror are parallel with the principal axis, they will meet approximately at the focal point where they will form an image of the object from which the light rays come. In this experiment, you will be placing an object at various distances from curved mirrors and observing the location, size, and orientation of the images.

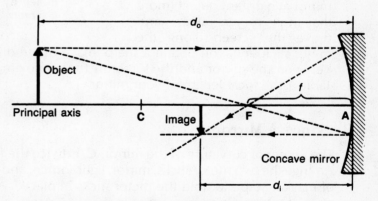

**Figure 1. The center of curvature C is located at a distance of twice the focal length $f$ from the center of the mirror, along the principal axis.**

## Materials

| | | |
|---|---|---|
| concave mirror | cardboard screen | masking tape |
| convex mirror | holders for mirror, screen, | metric ruler |
| 2 meter sticks | and light source | light source |

## Procedure

The laboratory room must be darkened so that images are easily observed. The electric light sources are sufficient to illuminate laboratory work areas while students set up equipment. One unshaded window should be available for student use.

### A. Focal Length of Concave Mirrors

1. Arrange your mirror, meter stick, and screen, as shown in Figure 2. If the sun is visible, the focal length can be determined by projecting a focused image of the sun onto a screen and measuring the distance from the mirror's vertex to the screen. CAUTION: *Do not look directly at the sun since this can cause severe damage to your eyes.* An alternate method is to point the mirror at a distant object (more than ten meters away) and move the screen along the

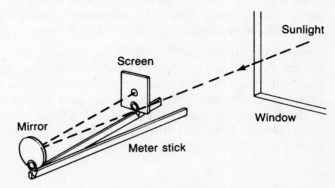

**Figure 2. The focal length of a mirror can be determined by projecting a focused image of the sun onto a screen.**

meter stick until you obtain a sharp image of the distant object on the screen. The distance between the mirror and the screen is the approximate focal length of the mirror. Record in Table 1 the focal length of your mirror.

### B. Concave Mirrors

1. The center of curvature of the mirror, **C**, is twice the focal length. Record this value in Table 1. Arrange the two meter sticks, mirror, light source, and screen, as shown in Figure 3. Use a piece of masking tape to hold the meter sticks in place.

2. Place the light source at a distance greater than **C** (beyond **C**) from the mirror. Measure the height of the light source and record this value in Table 1 as $h_o$. Move the screen back and forth along the meter stick until you obtain a sharp image of the light source. Determine the distance of the image from the mirror, $d_i$, by measuring from the mirror's vertex to the screen. Record in Table 2 your measurements of $d_i$, $d_o$, $h_i$, and your observations of the image.

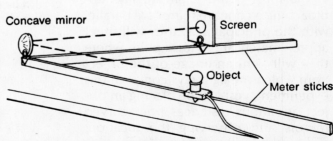

**Figure 3. Adjust the location of the screen until you obtain a clear, sharp image of the light source.**

# Concave and Convex Mirrors

3. Move the light source to the center of curvature, **C**. Move the screen back and forth to obtain a sharp image. Record in Table 2 your measurements of $d_i$, $d_o$, $h_i$ and your observations of the image.

4. Move the light source to a position that is between **F** and **C**. Move the screen back and forth to obtain a sharp image. Record in Table 2 your measurement of $d_i$, $d_o$, $h_i$ and your observations of the image.

5. Move the light source to a distance **F** from the mirror. Try to locate an image on the screen. Observe the light source in the mirror. Record your observations in Table 2.

6. Move the light source to a position between **F** and **A**. Try to locate an image on the screen. Observe the image in the mirror. Record your observations in Table 2.

## C. Convex Mirrors

Place the convex mirror in the holder. Place the light source anywhere along the meter stick. Try to obtain an image on the screen. Observe the image in the mirror. Move the light source to two additional positions along the meter stick and try each time to produce an image on the screen. Observe the image in the mirror. Record your observations in Table 3.

## Observations and Data

### Table 1

| | |
|---|---|
| Focal length of mirror, $f$ | |
| Center of curvature of mirror, **C** | |
| Height of light source, $h_o$ | |

### Table 2

| Position of object | Beyond **C** | At **C** | Between **C** and **F** | At **F** | Between **F** and **A** |
|---|---|---|---|---|---|
| $d_o$ | | | | | |
| $d_i$ | | | | | |
| $h_i$ | | | | | |
| Type of image: real, none, or virtual | | | | | |
| Direction of image: inverted or erect | | | | | |

### Table 3

| Trial | Position of object | Position of image | Type of image: real or virtual | Image size compared to object size | Direction of image: inverted or erect |
|---|---|---|---|---|---|
| 1 | | | | | |
| 2 | | | | | |
| 3 | | | | | |

# 25 Concave and Convex Mirrors

NAME —————————————————————————

## Analysis

1. Use your observations from Table 2 to summarize the characteristics of images formed by concave mirrors in each of the following situations.

   a. The object is located beyond the center of curvature.

   b. The object is located at the center of curvature.

   c. The object is located between the center of curvature and the focal point.

   d. The object is located at the focal point.

   e. The object is located between the focal point and the mirror.

2. Use your observations from Table 3 to summarize the characteristics of images formed by convex mirrors.

3. For each of the real images you observed, use the mirror equation to calculate $f$. Do your calculated values agree with each other?

4. Average the values of $f$ that you calculated for Question 3 and compute the relative error between the average and the measured value for $f$ recorded in Table 1.

## Application

The apparatus shown in Figure 4 can project the image of an illuminated light bulb onto an empty socket. Describe how to orient a large, spherical mirror to reflect the illuminated light bulb in the box onto the socket on top of the box.

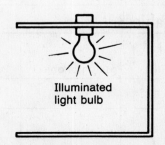

Illuminated light bulb

**Figure 4. Illusion apparatus for use with a concave mirror.**

# EXPERIMENT 26 : Snell's Law

## Purpose

Determine the index of refraction of glass using Snell's law.

## Concept and Skill Check

Light travels at different speeds in different media. As light rays pass at an angle from one medium to another, they are refracted or bent at the boundary between the two media. If a light ray enters an optically more dense medium at an angle, it is bent toward the normal. If a light ray enters an optically less dense medium, it is bent away from the normal. This change in direction or bending of light at the boundary at two media is called refraction.

The index of refraction of a substance, $n_s$, is the ratio of the speed of light in a vacuum, $c$, to its speed in the substance, $v_s$:

$$n_s = \frac{c}{v_s}.$$

All indices of refraction are greater than one, because light always travels slower in media other than a vacuum.

The index of refraction is also obtained from Snell's law, which states: a ray of light bends in such a way that the ratio of the sine of the angle of incidence to the sine of the angle of refraction is a constant. Snell's law can be written:

$$n = \frac{\sin \theta_i}{\sin \theta_r}.$$

For any light ray traveling between media, Snell's law, in a more general form, can be written as
$$n_i \sin \theta_i = n_r \sin \theta_r,$$

where $n_i$ is the index of refraction of the incident medium and $n_r$ is the index of refraction of the second medium. The angle of incidence is $\theta_i$, and the angle of refraction is $\theta_r$.

In this investigation, you will construct ray diagrams to analyze the path of light as it passes through plate glass. For each angle of incidence, you will measure the angle of refraction to find the index of refraction of plate glass.

## Materials

| | | |
|---|---|---|
| glass plate | 100-mL sample bottle of | 100-mL sample bottle of |
| metric ruler | mineral oil or oil of | olive oil (optional) |
| protractor | wintergreen (optional) | 100-mL sample bottle of |
| sheet plain white paper | 100-mL sample bottle of | cinnamon oil (optional) |
| small samples of glass, garnet, | clove oil (optional) | tweezers (optional) |
| tourmaline, beryl, topaz, | 100-mL sample bottle of | |
| and quartz (optional) | water (optional) | |

## Procedure

1. Place the glass plate in the center of a sheet of plain white paper. Use a pencil to trace an outline of the plate.
2. Remove the glass plate and construct a normal **N₁B** at the top left of the outline, as shown in Figure 1 on the next page.

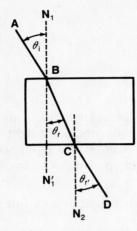

**Figure 1. Construct normals $N_1$ and $N_2$ to the surface of the glass at C and D, respectively.**

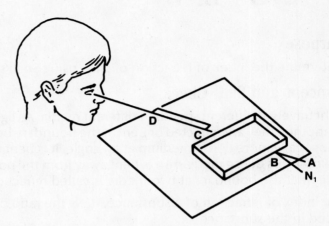

**Figure 2. Sight through the glass block and align the ruler with line AB.**

3. Use your ruler and protractor to draw a heavy line **AB** at an angle of 30° with the normal. Angle **ABN$_1$** is the angle of incidence, $\theta_i$.
4. Replace the glass plate over the outline on the paper. With your eyes on a level with the glass plate, sight along the edge of the glass plate opposite the line **AB** until you locate the heavy line through the glass as shown in Figure 2. Sight your ruler at the line until its edge appears to be a continuation of the line. Draw the line **CD** as shown in Figure 1.
5. Remove the glass plate and draw another line **CB** connecting lines **CD** and **AB**. Extend the normal **N$_1$B** through the rectangle, forming a new line **N$_1$CN$_1$'**.
6. Use a protractor to measure angle **CBN$_1$'**. This is the angle refraction, $\theta_r$. Record the value of this angle in Table 1. Record the values for the sines of the angles of incidence and refraction in Table 1. Determine the ratio of sin $\theta_i$/sin $\theta_r$ and record this value in Table 1 as the index of refraction, $n$.
7. Construct a normal **N$_2$** at point **C**. Measure angle **DCN$_2$**, which will be called $\theta_{r'}$, and record this value in Table 1.
8. Turn the paper over and repeat Steps 1 through 7, using an angle of incidence of 45°. Record all the data in Table 1. Again, determine the index of refraction from your data.

## Observations and Data
### Table 1

| $\theta_i$ | $\theta_r$ | sin $\theta_i$ | sin $\theta_r$ | $\theta_{r'}$ | Index of refraction, $n$ |
|---|---|---|---|---|---|
| 30° | | | | | |
| 45° | | | | | |

## Analysis

1. Is there good agreement between the two values for the index of refraction of plate glass?

2. According to your diagrams, are light rays refracted away from or toward the normal as they pass at an angle from an optically less dense medium into an optically more dense medium?

3. According to your diagrams, are light rays refracted away from or toward the normal as they pass from an optically more dense medium into an optically less dense medium?

4. Compare $\theta_i$ and $\theta_{r'}$. Is the measure of $\theta_{r'}$ what you should expect? Explain.

5. Use your results to determine the approximate speed of light as it travels through glass. By what percent is the speed of light traveling in a vacuum faster than the speed of light traveling in glass?

## Application

Will light be refracted more while passing from air into water or while passing from water into glass? Explain.

## Extension

Gemologists can identify an unknown gem by measuring its index of refraction, specific to each type of gemstone. The gemologist places the unknown gem in a series of liquids with known indices of refraction. When the gem becomes nearly invisible in a liquid, the gemologist concludes that the gem has an index of refraction corresponding to that of the liquid in which it is immersed. With this information, the gemologist consults a reference giving the indices of refraction of gemstones to identify the unknown gem. The following lists the indices of refraction of some liquids and materials.

| Liquid | n | Material | n |
|---|---|---|---|
| water | 1.34 | glass | 1.48–1.7 |
| olive oil | 1.47 | quartz | 1.54 |
| mineral oil | 1.48 | beryl | 1.58 |
| oil of wintergreen | 1.48 | topaz | 1.62 |
| clove oil | 1.54 | tourmaline | 1.63 |
| cinnamon oil | 1.60 | garnet | 1.75 |

Use these liquids to identify the unknown materials your teacher will give you. With tweezers, carefully place each sample material alternately into each bottle of liquid. Observe the material and the behavior of light as it passes through the liquid and the material. Before immersing the sample in the next bottle of liquid, rinse it off in water and dry it gently to prevent contamination of the oils. When the material seems to disappear, record the index of refraction for the liquid and use this value to identify the material.

EXPERIMENT

# 27 Convex and Concave Lenses

## Purpose

Observe the positions and characteristics of images produced by convex and concave lenses.

## Concept and Skill Check

A convex or converging lens is thicker in the middle of the lens than at the edges of the lens. A concave or diverging lens is thinner in the middle than at the edges. The principal axis of the lens is an imaginary line perpendicular to the plane of the lens that passes through its midpoint. It extends from both sides of the lens. At some distance from the lens along the principal axis is the focal point ($F$) of the lens. Light rays that strike a convex lens parallel to the principal axis come together or converge at this point. The focal length of the lens depends on both the shape and the index of refraction of the lens material. As with mirrors, an important point designated $2F$ is at twice the focal length. If the lens is symmetrical, the focal point $F$ and point $2F$ are located the same distances on either side of the lens, as shown in Figure 1.

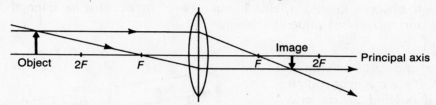

**Figure 1. Focal length, object distance, and image distance are measured from the lens along the principal axis.**

A concave lens causes all incident parallel light rays to diverge. Rays approaching a concave lens parallel to the principal axis appear to intersect on the near side of the lens. Thus, the focal length of a concave lens is negative. Figure 2 shows the relationship of the incoming and refracted rays passing through a concave lens.

The distance from the center of the lens to the object is designated $d_o$, while the distance from the center of the lens to the image is designated $d_i$. The lens equation is

$$\frac{1}{f} = \frac{1}{d_i} + \frac{1}{d_o}.$$

In this activity, you will measure the focal length, $f$, of a convex lens and place an object at various distances from the lens to observe the location, size, and orientation of the images. Find the focal length of a concave

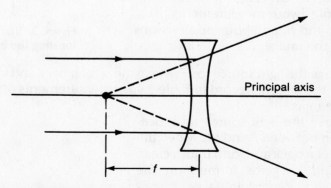

**Figure 2. The focal length of a concave lens is negative. All light rays passing through a concave lens diverge.**

lens by tracing diverging rays backwards to locate the focal point. Recall that real images can be projected onto a screen; virtual images cannot be projected.

## Materials

| | | |
|---|---|---|
| double convex lens | small cardboard screen | metric ruler |
| concave lens | light source | sunlight or He-Ne laser |
| 2 meter sticks | holders for screen, light source, | |
| 2 meter-stick supports | and lens | |

## Procedure

The laboratory room must be darkened so that images are easily observed. The electric light sources are sufficient to illuminate laboratory work areas while students set up equipment. One unshaded window should be available for student use.

### A. Focal Length of a Convex Lens

1. To find the focal length of the convex lens, arrange your lens, meter stick, and screen, as shown in Figure 3. Point the lens at a distant object and move the screen back and forth until you obtain a clear, sharp image of the object on the screen. A darkened room makes the image much easier to observe. Record in Table 1 your measurement of the focal length. Compute the distance *2F* and record this value in Table 1.

### B. Convex Lens

1. Arrange the apparatus as shown in Figure 4. Place the light source somewhere beyond *2F* on one side of the lens and place the screen on the opposite side of the lens. Move the screen back and forth until a clear, sharp image is formed on the screen. Record in Table 1 the height of the light source (object), $h_o$. Record in Table 2 your measurements of $d_o$, $d_i$, and $h_i$ and your observations of the image.

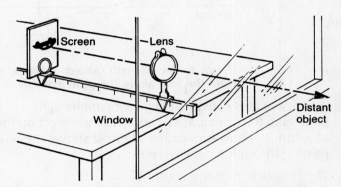

**Figure 3. The focal point of a convex lens is found by locating the image of a distant object.**

2. Move the light source to *2F*. Move the screen back and forth until a clear, sharp image is formed on the screen. Record in Table 2 your measurements of $d_o$, $d_i$, and $h_i$ and your observations of the image.

3. Move the light source to a location between *F* and *2F*. Move the screen back and forth until a clear, sharp image is formed on the screen. Record in Table 2 your measurements of $d_o$, $d_i$, and $h_i$ and your observations of the image.

4. Move the light source to a distance *F* from the lens. Try to locate an image on the screen. Record your observations in Table 2.

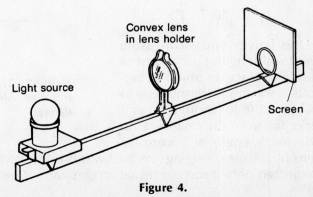

**Figure 4.**

# 27 Convex and Concave Lenses

NAME _____

5. Move the light source to a position between *F* and the lens. Try to locate an image on the screen. Look through the lens at the light source and observe the image. Record your observations in Table 2.

## C. Concave Lens

1. Place a concave lens in a lens holder and set it on the meter stick. Place a screen on one side of the lens. One of the following procedures can be used to determine the focal length.

Sunlight: Allow the parallel rays of the sun to strike the lens along the principal axis so that an image is formed on the screen. CAUTION: *Do not look directly at the sun since this can cause severe damage to your eyes.* The image should appear as a dark circle inside a larger, brighter circle. Place the screen close to the lens. Quickly measure the distance from the lens to the screen and the diameter of the bright circle. Record these data in Table 3. Move the screen and repeat the measurements for five additional sets of data.

He-Ne Laser: Shine a laser beam through the concave lens so that an image is formed on the screen. CAUTION: *Do not look directly at the laser source since this can cause severe damage to your eyes.* Measure the distance from the screen to the lens and the diameter of the circle of light projected onto the screen. Record these data in Table 3. Move the screen and repeat the measurements for five additional sets of data.

## Observations and Data

### Table 1

| | |
|---|---|
| Focal length | |
| 2F | |
| Height of light source, $h_o$ | |

### Table 2

| Position of object | Beyond 2F (cm) | At 2F (cm) | Between 2F and F (cm) | At F (cm) | Between F and lens (cm) |
|---|---|---|---|---|---|
| $d_o$ | | | | | |
| $d_i$ | | | | | |
| $h_i$ | | | | | |
| Type of image: real, none, or virtual | | | | | |
| Direction of image: inverted or erect | | | | | |

## Table 3

| Distance from lens (m) | Diameter of screen image (cm) |
|------------------------|-------------------------------|
|                        |                               |
|                        |                               |
|                        |                               |
|                        |                               |
|                        |                               |
|                        |                               |

## Analysis

1. Use the data from Table 2 to summarize the characteristics of images formed by convex lenses in each of the following situations.

   a. The object is located beyond 2F.

   b. The object is located at 2F.

   c. The object is located between 2F and F.

   d. The object is located at F.

   e. The object is located between F and the lens.

2. For each of the real images you observed, calculate the focal length of the lens using the lens equation. Do your values agree with each other?

3. Average the values for *f* found in Question 2 and calculate the relative error between this average and the value for *f* from Table 1.

4. Plot a graph of the image diameter on the vertical axis versus the distance from the lens on the horizontal axis. Allow room along the horizontal axis for negative distances. The rays expanding from the lens appear to originate from the focal point. Draw a smooth line that best connects the data points and extend the line until it intersects the horizontal axis. The negative distance along the horizontal axis at the intersection represents the value of the focal length. What is the focal length you derived from your graph? If your lens package includes an accepted focal length, calculate the relative error for the focal length you determined.

# 27 Convex and Concave Lenses

## Application

The following demonstration is best done when participants have removed their glasses or contact lenses. The relative direction of movement of reflected laser light with respect to an observer's moving head is a means to determine near- or farsightedness. In a darkened room, use a concave lens to expand the beam diameter of the laser so that a large spot is projected onto a screen. Observers should move their heads from side to side while looking at the spot. Each student should record the direction in which the reflected speckles of laser light appear to move and the direction in which his or her head is moving. Then ask observers to replace their glasses or contact lenses and to repeat the process and make their observations.

## Extension

Use two identical, clear watch glasses and a small fish aquarium to investigate an air lens. Carefully glue the edges of the watch glasses together with epoxy cement or a silicone sealant so that the unit is watertight. Attach the air lens to the bottom of an empty fish aquarium with a lump of clay and arrange an object nearby, as shown in Figure 5. Observe the object through the lens and record your observations. Predict what will happen when the aquarium is filled with water such that light is passing from a more dense to a less dense medium as it passes through the air lens. Fill the aquarium with water and repeat your observations. Compare the two sets of observations. Explain your results. What should you expect if you used a concave air lens? Design and construct a concave air lens to test your hypothesis.

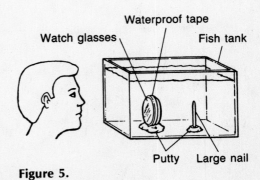

**Figure 5.**

EXPERIMENT
# 28 Double-Slit Interference

## Purpose

Use a double slit to determine the wavelength of light.

## Concept and Skill Check

As Thomas Young demonstrated in 1801, light falling on two closely spaced slits will pass through the slits and be diffracted. The spreading light from the two slits overlaps and produces an interference pattern, an alternating sequence of dark and light lines, on a screen. A bright band appears at the center of the screen. On either side of the central band, alternating bright and dark lines appear on the screen. Bright lines appear at points where constructive interference occurs, and dark lines appear where destructive interference occurs. The first bright line on either side of the central band is called the first-order line. This line is bright because, at this point on the screen, the path lengths of light waves arriving from each slit differ by one wavelength. The next line is the second-order line. Several lines are visible on either side of the central bright band. The equation that relates the wavelength of light, the separation of the slits, the distance from the central band to the first-order line, and the distance to the screen is

$$\lambda = xd/L.$$

The figure below shows this relationship.

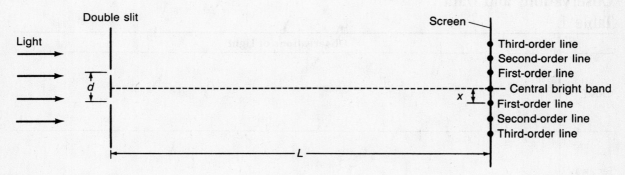

**Figure 1.**

In this experiment, you will use an arrangement similar to that employed in Young's experiment to determine the wavelengths of several different colors of light. As the colored light is diffracted through a double slit, you will observe an interference pattern and will measure the distance between the central bright line and the first-order line. Other distances can be measured, and these values can be used to calculate the wavelength of light.

## Materials

| | | |
|---|---|---|
| double slit | rubber band | masking tape |
| clear filament light with socket | meter stick | index cards, 5 × 7 inch |
| red, yellow, and violet transparent filter material | metric ruler | He-Ne laser (optional) |
| | magnifier with scale | |

# 28 Double-Slit Interference

## Procedure

1. Measure the slit separation of your double slit, using the magnifier and a graduated millimeter scale. If it is known, use the actual slit separation. Record in millimeters the width, *d*, in Table 2. This value will remain constant for all trials.
2. Place a red-colored filter over the light bulb and secure it with a rubber band. Turn on the light bulb. Turn off the classroom lights.
3. Stand 1 m from the light and observe the light bulb through the double slit. Record in Table 1 your observations of the light bands. Move several meters away from the light and repeat your observations, recording them in Table 1.
4. Move approximately 2–4 m from the light and mark the location with a small piece of tape. Measure the distance from the light bulb filament to the small piece of tape. Record the distance, *L*, in Table 2.
5. Make a dark line on a 5 × 7-inch index card. With your lab partner viewing the light through the double slit, place the card, as directed by your partner, so that the dark line is located over the central bright band, or light bulb filament.
6. While your lab partner observes the light bulb through the double slit, slowly move your pencil back and forth until your partner tells you that it is located over one of the first-order lines. Make a mark on the index card where the first-order line occurs. Measure the distance, *x*, from the central bright line to the first-order line. Record this value in Table 2.
7. Remove the red filter material from the light. Repeat Steps 4 through 6 with the other colored filters. Record your data in Table 2.

## Observations and Data
### Table 1

| | Observations of Light |
|---|---|
| Close | |
| Far away | |

### Table 2

| Color | *d* (mm) | *x* (m) | *L* (m) |
|---|---|---|---|
| Red | | | |
| Yellow | | | |
| Violet | | | |

# 28 Double-Slit Interference

NAME ————————————————————————

## Analysis

1. Describe how the double-slit pattern changes as you move farther away from the light source.

2. Calculate the wavelength, in nm, for each of the colors. Show your work.

3. Which color of light will be diffracted most as it passes through the double slit? Explain.

4. Predict where the first-order line for infrared light of wavelength 1000 nm would be located. How does this location compare with the other first-order lines observed in this experiment?

5. Predict where the first-order line for ultraviolet light of wavelength 375 nm would be located. How does this location compare with the other first-order lines observed in this experiment?

# ₄ENT  28 Double-Slit Interference

## Applications

1. A bright light, visible from 50.0 m away, produces an angle of 0.8° to the first-order line when viewed through a double slit having a width of 0.05 mm. What color is the light?

2. Blue-green light of wavelength 500 nm falls on two slits that are 0.0150 mm apart. A first-order line appears 14.7 mm from the central bright line. What is the distance between the slits and the screen? Show all your calculations.

3. The slit separation and the distance between the slits and the screen are the same as those in Question 2, but sodium light of wavelength 589 nm is used. What is the distance from the central line to the first-order line with this light? Show all your calculations.

## Extension

Place a double slit over the front of a He-Ne laser. Project the interference pattern onto a screen. Measure $L$ and $x$ for the first-order line and calculate the separation of the double slit using the value 632.8 nm for the wavelength of the laser light.

## EXPERIMENT
# 29 Semiconductor Properties

## Purpose

Investigate the voltage-current relationship in a semiconductor diode.

## Concept and Skill Check

The diode is the simplest type of semiconductor device. It is made of a semiconductor material, such as germanium or silicon, that is specially doped with two types of impurity. The n-type end of the diode is doped with a donor impurity, such as arsenic, that has loosely bound electrons. The p-type end is doped with an acceptor impurity, such as gallium, that has "holes" for electrons. Figure 1 shows a schematic representation of the diode.

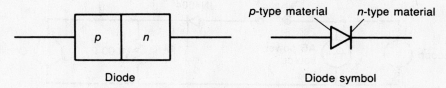

**Figure 1.**

When a direct current is applied in one direction, called reverse bias, the diode has a very high resistance and does not conduct current. However, when current is applied in the other direction, called forward bias, there is a very low resistance, and a large current flows through the diode. In this experiment, you will forward- and reverse-bias a diode and measure the current as a function of the applied voltage. Silicon diodes begin to conduct current in the forward-biased direction when the voltage reaches about 0.6 V. A germanium diode begins to conduct at about 0.2 V.

## Materials

1N4004 diode (rectifier) or
   equivalent; milliammeter,
   0–100 mA or 0–500 mA
voltmeter, 0–5 VDC
1.5–3-V battery or DC
   power supply, 0–6 V

10-Ω resistor
1000-Ω resistor
1000-Ω potentiometer
power supply, 6 VAC, or AC adapter,
   such as that used for a tape recorder

connecting
   wires
switch
oscilloscope

## Procedure

1. Connect the circuit as shown on the schematic diagram in Figure 2. The voltage is applied to the pn junction such that the p-type end is positive with respect to the n-type end. Note that a band or ring around one end of the diode corresponds to the n-type end.

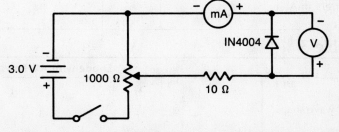

**Figure 2. Schematic diagram of a forward-biased circuit.**

2. Close the switch and slowly increase the voltage from 0.00 V to 0.80 V in 0.1-V increments, measuring the circuit current at each step. Record the current readings in Table 1. When you have taken the last reading, open the switch to turn off the current to the circuit.

3. Reverse the diode so that the positive end of the diode, the *p*-type end, is attached to the ammeter. Close the switch and slowly increase the voltage from 0.00 V to 0.80 V in 0.1-V increments, measuring the circuit current at each step. Record the current readings in Table 2. When you have taken the last reading, open the switch to turn off the current to the circuit.
4. Connect an alternating-current source—from a power supply or from a small adapter for a tape recorder—and a diode, as shown in Figure 3. CAUTION: *Under no circumstances should the diode be connected to the 120-V alternating current of a standard electrical outlet.* Attach an oscilloscope across the AC input. Close the switch to apply the AC power and observe the wave shape on the oscilloscope screen. In Table 3, draw the wave shape. Turn off the AC power. Connect the oscilloscope leads across the 1000-Ω resistor. Close the switch to apply current to the diode. Observe the wave shape on the oscilloscope screen. In Table 3, draw this other wave shape. Open the switch.

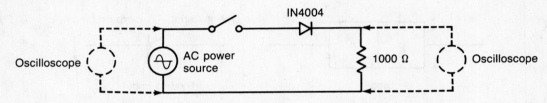

**Figure 3. Schematic diagram of an alternating current (AC) source and a diode.**

## Observations and Data
### Table 1

| Forward-Biased Diode | | | | | | | | |
|---|---|---|---|---|---|---|---|---|
| Voltage (V) | 0.00 | 0.10 | 0.20 | 0.30 | 0.40 | 0.50 | 0.60 | 0.70 | 0.80 |
| Current (mA) | | | | | | | | |

### Table 2

| Reverse-Biased Diode | | | | | | | | |
|---|---|---|---|---|---|---|---|---|
| Voltage (V) | 0.00 | 0.10 | 0.20 | 0.30 | 0.40 | 0.50 | 0.60 | 0.70 | 0.80 |
| Current (mA) | | | | | | | | |

### Table 3

| Diode with AC Power | |
|---|---|
| AC wave shape | |
| AC wave shape through a diode | |

## Analysis

1. Plot a graph of current versus voltage, with current on the vertical axis. Set up the horizontal axis with both negative and positive voltages and center the vertical axis. The negative voltage values represent the reverse-biased diode, while the positive voltage values represent the forward-biased diode.

2. Compare the graph of current versus voltage with a diode in the circuit to a similar graph with a resistor in the circuit.

3. What is unique about the current flow through a diode?

4. Is the diode you used a germanium or a silicon diode? What data support your answer?

5. Compare the two current wave shapes you observed. Explain any differences.

6. Describe the wave shape of the AC input across the 1000-$\Omega$ resistor if the diode is reversed.

# 29 Semiconductor Properties

## Application

The circuit that you constructed with the single diode and an alternating current input is called a half-wave rectifier. Conventional current flow is in the direction of the arrow on the diode symbol. A full-wave rectifier circuit, shown in Figure 4, is used in electronic equipment, such as computers, televisions, stereo receivers, and amplifiers, to provide a clean source of power. Draw the wave shape that you predict would appear across the output load resistor, $R_L$, for the circuit shown in Figure 4.

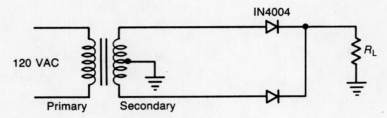

**Figure 4. Full-wave rectifier circuit.**

## Extension

Build the circuit shown in Figure 4 and attach the oscilloscope to the output load resistor. Compare your predicted wave shape to the one you observe.

# APPENDIX A

## A:1. Common Physical Constants

Absolute zero = $-273.15°C = 0$ K
Acceleration due to gravity at sea level, (Washington, D.C.): $g = 9.80$ m/s$^2$
Atmospheric pressure (standard): 1 atm = $1.013 \times 10^5$ Pa = 760 mm Hg
Avogadro's number: $N_A = 6.02 \times 10^{23}$ mol$^{-1}$
Charge of one electron (elementary charge): $e = -1.602 \times 10^{-19}$ C
Coulomb's law constant: $K = 9.0 \times 10^9$ N $\cdot$ m$^2$/C$^2$
Gas constant: $R = 8.31$ J/mol $\cdot$ K
Gravitational constant: $G = 6.67 \times 10^{-11}$ N $\cdot$ m$^2$/kg$^2$
Heat of fusion of ice: $3.34 \times 10^5$ J/kg
Heat of vaporization of water: $2.26 \times 10^6$ J/kg
Mass of electron: $m_e = 9.1 \times 10^{-31}$ kg = $5.5 \times 10^{-4}$ u
Mass of neutron: $m_n = 1.675 \times 10^{-27}$ kg = 1.00867 u
Mass of proton: $m_p = 1.673 \times 10^{-27}$ kg = 1.00728 u
Planck's constant: $h = 6.626 \times 10^{-34}$ J/Hz (J $\cdot$ s)
Velocity of light in a vacuum: $c = 2.99792458 \times 10^8$ m/s

## A:2. Conversion Factors

Mass:      1000 g    = 1 kg
           1000 mg   = 1 g

Volume:    1000 mL   = 1 L
              1 mL   = 1 cm$^3$

Length:    1000 mm   = 1 m
            100 cm   = 1 m
           1000 m    = 1 km

1 atomic mass unit (u) = $1.66 \times 10^{-27}$ kg = 931 MeV/c$^2$
1 electron volt (eV) = $1.602 \times 10^{-19}$ J
1 joule (J) = 1 N $\cdot$ m = 1 V-C
1 coulomb = $6.242 \times 10^{18}$ elementary charge units

## A:3. Color Response of Eye to Various Wavelengths of Light

| Color | Wavelength | |
|---|---|---|
| | in nm | in m |
| Ultraviolet | less than 380 | less than $3.8 \times 10^{-7}$ |
| Violet | 400–420 | $4.0$–$4.2 \times 10^{-7}$ |
| Blue | 440–480 | $4.4$–$4.8 \times 10^{-7}$ |
| Green | 500–560 | $5.0$–$5.6 \times 10^{-7}$ |
| Yellow | 580–600 | $5.8$–$6.0 \times 10^{-7}$ |
| Red | 620–700 | $6.2$–$7.0 \times 10^{-7}$ |
| Infrared | above 760 | above $7.6 \times 10^{-7}$ |

## A:4. Prefixes Used with SI Units

| Prefix | Symbol | Fraction | Prefix | Symbol | Multiplier |
|---|---|---|---|---|---|
| Pico | p | $10^{-12}$ | Tera | T | $10^{12}$ |
| Nano | n | $10^{-9}$ | Giga | G | $10^{9}$ |
| Micro | $\mu$ | $10^{-6}$ | Mega | M | $10^{6}$ |
| Milli | m | $10^{-5}$ | Kilo | k | $10^{3}$ |
| Centi | c | $10^{-2}$ | Hecto | h | $10^{2}$ |
| Deci | d | $10^{-1}$ | Deka | da | $10^{1}$ |

# APPENDIX B

## Properties of Common Substances
### B:1. Specific Heat and Density

| Substance | Specific heat (J/kg · K) | Density |
|---|---|---|
| Alcohol | 2450 | 0.8 |
| Aluminum | 903 | 2.7 |
| Brass | 376 | 8.5 varies by content |
| Carbon | 710 | 1.7–3.5 |
| Copper | 385 | 8.9 |
| Glass | 664 | 2.2–2.6 |
| Gold | 129 | 19.3 |
| Ice | 2060 | 0.92 |
| Iron (steel) | 450 | 7.1–7.8 |
| Lead | 130 | 11.3 |
| Mercury | 138 | 13.6 |
| Nickel | 444 | 8.8 |
| Platinum | 133 | 21.4 |
| Silver | 235 | 10.5 |
| Steam | 2020 | — |
| Tungsten | 133 | 19.3 |
| Water | 4180 | 1.0 at 4°C, 0.99 at 0°C |
| Zinc | 388 | 7.1 |

## B:2. Index of Refraction

| Substance | Index of refraction |
|---|---|
| Air | 1.00029 |
| Alcohol | 1.36 |
| Benzene | 1.50 |
| Beryl | 1.58 |
| Carbon dioxide | 1.00045 |
| Cinnamon oil | 1.6026 |
| Clove oil | 1.544 |
| Diamond | 2.42 |
| Garnet | 1.75 |
| Glass, crown | 1.52 |
| Glass, flint | 1.61 |
| Mineral oil | 1.48 |
| Oil of wintergreen | 1.48 |
| Olive oil | 1.47 |
| Quartz, fused | 1.46 |
| Quartz, mineral | 1.54 |
| Topaz | 1.62 |
| Tourmaline | 1.63 |
| Turpentine | 1.4721 |
| Water | 1.33 |
| Water vapor | 1.00025 |
| Zircon | 1.87 |

# B:3. Spectral Lines of Elements

| Element | Wavelength (nanometers) | Color |
|---------|------------------------|-------|
| | many close lines | |
| | 706.7 | red (strong) |
| | 696.5 | red (strong) |
| | 603.2 | orange |
| | 591.2 | orange |
| | 588.8 | orange |
| | 550.6 | yellow |
| Argon | 545.1 | green |
| | 525.2 | green |
| | 522.1 | green |
| | 518.7 | green |
| | 451.0 | purple |
| | 433.3 | purple |
| | 430.0 | purple |
| | 427.2 | purple |
| | 420.0 | purple |
| | 659.5 | red |
| | 614.1 | orange |
| Barium | 585.4 | yellow |
| | 577.7 | yellow |
| | 553.5 | green (strong) |
| | 455.4 | blue (strong) |
| | many green lines | |
| | 481.7 | green |
| Bromine | 478.6 | blue |
| | 470.5 | blue |
| | many purple lines | |
| | 445.4 | blue |
| | 443.4 | blue-violet |
| Calcium | 442.6 | violet (strong) |
| | 396.8 | violet (strong) |
| | 393.3 | violet (strong) |
| | 520.8 | green |
| | 520.6 | green |
| Chromium | 520.4 | green |
| | 428.9 | violet (strong) |
| | 427.4 | violet (strong) |
| | 425.4 | violet (strong) |
| | 521.8 | green |
| Copper | 515.3 | green |
| | 510.5 | green |
| | 656.2 | red |
| Hydrogen | 486.1 | green |
| | 434.0 | blue-violet |
| | 410.1 | violet |
| | 706.5 | red |
| | 667.8 | red |
| Helium | 587.5 | orange (strong) |
| | 501.5 | green |
| | 471.3 | blue |
| | 388.8 | violet (strong) |

| Element | Wavelength (nanometers) | Color |
|---------|------------------------|-------|
| | many lines | |
| Iodine | 546.4 | green (strong) |
| | 516.1 | green (strong) |
| | many faint close lines | |
| | 587.0 | orange (strong) |
| | 557.0 | yellow (strong) |
| Krypton | 455.0 | blue |
| | 442.5 | blue |
| | 441.0 | blue |
| | 430.2 | blue-violet |
| | 623.4 | red |
| | 579.0 | yellow (strong) |
| Mercury | 576.9 | yellow (strong) |
| | 546.0 | green (strong) |
| | 435.8 | blue-violet |
| | many lines in the violet and ultra violet | |
| | 567.6 | green (strong) |
| Nitrogen | 566.6 | green |
| | 410.9 | violet (strong) |
| | 409.9 | violet |
| Potassium | 404.7 | violet (strong) |
| | 404.4 | violet (strong) |
| | 670.7 | red (strong) |
| Lithium | 610.3 | orange |
| | 460.3 | violet |
| | 589.5 | yellow (strong) |
| Sodium | 588.9 | yellow (strong) |
| | 568.8 | green |
| | 568.2 | green |
| | many lines in the red | |
| | 640.2 | orange (strong) |
| Neon | 585.2 | yellow (strong) |
| | 583.2 | yellow (strong) |
| | 540.0 | green (strong) |
| | 496.2 | blue-green |
| | 487.2 | blue |
| | 483.2 | blue |
| Strontium | 460.7 | blue (strong) |
| | 430.5 | blue-violet |
| | 421.5 | violet |
| | 407.7 | violet |
| | 492.3 | blue-green |
| | 484.4 | blue |
| | 482.9 | blue |
| | 480.7 | blue |
| | 469.7 | blue |
| Xenon | 467.1 | blue (strong) |
| | 462.4 | blue (strong) |
| | 460.3 | blue |
| | 458.3 | blue |
| | 452.4 | blue |
| | 450.0 | blue (strong) |

# APPENDIX C
## Rules for the Use of Meters

### Introduction

Electric meters are precision instruments and must be handled with great care. They are easily damaged physically or electrically and are expensive to replace or repair. Meters are damaged physically by bumping or dropping and electrically by allowing excessive current to flow through the meter. The heating effect in a circuit increases with the square of the current. The wires inside the meter actually burn through if too much current flows through the meter. When possible, use a switch in the circuit to prevent having the circuit closed for extended periods of time. Note that meters are generally designed for use in either AC or DC circuits and are not interchangeable. In DC circuits, the polarity of the meter with respect to the power source in the circuit is critical. Be sure to use the proper meter for the type of circuit you are investigating. Always have your teacher check the circuit to be sure you have assembled it correctly.

### The Voltmeter

A voltmeter is used to determine the potential difference between two points in a circuit. It is always connected in parallel, never in series, with the element to be measured. If you can remove the voltmeter from the circuit without interrupting the circuit, you have connected it correctly.

On a DC voltmeter, the terminals are marked + or −. The positive terminal should be connected either directly or through components to the positive side of the power supply. The negative terminal must be connected either directly or through circuit components to the negative side of the power supply. After you have connected the voltmeter, close the switch for a moment to see if the polarity is correct.

Some meters have several ranges from which to choose. You may have a meter with ranges from 0–3 V, 0–15 V, and 0–300 V. If you do not know the potential difference across the circuit on which the voltmeter is to be used, choose the highest range initially, and then adjust to a range that gives readings in the middle of the scale (when possible).

### The Ammeter

An ammeter is used to measure the current in a circuit and must always be connected in series. Since the internal resistance of an ammeter is very small, the meter will be destroyed if it is connected in parallel. When you connect or disconnect the ammeter, the circuit must be interrupted. If the ammeter can be included or removed without breaking the circuit, the ammeter is incorrectly connected.

Like a voltmeter, an ammeter may have different ranges. Always protect the instrument by connecting it first to the highest range and then proceeding to a smaller scale until you obtain a reading in the middle of the scale (when possible).

On a DC ammeter, the polarity of the terminals is marked + or −. The positive terminal should be connected either directly or through components to the positive side of the power supply. The negative terminal must be connected either directly or through circuit components to the negative side of the power supply. After you have connected the ammeter, close the switch for a moment to see if the polarity is correct.